U0177437

人工智能
技术与应用（双语）

张 明 孙晓丽◎主编

中国铁道出版社有限公司
CHINA RAILWAY PUBLISHING HOUSE CO., LTD.

内 容 简 介

本书是中英文双语教材，首先通过介绍线性回归了解机器学习中的数据集、归一化、损失函数、梯度下降函数等基础知识，在此基础上通过多个典型案例，详细分析了逻辑回归、决策树、聚类、支持向量机等机器学习算法的实现流程和相关知识，并融入国家职业技能标准"人工智能训练师"和"大数据应用开发"1＋X证书等相关内容，具有较强的实践指导意义。

本书适合作为高职院校人工智能、大数据等相关专业课程、双语课程的教材，同时也可作为机器学习应用开发技术人员的指导教程。

图书在版编目（CIP）数据

人工智能技术与应用：汉文、英文/张明，孙晓丽

主编. —北京：中国铁道出版社有限公司，2023.12

ISBN 978-7-113-30417-1

Ⅰ. ①人… Ⅱ. ①张… ②孙… Ⅲ. ①人工智能-汉、英 Ⅳ. ①TP18

中国国家版本馆 CIP 数据核字（2023）第 133996 号

书　　名：人工智能技术与应用（双语）
RENGONG ZHINENG JISHU YU YINGYONG（SHUANGYU）

作　　者：张　明　孙晓丽

策　　划：潘晨曦　祁　云　　　　　　编辑部电话：(010)63551006

责任编辑：祁　云　绳　超

封面设计：刘　颖

责任校对：安海燕

责任印制：樊启鹏

出版发行：中国铁道出版社有限公司(100054,北京市西城区右安门西街 8 号)

网　　址：http://www.tdpress.com/51eds/

印　　刷：北京联兴盛业印刷股份有限公司

版　　次：2023 年 12 月第 1 版　　2023 年 12 月第 1 次印刷

开　　本：850 mm×1 168 mm　1/16　印张：21.75　字数：480 千

书　　号：ISBN 978-7-113-30417-1

定　　价：79.80 元

前 言

人工智能是计算机科学中最热门的领域之一,它的应用范围涵盖了很多领域,包括机器学习、自然语言处理、图像识别、语音识别等。对于大部分行业而言,人工智能应用、机器学习与大数据分析已经成为非常重要的一环,因此掌握人工智能技术将是一个极大的优势。无论是企业的管理层,还是行业中的研发人员,都需要掌握人工智能技术才能更好地应对未来的挑战。

本书站在学生的角度,以典型的人工智能机器学习案例为载体,以项目案例实现的过程为主线,将人工智能机器学习的理论知识融入项目实现过程,循序渐进地介绍了人工智能机器学习必备的基础知识,以及一些比较优秀的工具,帮助学生具备使用 sklearn 库实现机器学习任务的相关技能,能够独立编写程序完成相关项目,以胜任人工智能机器学习相关岗位的工作。

本书共 13 个单元,分为 3 个部分:单元 1~单元 5 介绍线性回归部分,包括简单线性回归、多元线性回归、多项式回归、岭回归等;单元 6~单元 8 介绍逻辑回归,包括逻辑回归、决策边界等;单元 9~单元 13 介绍经典且常用的机器学习方法,包括聚类算法、决策树、支持向量机、贝叶斯算法、Word2Vec 等。

本书不仅适合作为高职院校人工智能、大数据等相关专业课程、双语课程的教材,同时也可作为机器学习应用开发技术人员的指导教程。

在学习过程中,读者一定要亲自实践本书的案例代码。与本书相配套的在线课程资源可在"智慧职教"网站上搜索"人工智能应用技术"课程获取,或加入智慧职教 MOOC 学院中的"人工智能应用技术"课程,通过平台中提供的教学视频进行深入学习。

尽管编者付出了最大的努力,但书中仍难免会有不妥之处,欢迎各界专家和读者提出宝贵意见,以便本书进一步完善。

读者在阅读本书时,如果需要申请课程资源,或者如发现任何问题或有不认同之处,可以通过电子邮件与我们取得联系。编者邮箱:45655600@qq.com。

编 者
2023 年 4 月

目 录

单元1

使用简单线性回归模型
预测广告投入的收益

　　某公司统计了近期其在微信、微博、电视和其他广告媒体上的投入，现在需要预测在广告上投入多少资金，公司能获得多大的收益。

1.1　简单线性回归介绍

　　简单线性回归又称一元线性回归，也就是回归模型中只含一个自变量，否则称为多重线性回归。简单线性回归模型为

$$y = ax + b$$

式中，y 是因变量；x 是自变量；b 是常数项，是回归直线在纵坐标轴上的截距；a 是回归系数，是回归直线的斜率。

　　通常表述为 $y = wx + b$，其中，w 表示权重（weight），b 表示偏置值（bias）。

　　公司需要根据给定的企业的收入 y 和企业的广告投入 x，去预测 w 和 b，最终得到一个一元线性方程，通过这个方程，可以预测出企业投入的广告资金大概可以获得多大的收益，如图 1-1 所示。横轴代表企业投入，纵轴代表企业预期收益。直线是通过训练后得到的一元线性方程的点。

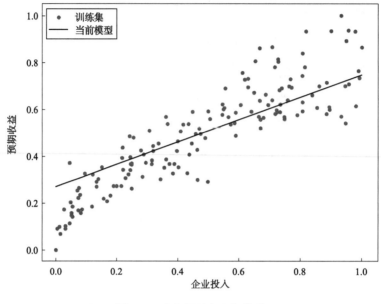

图 1-1   企业投入和预期收益

1.2 读取数据集

1. 导入相应的包

新建一个 Jupyter Notebook 文件，导入 pandas、matplotlib、numpy 包。导入这些包完成数据的处理和数据的可视化展示，如果需要在图标显示中文，还需要设置 matplotlib 的中文显示格式。代码如下：

```
1.  import pandas as pd
2.  import matplotlib.pyplot as plt
3.  import numpy as np
4.  #显示中文
5.  plt.rcParams['font.sans - serif'] = ['SimHei']
6.  plt.rcParams['axes.unicode_minus'] = False
```

2. 加载数据

使用 pandas 的 read_csv 方法打开数据集，使用 head( ) 函数查看前五条数据，同时可以使用 head( ) 函数查看数据的信息。代码如下：

```
1.  df_data =pd.read_csv("advertising.csv")
2.  df_data.head()
```

代码运行后,输出读 csv 文件的前五条数据,效果如图 1-2 所示。

	wechat	weibo	others	sales
0	304.4	93.6	294.4	9.7
1	1011.9	34.4	398.4	16.7
2	1091.1	32.8	295.2	17.3
3	85.5	173.6	403.2	7.0
4	1047.0	302.4	553.6	22.1

图 1-2　代码运行效果

1.3　数据相关性分析

在样本属性很多的数据集中,样本的特征(x)和标签(y)之间存在一些关系,有些特征与标签的相关性强,有些弱,可以通过画图、协方差、相关系数、信息熵来显示特征和标签之间的相关性。通常使用 seaborn 包的 heatmap( )函数绘制特征和标签之间的热力图,代码如下:

```
1.  import seaborn as sns
2.  sns.heatmap(df_data.corr(),cmap = 'YlGnBu',annot = True)
3.  plt.show()
```

代码运行后,效果如图 1-3 所示。通过可视化形式,很容易看到特征之间的相关性。根据图 1-3 可以轻易地看到特征之间的"影响因子",横纵坐标交叉的区域颜色越深,代表它们之间关系越深,区域块上面标识的数字同样显示着这个特征。

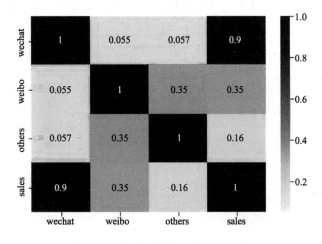

图 1-3　特征和标签之间的热力图

通过图 1-3 可以看到"wechat"的投入和公司收入之间存在很强的相关性,相关系数为 0.9。以双变量为例,变量 x 和变量 y 存在三种关系,即正线性相关、负线性相关、非线性相关,如

图1-4所示。横轴表示 x，纵轴表示 y。

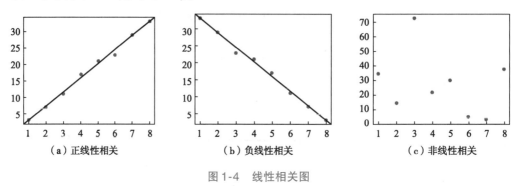

图1-4　线性相关图

同时也可以使用seaborn绘制pairplot图，也可以分析出"wechat"的投入基本和公司收入呈现一种线性关系，代码如下：

```
1.  sns.pairplot(df_data,x_vars = ['wechat','weibo','others'],
2.                y_vars = 'sales',
3.                height = 4,aspect = 1)
4.  plt.show()
```

代码运行结果如图1-5所示，从图中可以看出微信投入和收入呈现线性关系。

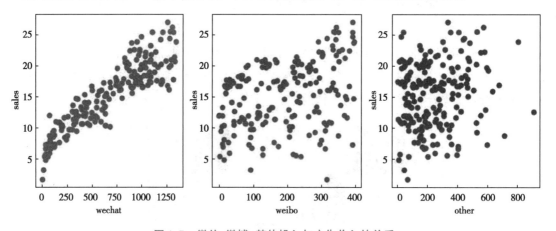

图1-5　微信、微博、其他投入与广告收入的关系

1.4　数据处理

1. 选取数据

建立线性模型的第一步是选取数据，通过对数据相关性的分析，可以从数据集中取出"wechat"数据作为 X，"sales"作为 y，代码如下：

```
1.  X_data = df_data['wechat']
2.  y_data = df_data['sales']
3.  print(type(X_data))
4.  print(type(y_data))
```

代码运行后,输出 X_data 的类型为 < class 'pandas. core. series. Series ' > , y_data 的类型为 < class 'pandas. core. series. Series ' > ,可以看出取出的 X 和 y 是一个 series 对象。

线性模型需要使用张量作为输入,需要将 series 类型数据转换为张量,使用 array( )函数,将 series 转换为 ndarry 类型,代码如下:

```
1.  #张量变换
2.  X_data = np.array(X_data)
3.  y_data = np.array(y_data)
4.  print(type(X_data))
5.  print(X_data.shape)
6.  print(X_data.ndim)
7.  X = X_data.reshape(len(X_data),1)
8.  print(type(X))
9.  print(X.shape)
10. print(X.ndim)
```

代码运行后,输出 X_data 的类型为 'numpy. ndarray ', X_data 的 shape 为(200,), X. ndim 的值为2。

2.拆分数据集为训练集和测试集

模型训练需要将数据集划分为训练集和测试集,使用 sklearn 的 model_selection( )函数对 X_data 和 y_data 按照80%的训练集和20%的测试集进行拆分,得到训练集和测试集。

```
1.  from sklearn.model_selection import train_test_split
2.  X_train,X_test,y_train,y_test = train_test_split(X_data,y_data,test
    _size = 0.2,random_state = 0)
```

同时对训练集和测试集的数据绘制 scatterplot 显示特征和标签的分布状况,代码如下:

```
1.  sns.scatterplot(X_data,y_data)
2.  plt.xlabel("weibo")
3.  plt.ylabel("sales")
4.  plt.show()
```

代码运行后,显示了训练集中微博(weibo)与收入(sales)的关系如图1-6所示,从图中可以看出两个数据之间存在线性关系。

使用 sns 的 scatterplot 方法绘制测试集中微博与收入散点图,代码如下:

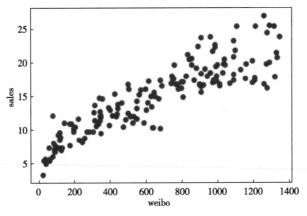

图 1-6   微博与收入的关系 (训练集)

```
1.  sns.scatterplot(X_test,y_test)
2.  plt.xlabel("weibo")
3.  plt.ylabel("sales")
4.  plt.show()
```

程序运行后显示测试集中微博与收入的关系,如图 1-7 所示。从两个图中可以看出,测试集和训练集的数据分布相同,不同之处在于训练集的数据要多一些。

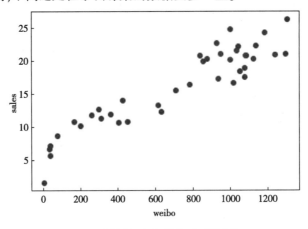

图 1-7   微博与收入的关系 (测试集)

1.5 数据归一化处理

在机器学习中,为何要经常对数据做归一化? 原因在于,归一化后加快了梯度下降求最优解的速度,归一化有可能提高精度。在这里使用线性归一化,方程如下:

$$X = \frac{x - \min(x)}{\max(x) - \min(x)}$$

式中，X 为归一化后的数据；x 为数据集中的数据；$\min(x)$ 为求出数据集中的最小值；$\max(x)$ 为求出数据集中的最大值。

这种归一化方法比较适用在数值比较集中的情况。这种方法有个缺陷，如果 max 和 min 不稳定，很容易使得归一化结果不稳定，使得后续使用效果也不稳定。实际使用中，可以用经验常量值来替代 max 和 min。

这里定义了一个函数 scalar( ) 来实现归一化操作，然后调用函数实现对 X_train，X_test，y_train，y_test 的归一化操作，同时绘制 scatterplot 图形，代码如下：

```
1.    def scalar(train,test):
2.        min = train.min(axis = 0)
3.        max = train.max(axis = 0)
4.        gap = max - min
5.        train - = min
6.        train/ = gap
7.        test - = min
8.        test/ = gap
9.        return train,test
10.
11.   X_data,X_test = scalar(X_train,X_test)
12.   y_train,y_test = scalar(y_train,y_test)
13.
14.   sns.scatterplot(X_train,y_train)
15.   plt.show()
```

代码运行后，输出归一化后的微信和收入之间的关系，如图 1-8 所示。通过对比归一化之后和归一化之前的数据图，可以看出微信(wechat)数据值的范围由 0 到 1400 通过归一化操作变成了 0~1；收入(sales)数据值的范围由 0 到 25 通过归一化操作变成了 0~1 之间。两幅图中微信投入和收入的取值范围发生了变化，但是两者之间的分布没有发生变化，如图 1-8 所示。

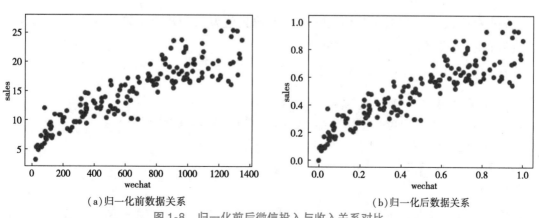

(a)归一化前数据关系　　　　　　　　(b)归一化后数据关系

图 1-8　归一化前后微信投入与收入关系对比

1.6　定义损失函数

机器学习中,目标函数是整个模型优化学习的核心导向,需要最小化目标函数时,目标函数也被称为损失函数或代价函数。对于有监督学习任务,通常目标都是使得预测值尽可能接近标签,即最小化目标函数。损失函数的选取主要取决于学习任务的要求。最常用的两个损失函数 MSE 与交叉熵分别用于回归与分类任务。MSE 均方差损失函数公式为

$$\text{MSE} = \frac{1}{2m} \sum_{i=1}^{n} (y_i - y_i')^2$$

式中,y_i'表示预测值;y_i 表示实际值。

实际值和预测值之间的距离表示为 $y_i - y_i'$,求出所有实际值和预测值距离的和,就是模型的整体损失。为了保证距离不会产生正负抵消,需要加上平方项,再除以 $2m$（m 表示样本数量）求平均值,就可以得到 MSE 均方差损失函数。

MSE 求导后计算量不大,最为常用,代码实现简单,可以自己编写代码实现,代码如下:

```
1.  def loss_function(X,y,weight,bias):
2.      y_hat =weight* X +bias    #y =a* x +b
3.      loss =y_hat - y
4.      cost =np.sum(loss* * 2)/(2* len(X))
5.      return cost
```

首先初始化权重 w 和偏置值 b,然后输出 MSE 的值,代码如下:

```
1.  print("权重为 5,偏置值为 3 时,损失函数为",loss_function(X_train,y_
    train,weight =5,bias =3))
2.  print("权重为 100,偏置值为 1 时,损失函数为",loss_function(X_train,y_
    train,weight =5,bias =3))
```

代码运行后,输出当权重为 5,偏置值为 3 时,损失值为 12.796 390 970 780 058;权重为 100,偏置值为 1 时,损失值为 12.796 390 970 780 058。

然后使用 scatterplot() 函数绘制模型预测值和实际值的图像,代码如下:

```
1.  sns.scatterplot(X_train,y_train)
2.  line_x =np.linspace(X_train.min(),X_train.max())
3.  line_y =[-5* xx +3 for xx in line_x]
4.  sns.scatterplot(line_x,line_y)
5.  plt.show()
```

代码运行后,结果如图 1-9 所示。从输出结果可以看出,当模型使用初始化参数时,模型预测

值和实际值之间的误差较大。

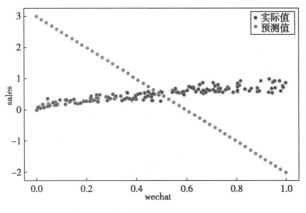

图 1-9　模型初始化参数运行结果

1.7　定义梯度下降函数

上一节定义了损失函数,把 x 和 y 看成已知量,可以将损失函数转换为 $l(w,b)$ 的形式。在这个函数中,w 和 b 是未知的值,MSE 函数可以用如下公式表示:

$$l(w,b) = \frac{1}{2m}\sum_{i=1}^{m}\big[(wx+b)-y_i'\big]^2$$

梯度下降实质上就是要求出当 $l(w,b)$ 的值最小时 w 和 b 的值。$l(w,b)$ 可以看成是一个二次函数,其中的 w 和 b 是变量。

可以先联想一下,函数 $y=x^2$,它的图象如图 1-10 所示,$l(w,b)$ 函数图象如图 1-11 所示。$l(w,b)$ 值可以看成是 z 轴的值。

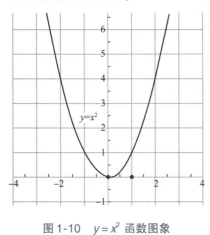

图 1-10　$y=x^2$ 函数图象　　　　　　　图 1-11　$l(w,b)$ 函数图象

梯度下降需要使用偏导数,求出 $l(w,b)$ 的梯度,根据梯度和学习率(步长)来确定下一个 (w,b) 的值。具体公式如下:

$$\frac{\partial l}{\partial b} = \frac{1}{n}\sum_{i=1}^{m}\left[(wx_i'+b)-y_i'\right]$$

$$= \frac{1}{m}\sum_{i=1}^{n}(\text{loss})$$

式中，loss 表示单个样本的损失，$\text{loss}=(wx_i'+b)-y_i'$。

$$\frac{\partial l}{\partial w} = \frac{1}{m}\sum_{i=1}^{n}\left[(wx_i'+b)-y_i'\right]x_i'$$

使用矩阵对上式做变换可得$\left[\boldsymbol{X}=(x_0,\cdots,x_n),\boldsymbol{y}=(y^1,\cdots,y^m)\right]$

$$\nabla J(w)=\frac{1}{n}\boldsymbol{X}^{\mathrm{T}}((\boldsymbol{X}w+b)-\boldsymbol{y})$$

$$\nabla J(w)=\frac{1}{n}\boldsymbol{X}^{\mathrm{T}}(\text{loss})$$

式中，$\boldsymbol{X}$ 表示数据集中的特征集合；$\boldsymbol{y}$ 表示是数据集中标签的集合。

　　使用代码实现梯度下降函数，传入 X 训练集、y 标签、w 权重、b 偏置值、alpha 学习率/步长、iter 迭代次数等参数，使用 l_history，w_history，b_history 列表保持每次迭代后的损失、权重和偏置值，并将结果返回，代码如下：

```
1.  def gradient_descent(X,y,w,b,lr,iter):
2.      l_history = np.zeros(iter)
3.      w_history = np.zeros(iter)
4.      b_history = np.zeros(iter)
5.      for i in range(iter):
6.          y_hat = w* X +b
7.          loss = y_hat - y
8.          derivative_w = X.T.dot(loss)/len(X)
9.          derivative_b = sum(loss)/len(X)
10.         w = w - derivative_w
11.         b = b - derivative_b
12.         l_history[i] = loss_function(X,y,w,b)
13.         w_history[i] = w
14.         b_history[i] = b
15.     return l_history,w_history,b_history
```

1.8 训练模型

模型训练首先给 iter 迭代次数、alpha 学习率/步长、weight 权重、hias 偏置值四个参数进行初始化,同时计算一个初始的损失值。

```
1.  #初始化参数
2.  iter = 100
3.  alpha = 0.5
4.  weight = -5
5.  bias = 3
6.  print("权重为 -5,偏置值为 3 时,损失函数为",loss_function(X_train,y_
    train,weight,bias))
```

代码运行后,输出权重为 −5,偏置值为 3 时,损失函数为 1.343 795 534 906 634。然后调用梯度下降函数开始训练。通过迭代得到 loss_history,weight_history,bias_history 的值,代码如下:

```
1.  #开始训练
2.  loss_history,weight_history,bias_history = gradient_descent(X_
    train,y_train,weight,bias,alpha,iterations)
```

代码运行后,打印输出 loss_history,因为设置的 iter = 100,所以输出 100 个 loss 结果,如图 1-12 所示。

```
array([1.16870167, 1.01830004, 0.88743937, 0.77347921, 0.67423079,
       0.5877945 , 0.5125164 , 0.44695606, 0.38985899, 0.34013266,
       0.29682557, 0.25910905, 0.22626142, 0.19765412, 0.17273979,
       0.15104168, 0.13214461, 0.11568699, 0.10135391, 0.0888711 ,
       0.07799972, 0.06853174, 0.06028599, 0.0531047 , 0.04685045,
       0.04140357, 0.03665984, 0.03252848, 0.02893044, 0.02579688,
       0.02306784, 0.02069109, 0.01862116, 0.01681844, 0.01524843,
       0.0138811 , 0.01269028, 0.01165319, 0.01074997, 0.00996335,
       0.00927828, 0.00868165, 0.00816203, 0.00770949, 0.00731538,
       0.00697214, 0.0066732 , 0.00641286, 0.00618613, 0.00598866,
       0.00581669, 0.00566692, 0.00553648, 0.00542288, 0.00532394,
       0.00523778, 0.00516274, 0.00509738, 0.00504046, 0.00499089,
       0.00494772, 0.00491013, 0.00487738, 0.00484887, 0.00482403,
       0.0048024 , 0.00478356, 0.00476716, 0.00475287, 0.00474043,
       0.00472959, 0.00472015, 0.00471193, 0.00470477, 0.00469854,
       0.00469311, 0.00468838, 0.00468426, 0.00468067, 0.00467755,
       0.00467483, 0.00467246, 0.0046704 , 0.0046686 , 0.00466703,
       0.00466567, 0.00466448, 0.00466345, 0.00466255, 0.00466177,
       0.00466108, 0.00466049, 0.00465997, 0.00465952, 0.00465913,
       0.00465878, 0.00465849, 0.00465823, 0.004658  , 0.0046578 ])
```

图 1-12　迭代损失结果

模型训练完毕得到 weight_history 和 bias_history 的集合,取出最后一个值,作为模型最优的 weight 和 bias 值,并绘制模型。

```
1.  #绘制最终得到的这个模型
2.  line_x = np.linspace(X_train.min(),X_train.max(),500)
3.  line_y = [weight_history[-1]* xx +bias_history[-1] for xx in line_x]
4.  sns.scatterplot(X_train,y_train)
5.  sns.scatterplot(line_x,line_y)
6.  plt.show()
```

代码运行后,结果如图 1-13 所示,图中点表示样本实际值,直线表示模型的预测值。

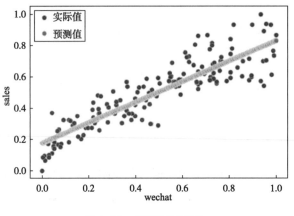

图 1-13　模型运行结果

weight_history[-1]值表示最后一次迭代后得到的 weight 值,这个值就是线性模型的权重 weight,同时也将 bias 输出,代码如下:

```
print("得到的参数","weight = ",weight_history[-1],"bias = ",bias_history
[-1])
```

代码运行结果:得到的参数 weight = 0.655 225 340 919 280 8,bias = 0.176 903 410 094 724 88,这两个值就是模型最优的 weight 和 bias 值。

1.9　简单线性模型损失

1. 绘制损失曲线

在模型训练过程中,iter = 100 是预先设定的参数,可以通过绘制损失曲线显示损失下降的过程来确定这个参数的范围。代码如下:

```
1.  line_x = np.linspace(0,iter,iter)
2.  sns.scatterplot(line_x,loss_history)
3.  plt.show()
```

代码运行后,输出结果如图 1-14 所示。图中横轴表示迭代的次数,纵轴表示损失。从图中可知,当迭代 40 次后损失的变化就很小了,所以迭代的参数可以设置为 40～50 之间的值。

2. 计算当前损失

根据得到的 weight 和 bias 可以计算出训练集的最终损失。代码如下:

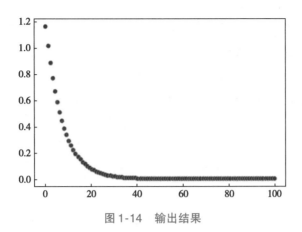

图 1-14　输出结果

```
1.  print('当前损失',loss_function(X_train,y_train,weight_history[-1],
    bias_history[-1]))
2.  print('当前权重',weight_history[-1])
3.  print('当前偏置',bias_history[-1])
```

代码运行后,输出:当前损失 0.004 657 804 055 314 04,当前权重 0.655 225 340 919 280 8,当前偏置 0.176 903 410 094 724 88。

3. 计算测试集的损失

根据得到的 weight 和 bias,使用 loss_function( )函数计算得到测试集的损失。代码如下:

```
1.  print('当前损失',loss_function(X_test,y_test,weight_history[-1],
    bias_history[-1]))
2.  print('当前权重',weight_history[-1])
3.  print('当前偏置',bias_history[-1])
```

代码运行后,输出:当前损失 0.004 581 809 380 247 21,当前权重 0.655 225 340 919 280 8,当前偏置 0.176 903 410 094 724 88。

1.10　绘制 *w*, *b* 的轮廓图形

在机器学习中,常使用轮廓图(contour plot)。这里使用轮廓图来表示偏置值的求解过程,也就是使用梯度下降函数 gradient_descent( )动态求得过程。使用 Matplotlib-Animation 生成一个动画,同时保存这个动画为一张 gif 图片。代码如下:

```
1.  # 设计 Contour Plot 动画
2.  import matplotlib.animation as animation
```

```
3.
4.   theta0_vals = np.linspace(-2, 3, 100)
5.   theta1_vals = np.linspace(-3, 3, 100)
6.   J_vals = np.zeros((theta0_vals.size, theta1_vals.size))
7.
8.   for t1, element in enumerate(theta0_vals):
9.       for t2, element2 in enumerate(theta1_vals):
10.
11.          weight = element
12.          bias = element2
13.          J_vals[t1, t2] = loss_function(X_train, y_train, weight, bias)
14.
15.  J_vals = J_vals.T
16.  A, B = np.meshgrid(theta0_vals, theta1_vals)
17.  C = J_vals
18.
19.  fig = plt.figure(figsize=(12,5))
20.  plt.subplot(121)
21.  plt.plot(X_train,y_train,'ro', label='Training data')
22.  plt.title('Sales Prediction')
23.  plt.axis([X_train.min()-X_train.std(),X_train.max()+X_train.std
     (),y_train.min()-y_train.std(),y_train.max()+y_train.std()])
24.  plt.grid(axis='both')
25.  plt.xlabel("WeChat Ads Volumn (X1) ")
26.  plt.ylabel("Sales Volumn (Y)")
27.  plt.legend(loc='lower right')
28.
29.  line, = plt.plot([], [], 'b-', label='Current Hypothesis')
30.  annotation = plt.text(-2, 3,'',fontsize=20,color='green')
31.  annotation.set_animated(True)
32.
33.  plt.subplot(122)
34.  cp = plt.contourf(A, B, C)
35.  plt.colorbar(cp)
```

```
36.  plt.title('Filled Contours Plot')
37.  plt.xlabel('Bias')
38.  plt.ylabel('Weight')
39.  track, = plt.plot([], [], 'r-')
40.  point, = plt.plot([], [], 'ro')
41.
42.  plt.tight_layout()
43.  plt.close()
44.
45.  def init():
46.      line.set_data([], [])
47.      track.set_data([], [])
48.      point.set_data([], [])
49.      annotation.set_text('')
50.      return line, track, point, annotation
51.
52.  def animate(i):
53.      fit1_X = np.linspace(X_train.min()-X_train.std(), X_train.max()+X_train.std(), 1000)
54.      fit1_y = b_history[i] + w_history[i]* fit1_X
55.
56.      fit2_X = b_history.T[:i]
57.      fit2_y = w_history.T[:i]
58.
59.      track.set_data(fit2_X, fit2_y)
60.      line.set_data(fit1_X, fit1_y)
61.      point.set_data(b_history.T[i], w_history.T[i])
62.
63.      annotation.set_text('Cost = % .4f' % (l_history[i]))
64.      return line, track, point, annotation
65.
66.  anim = animation.FuncAnimation(fig, animate, init_func=init,
67.                                frames=50, interval=0, blit=True)
68.
69.  anim.save('animation.gif', writer='imagemagick', fps=500)
```

加载显示 gif 图片代码：

```
1.  # 显示 Contour Plot 动画
```

```
2.   import io
3.   import base64
4.   from IPython.display import HTML
5.
6.   filename = 'animation.gif'
7.
8.   video = io.open(filename, 'r +b').read()
9.   encoded = base64.b64encode(video)
10.  HTML(data = '''< img src ="data:image/gif;base64,{0}" type ="gif" / >'
     ''.format(encoded.decode('ascii')))
```

设置 weight = −2，bias = −2 为初始化值，运行结果如图 1-15 所示。从图中可以看到，随着参数的拟合，损失越来越小，最终到达轮廓图的中心，也就是轮廓图颜色最深的部分，也就是最优解。最优点的状态如图 1-16 所示。

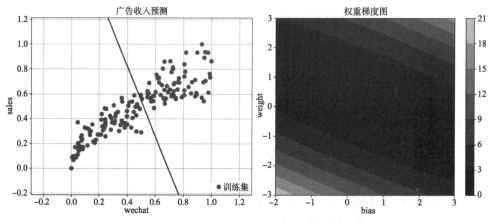

图 1-15　weight = −2，bias = −2 的初始状态图片

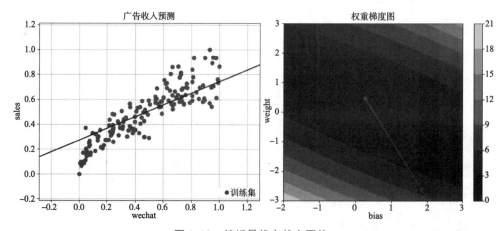

图 1-16　接近最优点状态图片

单元2

使用多元线性回归模型预测广告投入的收益

在单元1中使用微信的投入预测公司的预期收入,但是公司在微博和其他媒体中也投入了资金,现在需要预测在所有媒体上投入资金的收益,就需要使用多元线性回归模型。

2.1 多元线性回归介绍

多元线性回归,也就是回归模型中包含多个自变量,多元线性回归模型为

$$h(x) = w_1x_1 + w_2x_2 + w_3x_3 + w_4x_4 + w_5x_6 + \cdots + w_nx_n + b$$

式中,$x_1, x_2, x_3, \cdots, x_n$ 是自变量;b 是常数项。

在本单元中,有三个特征,"wechat"、"weibo"和"others",假设 $x_0 = 1$,就可得到 $w_0x_0 = w_0$,可以把 w_0 看成 b,所以可以使用如下的公式:

$$h(x) = w_0x_0 + w_1x_1 + w_2x_2 + w_3x_3$$

使用向量表示 $h(x)$,则权重可表示为 $\boldsymbol{\theta}^{\mathrm{T}}$,使用向量 $\boldsymbol{x}$ 表示特征,这个公式可以写成如下形式:

$$h_\theta(x) = \sum_{i=1}^{n} \theta_i x_i = \boldsymbol{\theta}^{\mathrm{T}}\boldsymbol{x}$$

2.2 读取数据集

1. 导入数据加载的包

在 Jupyter Notebook 中导入 numpy、pandas、matplotlib 三个包用来加载数据,同时对数据进行分析。代码如下:

```
1.  import pandas as pd
2.  import matplotlib.pyplot as plt
3.  import numpy as np
```

2.加载数据

使用 pandas 的 read_csv 方法打开数据集,使用 head( )函数查看前五条数据,同时可以使用 info( )函数查看数据的信息,代码如下:

```
1.  df_data =pd.read_csv("../data/advertising.csv")
2.  df_data.head()
```

2.3 生成训练集和测试集

1.选取数据

首先,保留所有的特征,然后使用 numpy 的 delete 删除标签字段"sales",得到特征集,单独取出"sales"得到标签。然后使用 np. array( )和 reshape( )函数将数据转为张量。

```
1.  X_data =np.array(df_data)
2.  X_data =np.delete(X_data,[3],axis =1)
3.  y_data =np.array(df_data.sales)
4.  #"the length of (% s) is % d" % ('runoob',len('runoob'))
5.  print("tensor X_data shape is :% s,dimension is % d"% (X_data.shape,
    X_data.ndim))
6.  print("tensor y_data shape is :% s,dimension is % d"% (y_data.shape,
    y_data.ndim))
7.  # print("张量 X_data 的形状是:% s,维度是:% d"% (X_data.shape,X_data.
    ndim))
8.  # print("张量 y_data 的形状是:% s,维度是:% d"% (y_data.shape,y_data.
    ndim))
```

代码运行输出结果为 tensor X_data shape is :(200,3),dimension is 2;tensor y_data shape is :(200,),dimension is 1,可以得到 X_data、y_data 转化为 tensor 后的形状。

这时的 y_data 的形状为向量,可以使用 reshape( )函数将向量变化为矩阵,也就是2D 张量。

```
1.  y_data =y_data.reshape(-1,1)
2.  print("tensor y_data shape is :% s,dimension is % d"% (y_data.shape,
    y_data.ndim))
```

代码运行输出结果为 tensor y_data shape is :(200,1),dimension is 2。

2. 拆分训练集和测试集

使用 sklearn 的 model_selection 函数对 x_data 和 y_data 按照 80% 的训练集 20% 的测试集进行拆分得到训练集和测试集。

```
1.  from sklearn.model_selection import train_test_split
2.  X_train,X_test,y_train,y_test =train_test_split(X_data,y_data,test_
    size =0.2,random_state =0)
```

2.4 反归一化

在机器学习中,模型训练完毕需要对测试集数据进行预测,这时就需要使用反归一化操作。反归一化函数用于数据的复原。还需要保存 y_min、y_max、y_gap,计算训练集最大、最小值以及它们的差,用于后面反归一化过程,同时保留一份原始的数据。代码如下:

```
1.  #反归一化
2.  def min_max_gap(train):
    #计算训练集最大、最小值以及它们的差,用于后面反归一化过程
3.      min = train.min(axis =0) #训练集最小值
4.      max = train.max(axis =0) #训练集最大值
5.      gap = max - min #最大值和最小值的差
6.      return min, max, gap
7.
8.  y_min, y_max, y_gap = min_max_gap(y_train)
```

保留原始数据,代码如下:

```
1.  #保留一份原始数据
2.  X_data_original =X_train.copy()
```

使用归一化函数对数据进行处理,代码如下:

```
1.  X_train,X_test = scaler(X_train,X_test) #对特征归一化
2.  y_train,y_test = scaler(y_train,y_test) #对标签也归一化
```

添加一个 x_0 的值,x_0 的值全部为 1,代码如下:

```
1.  _train0 =np.ones((len(X_train),1))
2.  print("张量 X_train 的形状是:% s,维度是:% d"% (X_train.shape,X_train.
    ndim))
```

```
3.  X_train = np.append(X_train0,X_train,axis =1)
4.  X_test0 = np.ones((len(X_test),1))
5.  X_test = np.append(X_test0,X_test,axis =1)
6.  print("张量 X_data 的形状是:% s,维度是:% d"% (X_train.shape,X_train.
    ndim))
```

代码运行后,输出张量 X_train 的形状是:(160, 3),维度是:2;张量 X_data 的形状是:(160, 4),维度是:2。

2.5 定义损失函数、MSE 均方差损失函数

多元线性回归和简单线性回归一样,使用 MSE 作为损失函数,公式如下:

$$\mathrm{MSE} = J(\theta) = \frac{1}{2m}\sum_{i=1}^{m}(h_\theta(x^{(i)}) - y^{(i)})^2$$

通常使用矩阵的方式来表示:

$$\mathrm{MSE} = J(\theta) = \frac{1}{2m}(X\theta^{\mathrm{T}} - y)^2$$

式中,$J(\theta)$ 表示损失函数;$h_\theta(x^{(i)}) = y' = X\theta^{\mathrm{T}}$ 表示模型的预测值;y 表示标签值。

可以使用向量内积来计算。在 Python 中使用 np. dot 来实现这个操作,X. dot(θ. T),T 的作用是把 θ 的 shape 由(1,4)变成(4,1),然后再减去对应的标签值 y,就可以最终得到 $h_\theta(x^{(i)}) - y^{(i)}$。

例如,当特征是一个 3×3 的矩阵,标签是一个 3×1 的矩阵时,计算 $J(\theta)$ 的值就可以使用下式:

$$y' - y = \begin{pmatrix} x_0^{(1)}\theta_0 + x_1^{(1)}\theta_1 + x_2^{(1)}\theta_2 \\ x_0^{(2)}\theta_0 + x_1^{(2)}\theta_1 + x_2^{(2)}\theta_2 \\ x_0^{(3)}\theta_0 + x_1^{(3)}\theta_1 + x_2^{(3)}\theta_2 \end{pmatrix} - \begin{pmatrix} y^{(1)} \\ y^{(2)} \\ y^{(3)} \end{pmatrix}$$

这时得到 $(y' - y)$ 的 shape 为 (len(x),1),然后计算 $(y' - y)^2$,最后使用 np. sum 对矩阵中的值进行求和,最终得到损失值。代码如下:

```
1.  def loss_function(X,y,w):
2.      y_hat = X.dot(w.T)
3.      loss = y_hat.reshape((len(y_hat)),1) - y
4.      cost = np.sum(loss* * 2)/(2* len(X))
5.      return cost
```

2.6 定义梯度下降函数

前面定义了损失函数：

$$J(\theta) = \frac{1}{2m} \sum_{i=1}^{m} (h_\theta(x^{(i)}) - y^{(i)})^2$$

式中，$h_\theta(x) = \theta_0 x_0 + \theta_1 x_1 + \cdots + \theta_{n-1} x_{n-1} + \theta_n x_n$ 是一个多元函数。按照梯度的求解法则，多元函数的梯度$\nabla j(\theta)$是一个由偏导数组成的列向量。

$$\nabla j(\theta) = \begin{pmatrix} \dfrac{\partial j(\theta)}{\partial \theta_0} \\[2mm] \dfrac{\partial j(\theta)}{\partial \theta_1} \\[2mm] \dfrac{\partial j(\theta)}{\partial \theta_2} \\[1mm] \vdots \\[1mm] \dfrac{\partial j(\theta)}{\partial \theta_n} \end{pmatrix}$$

按照链式求导法则计算偏导可得

$$\frac{\partial j(\theta)}{\partial \theta_0} = \frac{1}{m} \sum_{i=1}^{m} (h_\theta(x^{(i)}) - y^{(i)}) x_0^{(i)}$$

$$\frac{\partial j(\theta)}{\partial \theta_1} = \frac{1}{m} \sum_{i=1}^{m} (h_\theta(x^{(i)}) - y^{(i)}) x_1^{(i)}$$

$$\frac{\partial j(\theta)}{\partial j} = \frac{1}{m} \sum_{i=1}^{m} (h_\theta(x^{(i)}) - y^{(i)}) x_j^{(i)}$$

因为$(h_\theta(x^{(i)}) - y^{(i)}) = X\theta - y$，又因为可以将$x^{(i)}$部分转化为一个$1 \times m$的行向量（$m$为数据个数），那么上式可以写成

$$\frac{\partial j(\theta)}{\partial j} = \frac{1}{m} \sum_{i=1}^{m} (h_\theta(x^{(i)}) - y^{(i)}) x_j^{(i)} = x_j^{\mathrm{T}} (X\theta - y)$$

那么

$$\nabla j(\boldsymbol{\theta}) = \begin{pmatrix} \dfrac{\partial j(\theta)}{\partial \theta_0} \\ \dfrac{\partial j(\theta)}{\partial \theta_1} \\ \dfrac{\partial j(\theta)}{\partial \theta_2} \\ \vdots \\ \dfrac{\partial j(\theta)}{\partial \theta_n} \end{pmatrix} = \begin{pmatrix} \boldsymbol{x}_1^{\mathrm{T}}(\boldsymbol{X\theta}-\boldsymbol{y}) \\ \boldsymbol{x}_2^{\mathrm{T}}(\boldsymbol{X\theta}-\boldsymbol{y}) \\ \boldsymbol{x}_3^{\mathrm{T}}(\boldsymbol{X\theta}-\boldsymbol{y}) \\ \vdots \\ \boldsymbol{x}_n^{\mathrm{T}}(\boldsymbol{X\theta}-\boldsymbol{y}) \end{pmatrix} = \begin{pmatrix} \boldsymbol{x}_1^{\mathrm{T}} \\ \boldsymbol{x}_2^{\mathrm{T}} \\ \boldsymbol{x}_3^{\mathrm{T}} \\ \vdots \\ \boldsymbol{x}_n^{\mathrm{T}} \end{pmatrix} \times (\boldsymbol{X\theta}-\boldsymbol{y})$$

$$= (\boldsymbol{x}_1,\boldsymbol{x}_2,\cdots,\boldsymbol{x}_n)^{\mathrm{T}}(\boldsymbol{X\theta}-\boldsymbol{y})$$

式中，$(\boldsymbol{x}_1,\boldsymbol{x}_2,\cdots,\boldsymbol{x}_n)$ 为 $m \times n$ 的矩阵，$(\boldsymbol{X\theta}-\boldsymbol{y})$ 为 $m \times 1$ 的列向量。所以，多元线性模型的梯度函数为

$$\nabla j(\boldsymbol{\theta}) = \frac{1}{2m}\boldsymbol{X}^{\mathrm{T}}(\boldsymbol{X\theta}-\boldsymbol{y})$$

在这个公式中，可以分步实现：

第一步：使用 numpy 的 dot 方法实现 $(wx_i'+b)$。代码如下：

```
y_hat = X.dot(w)
```

在这里 X 的形状 $(160,4)$，w 的形状是 $(4,)$，进行 dot 运算后 y_hat 的形状是 $(160,)$。

第二步：实现 $(wx_i'+b)-y_i'$。代码如下：

```
loss =y_hat.reshape((len(y_hat),1)) - y
```

对 y_hat.reshape 后，y_hat 的形状是 $(160,1)$，y 的形状是 $(160,1)$，然后再和 y 进行减法得到 loss 的形状是 $(160,1)$。

第三步：实现 $\frac{1}{2m}\sum_{i=1}^{m}[(wx_i'+b)-y_i']x_i'$。代码如下：

```
derivative =X.T.dot(loss)/2(len(X))
```

X 的形状为 $(160,4)$，对 X 做矩阵转置操作，X 的形状变为 $(4,160)$，然后和 loss$(160,1)$ 进行 dot 操作，可以得到一个形状为 $(4,1)$ 的矩阵，这个矩阵就是求的梯度矩阵（偏导矩阵）。

第四步：derivative_w = derivative_w. reshape$(len(w))$。

进行 derivative_w. reshape(4) 操作，derivative_w 的形状变为 $(4,)$，然后进行 w = w − lr * derivative_w 得到更新后的 w 值。梯度下降函数定义代码如下：

```
1.  def gradient_descent(X,y,w,lr,iter):
2.      l_history =np.zeros(iter)
3.      w_history =np.zeros((iter,len(w)))
```

```
4.      for iter in range(iter):
5.          y_hat = X.dot(w)
6.          loss = y_hat.reshape((len(y_hat),1)) - y
7.          derivative_w = X.T.dot(loss)/(2* len(X))
8.          derivative_a = derivative_w
9.          derivative_w = derivative_w.reshape(len(w))
10.         w = w - lr* derivative_w
11.         l_history[iter] = loss_function(X,y,w)
12.         w_history[iter] = w
13.         #print(X.shape,y_hat.shape,loss.shape,derivative_a.shape)
14.     return l_history,w_history
```

2.7 定义模型并训练

1. 初始化参数

对 iter、lr、weight 给一个初始值,同时计算当前值的损失值。代码如下:

```
1.  iter = 300
2.  lr = 0.15
3.  weight = np.array([0.5,1,1,1]) # weight[0] = bias
4.  print ('当前损失:',loss_function(X_train, y_train, weight))
5.  print(weight.shape)
```

代码运行后,输出当前损失: 0.803 918 373 360 485 7。

2. 定义线性模型

它主要包括两部分内容,调用梯度下降函数开始迭代运算,同时计算准确率。代码如下:

```
1.  def linear_regression(X,y,weight,lr,iter):
2.      loss_history,weight_history = gradient_descent(X,y,weight,lr,iter)
3.      print("训练最终损失:",loss_history[-1])
4.      y_pred = X.dot(weight_history[-1])
5.      training_acc = 100 - np.mean(np.abs(y_pred - y)) * 100
6.      print("线性回归训练准确率:{:.2f}% ".format(training_acc))
7.      return loss_history,weight_history
8.
9.  loss_history, weight_history = linear_regression(X_train, y_train,
    weight,lr,iter)
```

先计算预测值 y_pred 和实际值 y 的差值，并使用 np. abs( )求绝对值，然后使用 np. mean( )求全部距离的平均值，这个值是 y_pred 和实际值 y 的"距离"，可以通过"距离"的远近作为模型的准确率。

3. 训练模型

调用模型函数并运行。代码如下：

```
loss_history, weight_history = linear_regression (X_train, y_train,
weight, lr, iter)
```

代码运行后，输出训练最终损失：0.004 334 018 335 12，线性回归训练准确率为 74.52%。

4. 使用模型预测

需要对预测到的结果进行反向归一化操作。代码如下：

```
1.  print(y_min, y_max, y_gap)
2.  X_plan = [250, 50, 50]
3.  X_train, X_plan = scaler(X_data_original, X_plan)
4.  X_plan = np.append([1], X_plan)
5.  y_plan = np.dot(weight_history[-1], X_plan)
6.  y_value = y_plan* y_gap + y_min
7.  print ("预计商品销售额：", y_value, "千元")
```

代码运行后，输出预计商品销售额：[6.08890958] 千元。

2.8　计算损失

1. 绘制损失曲线，并输出线性方程

使用 np. linspace( )函数，生成 x 坐标点，y 坐标为 loss_history。代码如下：

```
1.  line_x = np.linspace(0, iter, iter)
2.  sns.scatterplot(line_x, loss_history)
3.  plt.xlabel("迭代次数")
4.  plt.ylabel("损失")
5.  plt.show()
```

代码运行后，结果如图 2-1 所示，可以看到 iter 的值大于 25 时，损失的值基本达到了一个极小值。

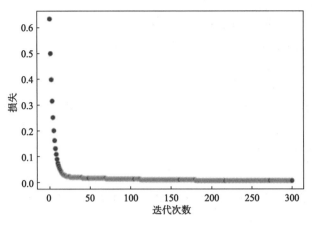

图 2-1　损失曲线

模型训练完毕后,可以将训练好的 θ 值,也就是 weight 输出,代码如下:

```
1.  print(weight_history[-1])
2.  print("预测线性模型为:y =% .2f * x1 +% .2f * x2 +% .2f * x3% .2f"%
    (weight_history[-1][1],weight_history[-1][2], weight_history[-1]
    [3],weight_history[-1][0]))
```

代码运行后,输出 weight 的值为[−0.041 612 05　0.652 300 9　0.246 867 67　0.377 415 12],同时输出预测线性模型为 y = 0.65 * x1 + 0.25 * x2 + 0.38 * x3 − 0.04。

2. 绘制拟合曲线

得到模型后可以进行预测。为了更加直观地观察模型的效果,可以将预测值和实际值绘制到同一张图中进行直观对比。首先获取 X_train 的长度并输出,然后使用 np. ones 生成模型中 $\theta_0 x_0$ 的权重 θ_0,然后将 θ_0 和 X_train 进行拼接,最终得到训练集数据。代码如下:

```
1.  x = np.arange(len(X_train))
2.  print(x.shape)
3.  y_train = y_train.reshape(len(y_train))
4.  X_train0 = np.ones((len(X_train),1))
5.  X_train = np.append(X_train0,X_train,axis =1)
```

使用 weight_history[−1]参数,也就是最终的模型权重系数,然后使用 dot 计算 y_pred,得到模型的预测值。代码如下:

```
1.  y_pred = X_train.dot(weight_history[-1])
2.  y_pred
```

最后使用 lineplot 绘制拟合曲线,将标签值(y)和预测值(y_pred)在同一张图中输出。代码如下:

```
1.  print(y_train.shape,y_pred.shape)
2.  data=np.append(y_train,y_pred,axis=0)
3.  data=data.reshape(160,2)
4.  print(data.shape)
5.  w_d=pd.DataFrame(data,x,["实际值","预测值"])
6.  plt.figure(figsize=(15,5))
7.  sns.lineplot(data=w_d)
8.  plt.show()
```

代码运行后，结果如图2-2所示。可以看出预测值和实际值还是有一定的差值。如果要减小误差，可以采用增加高次项等方法。

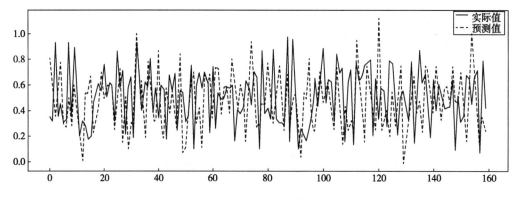

图2-2　拟合曲线

单元3

使用Scikit-learn库来实现回归

前面的例子中使用简单线性回归和多元线性回归预测广告投入时,在 Jupyter Notebook 中编写了模型的代码、损失函数、梯度下降函数等。在实际项目中,通常借助开源的 Scikit-learn 库来完成上述工作。本单元主要讲述如何使用 Scikit-learn 库来搭建模型完成广告投入的预测。

3.1 Scikit-learn 基本知识

Scikit-learn(曾称 scikits learn,简称 sklearn)是针对 Python 编程语言的免费软件机器学习库,是一个 Python 的机器学习项目,是一个简单高效的数据挖掘和数据分析工具。基于 NumPy、SciPy 和 matplotlib 构建。它具有各种分类、回归和聚类、降维、模型选择、预处理等功能, 它是开源的,可商业使用。

3.2 加载数据集

1. 导入数据加载的包

在 Jupyter Notebook 中导入 numpy、pandas、matplotlib 三个包用来加载数据,同时对数据进行分析,代码如下:

```
1.  #导入数据处理库
2.  import numpy as np
3.  import pandas as pd
4.  from matplotlib import font_manager as fm, rcParams
5.  import matplotlib.pyplot as plt
6.  from sklearn.model_selection import train_test_split
7.  from sklearn import datasets
8.  plt.rcParams['font.sans-serif'] = ['SimHei']
9.  plt.rcParams['axes.unicode_minus'] = False
```

2. 数据集可视化展示

seaborn 中的 distplot 主要功能是绘制单变量的直方图，还可以在直方图的基础上增加 kdeplot 和 rugplot 的部分内容（直方图＋核密度估计），是一个功能非常强大且实用的函数。下面的代码演示了如何使用 seaborn 绘制一个 distplot 图像。

```
1.  import seaborn as sns
2.  f, ax = plt.subplots(1, 1)
3.  sns.distplot(data['weibo'],color = 'r',label = "weibo",ax = ax)
4.  sns.distplot(data['wechat'],color = 'b',label = "wechat",ax = ax)
5.  sns.distplot(data['others'],color = 'g',label = "others",ax = ax)
6.  plt.ylabel('核密度')
7.  plt.xlabel('广告投入')
```

代码运行后绘制了 weibo（虚线）、wechat（实线）、others（点画线）的核密度函数，可以看出微信和微博都存在左偏态分布，如图 3-1 所示。

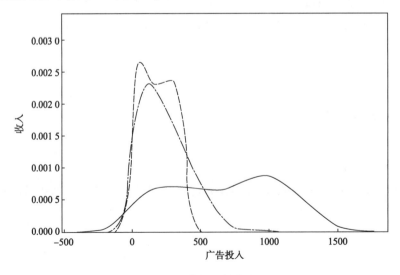

图 3-1　广告投入核密度图

同时也使用了 pairplot，对数据进行可视化展示。pairplot 中 pair 是成对的意思，pairplot 主要展现的是变量两两之间的关系(线性或非线性，有无较为明显的相关关系)，代码如下：

```
1.  #显示销量和各种广告投放量的散点图
2.  sns.pairplot(data, x_vars = ['wechat', 'weibo', 'others'],
3.                             y_vars = 'sales',
4.                             height =4, aspect =1, kind = 'scatter')
5.  plt.show()
```

3.3　数据归一化

1. Z-score 标准化

Z-score 标准化是数据处理的一种常用方法，这种方法使用原始数据的均值 x_{mean} 和标准差 σ (standard deviation)对数据做标准化。经过处理的数据符合标准正态分布，即均值为 0，标准差为 1。转换函数为

$$x_{scale} = \frac{x - x_{mean}}{\sigma}$$

式中，x_{mean} 为所有样本数据的均值；σ 是所有样本数据的标准差；x_{scale} 是标准化后的值。

标准化的代码如下：

```
1.  #特征缩放 (x -data.mean())/data.std()
2.  data = (data -data.mean())/data.std()
3.  #查看缩放后的数据
4.  data.head(10)
```

代码运行后，输出数据标准归一化后的值，如图 3-2 所示。

2.归一化、标准化、中心化

在机器学习中，通常会使用到归一化、标准化、中心化对数据进行预处理。

(1)归一化：把数据变成(0,1)或者(-1,1)之间的小数，主要是为了数据处理方便。把数据映射到 0~1 范围之内处理，更加便捷快速。归一化是一种简化计算的方式，即将有量纲的表达式，经过变换，化为无量纲的表达式，成为纯量，便于不同单位或量级的指标能够进行比较和加权。

	wechat	weibo	others	sales
0	-0.933464	-0.778888	0.286795	-1.027746
1	0.877676	-1.277311	0.883711	0.297035
2	1.080421	-1.290782	0.291387	0.410587
3	-1.493829	-0.105343	0.911261	-1.538733
4	0.967529	0.979066	1.774493	1.319009
5	0.695922	-1.216692	-0.512154	0.391662
6	1.556821	-0.630708	0.295978	0.183482
7	-1.614913	0.265107	-1.306511	-1.784764
8	-0.835675	-0.199639	0.089354	-0.724939
9	-0.822876	-1.513052	-0.723370	-1.084522

图 3-2　数据归一化后的结果

（2）标准化：在机器学习中，需要处理不同种类的资料，对不同特征维度的数据做伸缩变换（数据按比例缩放，使之落入一个小的特定区间），其目的是使不同度量之间的特征具有可比性。并且同时不改变原始数据的分布状态。转化为无量纲的纯数值后，便于不同单位或量级的指标能够进行比较和加权。

标准化被广泛使用在许多机器学习算法中（例如：支持向量机、逻辑回归和类神经网络）。

基本原理：样本值减去样本平均值，再除以其样本标准差，得到均值为0，标准差为1的服从标准正态分布的数据。

举例：根据人的身高和体重预测人的健康指数。假设有如下数据，见表3-1。

表3-1　身高-年龄-体重数据

样本	身高/m	年龄	体重/g
样本1	1.7	20	70 000
样本2	1.8	18	80 000
样本3	1.6	13	60 000

例中身高、年龄、体重三个权重是一样的，在做特征分析的时候会明显发现体重对计算结果的影响是最大的。因此，使用标准化的方法将数值处理到0~1的范围内。Python中提供了standardscaler类可以完成数据样本的标准化操作。

（3）中心化：平均值为0，对标准差无要求。

归一化和标准化的区别：归一化是将样本的特征值转换到同一量纲下，把数据映射到（0,1）或者（-1,1）区间内。标准化是依照特征矩阵的列处理数据，其通过求标准化计算方法，将数据转换为标准正态分布。标准化和整体样本分布相关，每个样本点都能对标准化产生影响。它们的相同点在于都能取消由于量纲不同引起的误差；都是一种线性变换，都是对向量 X 按照比例压缩再进行平移。

标准化和中心化的区别：标准化是原始数据减去平均数，然后除以标准差。中心化是原始数据减去平均数。所以，一般流程为先中心化再标准化。

3. 什么时候用归一化？什么时候用标准化？

如果对输出结果范围有要求，用归一化；如果数据较为稳定，不存在极端的最大最小值，用归一化；如果数据存在异常值和较多噪声，用标准化，可以间接通过中心化避免异常值和极端值的影响。

3.4　数据预处理

1. 选取数据

这里使用 iloc 来选取数据。iloc 通过使用索引来取值（loc 通过列名来取值，前后都要取值）。

这里用到了切片操作,是一种前闭后开的操作。

首先使用 shape[1],返回列的数量;然后使用 data. iloc[:,0:cols − 1],取前 cols − 1 列,即输入向量。data. iloc[:,cols − 1:cols] ,取最后一列,即目标变量也即标签 y 值,使用 iloc 函数截取数据的代码如下:

```
1.  #变量初始化
2.  #最后一列为 y,其他为 x
3.  cols = data.shape[1]
4.  X = data.iloc[:,0:cols −1]      #取前 cols −1 列,即输入向量
5.  y = data.iloc[:,cols −1:cols]    #取最后一列,即目标变量
6.  X.head(10)
```

代码运行后,输出使用 iloc 截取的数据。Python 中还有一个函数 loc 也可用来选取行和列的数据,它和 iloc 的区别如下:

loc 函数通过行索引 Index 中的具体值来取行数据(如取 Index 为 A 的行);iloc 函数通过行号来取行数据(如取第二行的数据)。iloc 根据标签的所在位置,从 0 开始计数,先选取行再选取列;loc 根据 DataFrame 的具体标签选取行列,同样是先行标签,后列标签。

2. 划分训练集和测试集

使用 sklearn 的 model_selection 函数对 x_data 和 y_data 按照 80% 的训练集 20% 的测试集进行拆分得到训练集和测试集,划分训练集和测试集代码如下:

```
1.  # 划分训练集和测试集
2.  X_train,X_test,y_train,y_test = train_test_split(X,y,test_size =0.2)
```

使用 np. matrix( )函数将数据转换为 numpy 的矩阵,转换代码如下:

```
1.  # 将数据转换成 numpy 矩阵
2.  X_train =np.matrix(X_train.values)
3.  y_train =np.matrix(y_train.values)
4.  X_test =np.matrix(X_test.values)
5.  y_test =np.matrix(y_test.values)
6.  theta =np.matrix([0,0,0,0])
7.  X_train.shape,X_test.shape,y_train.shape,y_test.shape
```

3.5 使用 sklearn 建立 LinearRegression 模型

1. 导入 sklearn 库使用 LinearRegression 来进行线性回归

导入 sklearn 中的线性模型 linear_model 库,linear_model 包含了多种多样的类和函数,Linear-

Regression 是 linear_model 库中用来完成普通线性回归的模型。

线性回归的性能往往取决于数据本身,而并非调参能力,线性回归也因此对数据有着很高的要求。现实中,大部分连续型变量之间,都存在着或多或少的线性联系。所以,线性回归虽然简单,功能却很强大。sklearn 中的线性回归可以处理多标签问题,只需要在训练模型的时候输入多维度标签即可,导入 sklearn 库建立线性回归模型的代码如下:

```
1.  from sklearn.linear_model import LinearRegression
2.  reg = LinearRegression()
3.  model = reg.fit(X_train, y_train)
4.  print(model)
5.  print(reg.coef_)
6.  print(reg.intercept_)
```

代码运行后,输出使用 sklearn 建立的 LinearRegression 的信息,同时输出了 coef_,也就是模型的 weight 值和 intercept_ 即 bias 值,如图 3-3 所示。

```
LinearRegression(copy_X=True, fit_intercept=True, n_jobs=None, normalize=False)
[[0.88775328 0.30866359 0.00554074]]
[0.01322183]
```

图 3-3　代码运行结果

2. linear_model. LinearRegression( ) 的用法

LinearRegression( ) 用来建立一个线性模型,函数及参数代码如下:

```
1.  LinearRegression(
2.      fit_intercept = True,
3.      normalize = False,
4.      copy_X = True,
5.      n_jobs = None,
6.  )
```

LinearRegression( ) 函数主要参数及用法如下:

fit_intercept:boolean,optional,default True。是否计算截距,默认为计算。如果使用中心化的数据,可以考虑设置为 False,不考虑截距。注意这里是考虑,一般要考虑截距。

normalize:boolean,optional,default False。标准化开关,默认关闭;该参数在 fit_intercept 设置为 False 时自动忽略。如果为 True,回归会标准化输入参数。这里建议将标准化工作放在训练模型之前。若为 False,在训练模型前,可使用 sklearn. preprocessing. StandardScaler 进行标准化处理。

copy_X:boolean,optional,default True。默认为 True, 否则 X 会被改写。

n_jobs:int,默认为 1。当为 −1 时,默认使用全部 CPU。

LinearRegression( ) 函数返回值及属性如下:

coef_ 数组型变量,形状为(n_features,)或(n_targets, n_features)。在本例中,返回的是权重系数($w_1 - w_n$)。对于线性回归问题,计算得到的是 feature 的系数。如果输入的是多目标问题,则返回一个二维数组(n_targets, n_features);如果是单目标问题,返回一个一维数组 (n_features,)。

intercept_ 数组型变量,本例中返回的就是 bias/w_0 的值,但是如果在 LinearRegression 设置 fit_intercept = False,则 intercept_为 0.0。

LinearRegression()函数的使用方法:

fit(self, X, y[,sample_weight])方法,训练模型,sample_weight 为每个样本权重值,默认 None。

predict(self, X)方法,模型预测,返回预测值。

score(self, X, y[, sample_weight])方法,评估模型并返回模型的评估分数,最优值为 1,说明所有数据都预测正确。

3.6　LinearRegression 模型预测及误差

1. 使用 LinearRegression 的 predict( )函数预测 y_test 值

定义线性回归模型后,调用 predict()函数预测 y_test 值,然后使用 mean_squared_error()函数计算均方误差 MSE,计算 MSE 代码如下:

```
1.  from sklearn.metrics import mean_squared_error
2.  y_hat = reg.predict(X_test)
3.  mean_squared_error(y_test,y_hat)
```

2. 线性回归算法评价指标

线性回归算法评价指标有 MSE、RMSE、MAE、R2_score。指标含义如下:

(1)均方误差 MSE(mean squared error)

举例:假设建立了两个模型,一个模型使用了 10 个样本,计算出的损失值是 100;另一个模型使用了 50 个样本,计算出的损失值是 200,很难判断两个模型哪个更好。那怎么办呢? 其实很好解决,将损失值除以样本数量就可以去除样本数量的影响。

针对上面举例的两个模型的 MSE 分别是 10(100/10)和 4(200/50),所以后者效果更好。

(2) 均方根误差 RMSE(root mean squared error)

MSE 公式(见 1.6 节)有一个问题,因为公式做了平方,所以会改变量纲。例如,y 值的单位是万元,MSE 计算出来的是万元的平方,对于这个值难以解释它的含义。所以为了消除量纲的影响,可以对这个 MSE 开方。可以看到,MSE 和 RMSE 二者是呈正相关的,MSE 值大,RMSE 值也大。

RMSE 计算公式如下:

$$RMSE = \sqrt{\frac{1}{2n} \sum_{i=1}^{n} (y_i - y_i')^2}$$

式中，y_i 表示实际值；y_i' 表示预测值；n 表示样本数量。

例如：要做房价预测，每平方米售价是万元，预测结果也是万元。那么差值的平方单位应该是千万级别。这时不太好描述所做模型效果。于是将结果开方。这时误差的结果就和数据是一个级别，在描述模型的时候，就说模型的误差是多少万元。

（3）平均绝对误差 MAE（Mean Absolute Error）

在 MSE 和 RMSE 中，为了避免误差出现正负抵消的情况，采用计算差值的平方。还有一种公式也可以起到同样效果，就是计算差值的绝对值。这时可以得到平均绝对误差 MAE。MAE 计算公式如下：

$$MAE = \frac{1}{n} \sum_{i=1}^{n} |(y_i - y_i')|$$

式中，y_i 表示实际值；y_i' 表示预测值；n 表示样本数量。

上面三个方法消除了样本数量 m 和量纲的影响。但是它们都存在一个相同的问题：当量纲不同时，难以衡量模型效果好坏。

举例：模型在一份房价数据集上预测得到的误差 RMSE 是 5 万元，在另一份学生成绩数据集上得到的误差是 10 分。凭这两个值，很难知道模型到底在哪个数据集上效果好。

（4）线性回归决定系数 R 方值（R2_score）

决定系数（coefficient of determination）又称判定系数、拟合优度。它是表征回归方程在多大程度上解释了因变量的变化，或者说方程对观测值的拟合程度如何。

如果单纯用残差平方和会受到因变量和自变量绝对值大小的影响，不利于在不同模型之间进行相对比较，而用拟合优度就可以解决这个问题。例如：一个模型中的因变量为 10 000、20 000……，而另一个模型中因变量为 1、2……，这两个模型中第一个模型的残差平方和可能会很大，而第二个会很小，但是这不能说明第一个模型就比第二个模型差，这时可以使用 R2_score 来判定模型的好坏。R2_score 计算公式如下：

$$R2\_score = 1 - \frac{MSE(\hat{y}, y)}{Var(y)}$$

式中，分子是均方误差，分母是方差，都能直接计算得到，从而能快速计算出 R2_score 值。R2_score 的取值可以分为三种情况：

R2_score = 1，达到最大值。即分子为 0，意味着样本中预测值和真实值完全相等，没有任何误差。也就是说，建立的模型完美拟合了所有真实数据，是效果最好的模型。但通常模型不会这么完美，总会有误差存在，当误差很小的时候，分子小于分母，模型会趋近 1，仍然是好的模型，随着误差越来越大，R2_score 也会离最大值 1 越来越远，直到出现第二种情况。

R2_score = 0。此时分子等于分母，样本的每项预测值都等于均值。也就是说，训练出来的模型和前面说的均值模型完全一样。当误差越来越大的时候就出现了第三种情况。

R2_score <0。分子大于分母,训练模型产生的误差比使用均值产生的还要大,也就是训练模型反而不如直接取均值效果好。出现这种情况,通常是模型本身不是线性关系的,而误使用了线性模型,导致误差很大。

可以直接使用 sklearn 中的函数计算模型的 MSE、MAE、R2 的值,代码如下:

```
1.  from sklearn.metrics import mean_squared_error
2.  from sklearn.metrics import mean_absolute_error
3.  from sklearn.metrics import r2_score
4.  y_hat = reg.predict(X_test)
5.  print(mean_squared_error(y_test,y_hat))#均方误差
6.  print(mean_absolute_error(y_test,y_hat))#平方误差
7.  print(r2_score(y_test,y_hat))#R2 值
```

程序运行后输出 MSE、MAE、R2 的值分别为 0.085 430 660 95、0.226 210 669 0、0.903 921 966 4。

3.7　绘制拟合曲线

取出 X_test 的值,利用模型预测得到 y_hat,在同一个 plt 中绘制 y_hat 和 y_test,直观地显示预测值和实际值之间的关系。

这里使用了 lineplot() 函数来绘制折线图,先生成 x 轴的数据,然后将 y_hat 和 y_test 的数据进行拼接,最后将拼接后的数据转换为 dataframe,拼接 dataframe 的代码如下:

```
1.  # y = y_test.reshape((40,1)).ravel()
2.  x = np.arange(len(X_test))#生成 x 轴数据
3.  print(y_test.shape,y_hat.shape)
4.  df = np.append(y_test,y_hat,axis =1)#拼接生成 y 轴数据
5.  df = df.reshape(40,2)
6.  print(df.shape)
7.  w_d = pd.DataFrame(df,x,["实际值","预测值"])#生成 dataframe 数据
8.  plt.figure(figsize = (15,5))#设置图像大小
9.  sns.lineplot(data =w_d)#绘制折线图
10. plt.show()
```

代码运行后,绘制出了预测值和实际值,如图 3-4 所示,可以从图中看出,预测值和实际值重合度较高,模型的精确度较高。

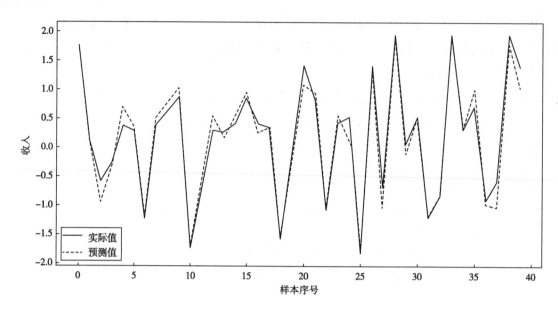

图 3-4　预测值和实际值的拟合曲线

单元4

使用Scikit-learn库来实现多项式回归

前面使用简单线性回归和多元线性回归来拟合广告投入,但是在实际生活中,很多数据之间是非线性关系,虽然也可以用线性回归拟合非线性回归,但是效果很差,这时候就需要对线性回归模型进行改进,在模型中增加高阶项,也就是增加非线性的量,使之能够拟合非线性数据。

4.1 多项式回归简介

在实际项目中通常会遇到非线性问题。图 4-1 所示数据呈现出线性关系,用线性回归可以得到较好的拟合效果。图 4-2 所示数据呈现非线性关系,需要多项式回归模型。多项式回归是在线性回归基础上进行改进,相当于为样本再添加特征项。如图 4-2 所示,为样本添加了一个 x^2 的特征项,可以较好地拟合非线性的数据。

图 4-2 所示数据集可以使用一个二次曲线来拟合,相应的方程式为 $y = ax^2 + bx + c$,这时数据模型就是一个多项式,其中的 x^2 可以看作添加的一个特征。

多项式回归解决非线性的问题,拟合不是直线关系而是其他曲线关系的数据。与线性回归类似,假设特征(x)与值(y)呈现多项式的数学模型的关系,具体操作时可以看成在线性关系($y = bx + c$)的基础上添加了多项式 ax^2。如果从回归的角度来看,这个式子是一个线性回归的式子;从 x 表达式的角度来看,它就是一个二次方程。

一个因变量与一个或多个自变量间多项式的回归分析方法,称为多项式回归。如果自变量只有一个时,称为一元多项式回归;如果有多个自变量时,称为多元多项式回归。由于任何函数都可以使用多项式逼近,因此多项式回归有着广泛的应用。

一元 n 次多项式回归:$y = b_0 + b_1 x + b_2 x^2 + \cdots + b_m x^m$。

二元二次多项式回归：$y = b_0 + b_1x_1 + b_2x_2 + b_3x_1^2 + b_4x_2^2 + b_5x_1x_2$。

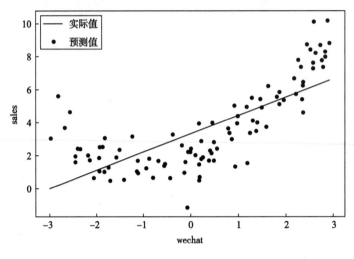

图 4-1　线性回归

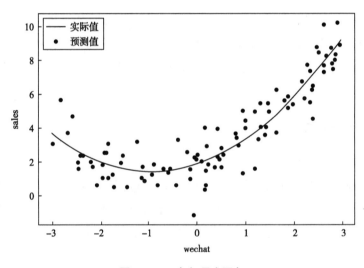

图 4-2　二次多项式回归

4.2　数据预处理

①首先在 Jupyter Notebook 中导入包，代码如下：

```
1.  # 导入数据处理库
2.  import numpy as np
```

```
3.   import pandas as pd
4.   from matplotlib import font_manager as fm, rcParams
5.   import matplotlib.pyplot as plt
6.   from sklearn.model_selection import train_test_split
7.   from sklearn import datasets
8.   plt.rcParams['font.sans-serif'] = ['SimHei']
9.   plt.rcParams['axes.unicode_minus'] = False
```

②读取数据，代码如下：

```
1.   data = pd.read_csv('../data/advertising.csv')
2.   data.head()
```

③生成特征和标签，代码如下：

```
1.   #变量初始化
2.   #最后一列为 y,其他为 x
3.   cols = data.shape[1]
4.   X = data.iloc[:,0:cols-1]         #取前 cols-1 列,即输入向量
5.   y = data.iloc[:,cols-1:cols]      #取最后一列,即目标变量
6.   X.head(10)
```

④拆分得到训练集和测试集，代码如下：

```
1.   from sklearn.preprocessing import PolynomialFeatures
2.   from sklearn.linear_model import LinearRegression
3.   from sklearn.metrics import r2_score
4.   from sklearn.pipeline import Pipeline
5.   from sklearn.model_selection import train_test_split
6.
7.   X_train,X_test,y_train,y_test = train_test_split(X,y,test_size=0.2)
8.
9.   X_train = np.array(X_train)
10.  X_test = np.array(X_test)
11.  y_train = np.array(y_train)
12.  y_test = np.array(y_test)
```

4.3　使用多项式线性回归预测广告投入的收益

1. 使用 PolynomialFeatures( ) 实现特征增扩

可以使用 sklearn 库中的 PolynomialFeatures( ) 函数来对特征进行增扩，生成 x^m 特征，然后使用

pipeline 来实现多项式回归。

（1）Polynomial Features( )函数

函数格式如下：

```
1.  sklearn.preprocessing.PolynomialFeatures(
2.          degree=2, * , interaction_only=False,
3.          include_bias=True, order='C')
```

（2）PolynomialFeatures( )的常用参数及用法

degree：控制多项式的次数。

interaction_only：默认为 False。如果指定为 True，那么就不会出现 x^2 项，如果设置 degree=2，那么组合的特征中没有 x^2。

include_bias：默认为 True，表示结果中就会有 0 次幂项，即全为 1 这一列。

这里使用 PolynomialFeatures( )进行数据增扩。设置 degree=2 对特征 $x_1 x_2 x_3$ 进行增扩，增扩后生成 10 个特征如下：

$$x_0, x_1, x_2, x_3, x_1 x_2, x_1 x_3, x_2 x_3, x_1^2, x_2^2, x_3^2$$

增扩代码如下：

```
1.  polyCoder = PolynomialFeatures(degree=2,include_bias=True,inter-
    action_only=False)
2.  df=polyCoder.fit_transform(X)
3.  print(df)
4.  print(df.shape)
```

上面代码中的 fit_transform( )函数的作用是对数据进行特征增扩。代码运行后，输出新生成的 10 个增扩特征，如图 4-3 所示。

```
[[1.0000000e+00 3.0440000e+02 9.3600000e+01 ... 8.7609600e+03
  2.7555840e+04 8.6671360e+04]
 [1.0000000e+00 1.0119000e+03 3.4400000e+01 ... 1.1833600e+03
  1.3704960e+04 1.5872256e+05]
 [1.0000000e+00 1.0911000e+03 3.2800000e+01 ... 1.0758400e+03
  9.6825600e+03 8.7143040e+04]
 ...
 [1.0000000e+00 1.2190000e+02 2.6400000e+02 ... 6.9696000e+04
  4.0761600e+04 2.3839360e+04]
 [1.0000000e+00 3.4350000e+02 8.6400000e+01 ... 7.4649600e+03
  4.1472000e+03 2.3040000e+03]
 [1.0000000e+00 7.9670000e+02 1.8000000e+02 ... 3.2400000e+04
  4.5360000e+04 6.3504000e+04]]
(200, 10)
```

图4-3　增扩后的特征结果

2. 使用 pipeline 实现多项式线性回归

定义 polynomial_regression( )函数，设置 pipeline 快速实现多项式回归。使用 pipeline 实现特征

增扩的代码如下：

```
1.  def polynomial_regression(degree=1):
2.      polyCoder = PolynomialFeatures(degree=degree,include_bias=
        True,interaction_only=False)
3.      linear_regression_model=LinearRegression(normalize=True)
        #数据归一化
4.      new_model=Pipeline([("polynomial_features",polyCoder),
5.                          ("linear_regression",linear_regression_model)])
6.      return new_model
```

polynomial_regression( )函数中，传入参数 degree，默认值为1，可以在调用函数时，传入不同 degree 实现 n 次多项式，最后返回生成多项式模型。

3. pipeline 函数的用法

Python 中的 pipeline 和 Linux 中的 pipeline 类似。把若干个命令连接起来，前一个命令的输出是后一个命令的输入，最终完成一个类似于流水线的功能。

（1）可以使用 sklearn. pipeline. Pipeline 来实现管道

参数：steps 为一个列表，列表的元素为(name,transform)元组，其中 name 是学习器的名字，用于输出和日志；transform 是学习器，之所以叫 transform，是因为这个学习器(除了最后一个)必须提供 transform 方法。

函数的属性：named_steps 为一个字典，字典的键就是 steps 中各元组的 name 元素，字典的值就是 steps 中各元组的 transform 元素。

（2）常用方法

fit(X[,y])：启动流水线，依次对各个学习器(除了最后一个学习器)执行 fit 方法和 transform 方法转换数据，对最后一个学习器执行 fit 方法训练学习器。

transform(X)：启动流水线，依次对各个学习器执行 fit 方法和 transform 方法转换数据。要求每个学习器都实现了 transform 方法。

fit_transform(X[,y])：启动流水线，依次对各个学习器(除了最后一个学习器)执行 fit 方法和 transform 方法转换数据，对最后一个学习器执行 fit_transform 方法转换数据。

inverse_transform(X)：将转换后的数据逆转换成原始数据，要求每个学习器都实现 inverse_transform 方法。

predict(X)/predict_log_proba(X)/predict_proba(X)：将 X 进行数据转换后，用最后一个学习器来预测。

score(X,y)：将 X 进行数据转换后，用最后一个学习器来给出预测评分。

当执行 pipe. fit(X_train, y_train)时，首先由 StandardScaler 在训练集上执行 fit 和 transform 方法，transformed 后的数据又被传递给 pipeline 对象的下一步，即 PCA( )。和 StandardScaler 一样，

PCA 也是执行 fit 和 transform 方法，最终将转换后的数据传递给 LogisticRegression。代码如下：

```
1.  #使用 pipeline 管道机制
2.  from sklearn.preprocessing import StandardScaler        #规范化,使各特征
    的均值为 1,方差为 0
3.  from sklearn.decomposition import PCA
4.  from sklearn.linear_model import LogisticRegression
5.
6.  from sklearn.pipeline import Pipeline
7.  pipe = Pipeline([('sc',StandardScaler()),        #数据标准化
8.                   ('pca',PCA(n_components=2)),     #PAC 数据降维
9.                   ('clf',LogisticRegression(random_state=666))
10.
11. pipe.fit(x_train, y_train)
12. print('Test accuracy is % .3f' % pipe.score(x_test, y_test))
```

4.4　训练模型

使用 fit( ) 函数训练模型，调用 score( ) 函数输出模型的决定系数 R2，代码如下：

```
1.  model = polynomial_regression(2)
2.  model.fit(X_train, y_train)
3.  train_score = model.score(X_train, y_train)
4.  test_score = model.score(X_test, y_test)
5.  print(model.named_steps['linear_regression'].coef_)
6.  print(train_score)
7.  print(train_score)
8.  print(model)
```

代码运行后，输出模型的 weight、训练集的 score 值、测试集的 score 值，如图 4-4 所示。

```
[[ 0.00000000e+00  1.66223817e-02 -3.03453877e-03 -2.56333282e-03
  -5.34773625e-06  1.15973927e-05  3.10131074e-08  2.24489815e-05
   4.97254102e-06  2.16733893e-06]]
0.9304880804198212
0.9304880804198212
Pipeline(memory=None,
     steps=[('polynomial_features', PolynomialFeatures(degree=2, include_bias=True, interaction_only=False)),
('linear_regression', LinearRegression(copy_X=True, fit_intercept=True, n_jobs=1, normalize=True))])
```

图 4-4　代码运行结果

<table>
<tr><td>4.5</td><td>绘制拟合曲线</td></tr>
</table>

使用 lineplot 绘制预测值和实际值的拟合曲线,代码如下:

```
1.  x = np.arange(len(X_test))#生成 X 轴数据
2.  print(y_test.shape,y_hat.shape)
3.  df = np.append(y_test,y_hat,axis =1)#拼接生成 y 轴数据
4.  df = df.reshape(40,2)
5.  print(df.shape)
6.  w_d = pd.DataFrame(df,x,["实际值","预测值"])#生成 dataframe 数据
7.  plt.figure(figsize = (15,5))#设置图像大小
8.  sns.lineplot(data = w_d)#绘制折线图
9.  plt.title("多元线性回归预测广告收入",fontsize =18)
10.  plt.show()
```

代码运行后,输出结果如图 4-5 所示。可以看出多项式模型中预测值和实际值之间的拟合度要高于多元线性回归。

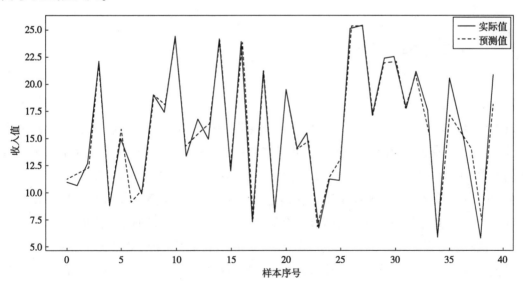

图 4-5 拟合运行结果

使用岭回归预测房价

房价预测数据集是一个高维的数据(有 80 列),在线性回归中随着特征的增加会出现过拟合现象。为了防止过拟合,在标准线性方程损失函数的基础上增加 L1 和 L2 正则化项,加入 L1 正则化项的为拉索(Lasso)回归,加入 L2 正则化项的为岭(Ridge)回归。

5.1 过拟合和欠拟合

1. 过拟合和欠拟合简介

过拟合和欠拟合是人工智能算法训练中经常会遇到的问题。"欠拟合"是指模型学习能力较弱,无法学习到样本数据中的"一般规律",因此导致模型泛化能力较弱。而"过拟合"则恰好相反,是指模型学习能力太强,以至于将样本数据中的"个别特点"也当成了"一般规律"。

欠拟合:一个假设在训练数据上不能获得更好的拟合,并且在测试数据集上也不能很好地拟合数据,此时认为这个假设出现了欠拟合的现象,如图 5-1 所示。(模型过于简单)

因为机器学习到的天鹅特征太少了,导致区分标准太粗糙,不能准确识别出天鹅。

过拟合:一个假设在训练数据上能够获得比其他假设更好的拟合,但是在测试数据集上却不能很好地拟合数据,此时认为这个假设出现了过拟合的现象,如图 5-2 所示。(模型过于复杂)

2. 数据集在过拟合和欠拟合状态下的准确率

对于训练数据集来说,模型越复杂,模型准确率越高,对训练数据集的拟合就越好。对于测试数据集来说,在模型很简单的时候,模型的准确率比较低,随着模型逐渐变复杂,测试数据集的准确率在逐渐提升,提升到一定程度后,如果模型继续变复杂,随着模型的复杂度的增加,模型在测试集上的准确率下降,那么就说明模型处于过拟合状态。

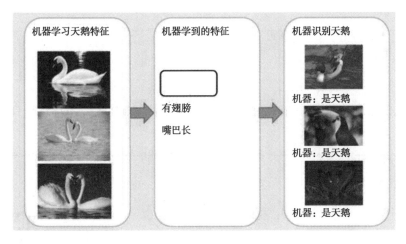

图 5-1　欠拟合示意图

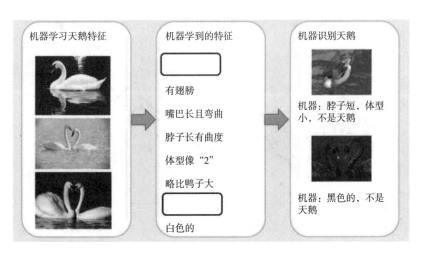

图 5-2　过拟合示意图

3. 模型泛化能力

人工智能算法的本质是利用算法模型对样本数据进行拟合,从而对未知的新数据进行有效预测。算法模型对训练集以外数据的预测越准确,模型的泛化能力越强。这就要求从训练样本中尽量学到适用于所有潜在样本的"普遍规律"。过拟合和欠拟合都会导致模型泛化能力较弱。过拟合和欠拟合是导致模型泛化能力不高的两种常见原因。

4. 过拟合和欠拟合的解决办法

出现欠拟合是因为模型学习的数据特征过少,可以增加数据的特征数量。过拟合是因为原始特征过多,存在一些噪声特征。模型过于复杂是因为模型尝试去兼顾各个测试数据点,从而学习了一些"普遍规律"之外的特征。

在这里针对回归,选择了正则化。但是,对于其他机器学习算法如分类算法来说也会出现这样的问题,除了一些算法本身作用之外(决策树、神经网络),更多的是做特征选择。

5.2 L1、L2 正则化

对于多项式模型来说，回归的阶数越高，模型会越复杂，这时就会出现过拟合现象，如图 5-3 所示。

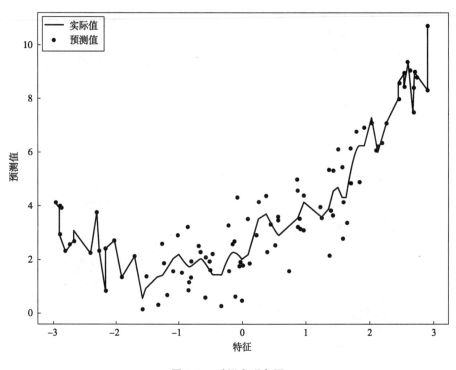

图 5-3   过拟合现象图

出现过拟合的原因是因为高次项的值对预测值的影响特别大，这就需要尽量减少高次项对特征的影响。于是，在损失函数后面加上惩罚项。惩罚项是指对损失函数中的某些参数做一些限制。常用的做法是在模型后面加上 L1、L2 惩罚项。

L1 正则化（拉索回归）是指在原来的损失函数基础上，加上权重参数的绝对值，公式如下：

$$J(\theta) = \frac{1}{2m}\sum_{i=1}^{m}(y_i - y_i')^2 + \frac{\lambda}{2m}\sum_{j=1}^{n}|\theta_j|$$

式中，y_i 表示实际标签值；y_i' 表示预测值；θ_j 表示权重系数；m 表示样本数量；$\frac{\lambda}{2m}\sum_{j=1}^{n}|\theta_j|$ 表示 L1 惩罚项。

L2 正则化（岭回归）是指在原来的损失函数基础上，加上权重参数的平方和，公式如下：

$$J(\theta) = \frac{1}{2m}\sum_{i=1}^{m}(y_i - y_i')^2 + \frac{\lambda}{2m}\sum_{j=1}^{n}\theta_j^2$$

惩罚项 $\frac{\lambda}{2m}\sum_{j=1}^{n}\theta_j^2$ 的加入保证了矩阵的可逆,单位矩阵 $\boldsymbol{I}$ 的对角线上全是 1,像一条山岭一样,这也是岭回归名称的由来。

5.3 房价数据集的读取

房价预测数据集是一个高维的数据(有 80 列),可以建立岭回归模型预测房价。

1. 导入数据加载的包,打开 csv 数据文件

导入 numpy、pandas、matplotlib 三个包来加载数据,同时对数据进行分析。源数据的 index 可以用来作为 pandas dataframe 的 index。在使用 read_csv() 函数读取数据时,加入 index_col = 0 参数,可以把数据的第一列作为编号,代码如下:

```
1.  train_df = pd.read_csv('../input/train.csv', index_col = 0)
2.  test_df = pd.read_csv('../input/test.csv', index_col = 0)
```

代码运行后,分别打开房价测试集和训练集,并且显示训练集,数据一共有 80 列,输出结果如图 5-4 所示。

Id	MSSubClass	MSZoning	LotFrontage	LotArea	Street	Alley	LotShape	LandContour	Utilities	LotConfig	...	PoolArea	PoolQC	Fence	MiscFeature	Misc
1	60	RL	65.0	8450	Pave	NaN	Reg	Lvl	AllPub	Inside	...	0	NaN	NaN	NaN	
2	20	RL	80.0	9600	Pave	NaN	Reg	Lvl	AllPub	FR2	...	0	NaN	NaN	NaN	
3	60	RL	68.0	11250	Pave	NaN	IR1	Lvl	AllPub	Inside	...	0	NaN	NaN	NaN	
4	70	RL	60.0	9550	Pave	NaN	IR1	Lvl	AllPub	Corner	...	0	NaN	NaN	NaN	
5	60	RL	84.0	14260	Pave	NaN	IR1	Lvl	AllPub	FR2	...	0	NaN	NaN	NaN	

5 rows × 80 columns

图 5-4　数据显示结果

接下来需要观察数据,需要确定数据中是否有"NaN",哪些字段是数值型的,哪些是类别型的,这些数据需要分别进行处理。

2. 使用 info() 函数查看数据信息

运行结果如图 5-5 所示。

从运行结果中可以看出所有的特征的类型以及每个特征的类型,例如,MSSubClass-表示住所类型,有 1460 个值,类型是 int64。

```
train_df.info()

<class 'pandas.core.frame.DataFrame'>
Int64Index: 1460 entries, 1 to 1460
Data columns (total 80 columns):
 #   Column        Non-Null Count   Dtype
---  ------        --------------   -----
 0   MSSubClass    1460 non-null    int64
 1   MSZoning      1460 non-null    object
 2   LotFrontage   1201 non-null    float64
 3   LotArea       1460 non-null    int64
 4   Street        1460 non-null    object
 5   Alley         91 non-null      object
 6   LotShape      1460 non-null    object
 7   LandContour   1460 non-null    object
 8   Utilities     1460 non-null    object
 9   LotConfig     1460 non-null    object
 10  LandSlope     1460 non-null    object
 11  Neighborhood  1460 non-null    object
 12  Condition1    1460 non-null    object
```

图 5-5　运行结果

5.4　处理标签值

1. 标签值分布可视化

首先查看标签值房价 SalePrice，取出 train 中的 SalePrice，绘制图形，代码如下：

```
1.  import seaborn as sns
2.  sns.histplot(train_df['SalePrice'])
```

程序运行后，横轴 SalePrice 表示房价，纵轴 Count 表示对应房价的统计数量，可以看出数据的分布存在右偏态分布，如图 5-6 所示。

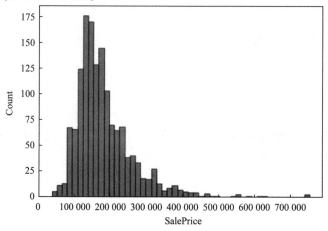

图 5-6　房价直方图

2. 使用 log1p 处理标签

在机器学习中用到了很多概率正态分布,所以首先需要把数据调整为正态分布,这样处理后的输出是一个正态分布。图中 SalePrice 本身并不平滑。为了使分类器的学习更加准确,会首先把 label"平滑化"(正态化)。这里使用的 log1p( ),也就是 $\log(x+1)$,避免了复值的问题。

如果这里把房价 SalePrice 数据做了 log1p 平滑处理,那么最后计算结果的时候,要记得把预测到的平滑数据给变回去。按照"怎么来的怎么去"原则,log1p( )就需要 expm1( );同理,log( )就需要 exp( )。使用 log1p( )处理的代码如下:

```
sns.histplot(np.log1p(train_df['SalePrice']))
```

代码运行后,可以看出使用 log1p( )处理后的数据分布基本符合正态分布,如图 5-7 所示。

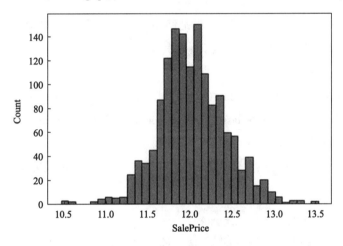

图 5-7　平滑处理后的房价直方图

5.5　类型特征 one-hot 编码

1. 合并训练集和测试集

项目提供了 train 和 test 数据集,test 是测试用的数据,训练完模型后需要用模型去预测 test 中的房价。因此需要先把 train 和 test 合起来,这么做主要是为了进行数据预处理的时候更加方便。等所有需要的预处理进行完之后,再把它们分隔开,代码如下:

```
1.  y_train = np.log1p(train_df.pop('SalePrice'))
2.  all_df = pd.concat((train_df,test_df),axis = 0)
3.  all_df.shape
4.  y_train.head()
```

2. one-hot 编码处理

（1）one-hot 编码

在机器学习算法中，经常会遇到分类特征，例如：人的性别有男女，国籍有中国、美国、法国等。这些特征值并不是连续的，而是离散的、无序的。这类数据是类别/标称（category）数据。如果 category 离散类型数据之间没有大小的意义，可以使用 one-hot 编码转换，如果 category 离散类型数据之间存在大小的意义，可以使用 LableEncoder 标签编码进行转换。

例如：["中国","美国","法国"]：

中国 = >100   美国 = >010   法国 = >001

上例中将分类值映射到整数值；然后，每个整数值被表示为二进制向量。这种处理方式就是one-hot 编码。

one-hot 编码又称一位有效编码。其方法是使用 N 位状态寄存器来对 N 个状态进行编码，每个状态都有它独立的寄存器位，并且在任意时候，其中只有一位有效。

（2）MSSubClass one-hot 编码

例如房价预测数据集中的 MSSubClass（住所类型）的值是一个 category 类型，在 DF 中它是一个 int 类型的值，使用 all_df['MSSubClass']. value_counts( )查看，MSSubClass 一共有 16 个值，这些值没有具体的含义，需要使用 one-hot 编码对它处理。

首先需要把它变回成 string，然后使用 pandas 自带的 get_dummies 方法转换为 one-hot 编码，代码如下：

```
1.  all_df['MSSubClass'].dtypes
2.  all_df['MSSubClass'] = all_df['MSSubClass'].astype(str)
```

代码运行后，按照 MSSubClass 的类型进行 one-hot 编码，结果如图 5-8 所示。

Id	MSSubClass_120	MSSubClass_150	MSSubClass_160	MSSubClass_180	MSSubClass_190	MSSubClass_20	MSSubClass_30	MSSubClass_40	MSSubClass_
1	0	0	0	0	0	0	0	0	
2	0	0	0	0	0	1	0	0	
3	0	0	0	0	0	0	0	0	
4	0	0	0	0	0	0	0	0	
5	0	0	0	0	0	0	0	0	
6	0	0	0	0	0	0	0	0	
7	0	0	0	0	0	1	0	0	
8	0	0	0	0	0	0	0	0	
9	0	0	0	0	0	0	0	0	
10	0	0	0	0	1	0	0	0	

图 5-8　转换为字符后的结果

由行结果可以看到，此刻 MSSubClass 被分成了 16 个 column，每一个代表一个 category。如果是对应的类别就用 1 表示，不是该类别就用 0 表示。可以用一个向量表示类别，如果是第一类，则可以表示为[1,0,0,0,0,0,0,0,0,0,0,0,0,0,0,0,0,0,0,0,0,0,0,0,0,0,0,0,0,0,0,0,0,0,0,0,0]，对应的值为1，其他为0。

3. 将 object 中的数据转换为 one-hot 编码

把所有的 category 数据,都进行 one-hot 编码处理。这样可以把分类数据转化为数值数据。这里使用 get_dummies( ) 来将数据转化为 one-hot 编码的格式,也就是对 object 类型和 str 类型的数据进行 one-hot 编码,代码如下:

```
1.  all_dummy_df =pd.get_dummies(all_df,drop_first =True)
2.  all_dummy_df.head()
```

代码运行后,可以看出 one-hot 编码后,数据集变成了 259 个 column,输出结果如图 5-9 所示。

Id	LotFrontage	LotArea	OverallQual	OverallCond	YearBuilt	YearRemodAdd	MasVnrArea	BsmtFinSF1	BsmtFinSF2	BsmtUn
1	65.0	8450	7	5	2003	2003	196.0	706.0	0.0	15
2	80.0	9600	6	8	1976	1976	0.0	978.0	0.0	28
3	68.0	11250	7	5	2001	2002	162.0	486.0	0.0	43
4	60.0	9550	7	5	1915	1970	0.0	216.0	0.0	54
5	84.0	14260	8	5	2000	2000	350.0	655.0	0.0	49

5 rows × 259 columns

图 5-9　one-hot 编码运行结果

5.6　房价数据集缺失数据处理

1. 统计数据缺失值

数据中如果存在缺失值,在模型训练时会产生异常,所以需要对缺失数据进行填充处理。对于 numerical 类型的数据,需要先查看缺失值,然后再对缺失值进行填充,最后做归一化操作。

这里首先使用 isnull( ) 方法检测数据集中哪些特征有 NULL 值,然后再使用 sort_values 排序,并计算出缺失的比例。这里有 35 个特征存在缺失值,代码如下:

```
1.  conut =all_dummy_df.isnull().sum().sort_values(ascending =False)
2.  percent =(all_dummy_df.isnull().sum()/all_df.isnull().count()).
    sort_values(ascending =False)
3.  miss_df =pd.concat([conut,percent],axis =1,keys =['Total','precent'],
    sort =False)
4.  miss_df.head(40)
```

处理这些缺失的数据需要根据题目提供的特征描述文件来进行分析,可以根据描述知道这些缺失代表的意思。

2. 填充缺失数据

找到缺失的数据后，需要对数据进行填充。数据填充有很多种方法，可以使用中位数、均值填充，随机森林填充，如果某个特征的缺失值太多也可以删除这个特征，这里使用均值填充，代码如下：

```
1.  mean_clos = all_dummy_df.mean()
2.  mean_clos.head()
3.  mean_clos.shape
```

使用 fillna 填充均值，然后再统计，代码如下：

```
1.  all_dummy_df = all_dummy_df.fillna(mean_clos)all_dummy_df.isnull().
    sum().sum()
2.  all_dummy_df.isnull().sum().sum()
```

5.7　房价数据集归一化处理

1. 数据归一化处理

这里使用标准分布对数据做归一化处理，可以使数据更平滑，更便于计算。这里需要分两步实现：首先找出数值型的特征；其次使用标准差归一化公式对数据归一化。先来看看哪些是 numerical 的，代码如下：

```
1.  numeric_cols = all_df.columns[all_df.dtypes! ='object']
2.  numeric_cols
```

使用标准差归一化对 numerical 做归一化处理，代码如下：

```
1.  numeric_cols_means = all_dummy_df.loc[:,numeric_cols].mean()
2.  numeric_cols_std = all_dummy_df.loc[:,numeric_cols].std()
3.  all_dummy_df.loc[:,numeric_cols] = (all_dummy_df.loc[:,numeric_
    cols] -numeric_cols_means) / numeric_cols_std
```

2. 拆分得到训练集和测试集

使用 loc( ) 函数，从数据中获得训练集和测试集，同时输出 train、test 的形状，代码如下：

```
1.  dummy_train_df = all_dummy_df.loc[train_df.index]
2.  dummy_test_df = all_dummy_df.loc[test_df.index]
3.  dummy_train_df.shape, dummy_test_df.shape
```

建立岭回归房价预测模型

1. 导入 sklearn 包建立岭回归模型

导入 sklearn 包来实现模型的建立和超参数的选择。因为是高维数据,所以选择岭回归作为基本模型。

2. 使用交叉验证集来获取岭回归的最优化参数

np. logspace(−2,2,50)构造的是一个从 10 的 −2 次方到 10 的 2 次方的等比数列,这个等比数列的长度是 50 个元素,这个值对应的岭回归中的 L2 正则化系数 λ。

然后通过循环,保存每个对应数值的 test_score 的值,并且保存每个 λ 对应的 test_score,代码如下:

```
1.  test_scores = []
2.  for alpha in alphas:
3.      clf = Ridge(alpha)
4.      test_score = np.sqrt( - cross_val_score(clf,X_train,y_train,cv =
    10,scoring = 'neg_mean_squared_error'))
5.      test_scores.append(np.mean(test_score))
6.  print(test_score)
```

然后使用 sns. lineplot 绘制 test_score 图形,代码如下:

```
1.  import seaborn as sns
2.  sns.lineplot(alphas,test_scores)
3.  plt.xlabel("迭代次数")
4.  plt.ylabel("损失")
5.  ax.legend()
```

运行结果如图 5-10 所示。由图 5-10 可知,迭代次数(alpha)的值在 10 ~ 20 的时候,损失(R_score)达到 0. 135 5 左右。可以选取 λ 的值为 15。

3. 交叉验证

将全部样本划分成 k 个大小相等的样本子集;依次遍历这 k 个子集,每次把当前子集作为验证集,其余所有样本作为训练集,进行模型的训练和评估;最后把 k 次评估指标的平均值作为最终的评估指标。在实际实验中,k 通常取 10。

举例:这里取 $k = 10$,如图 5-11 所示。

①先将原数据集分成 10 份。

②将其中的一份作为测试集,剩下的 9 份作为训练集,此时训练集就变成了 $k \times D(D$ 表示每一份中包含的数据样本数)。

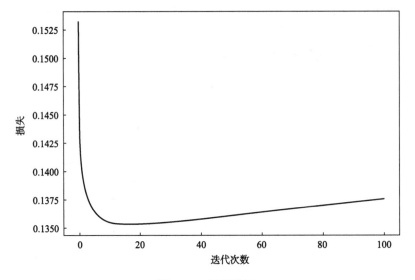

图 5-10　运行结果

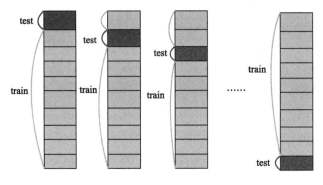

图 5-11　交叉验证

　　如图 5-11 所示，每次取出一份作为测试集，其他数据作为训练集，训练模型可以得到 10 个模型评估指标，最后将平均值作为最终的模型评估指标。

　　可以使用 cross_val_score( ) 函数实现交叉的功能，函数的用法如下：

```
sklearn.model_selection.cross_val_score(estimator, X, y = None, groups =
None, scoring = None, cv = 'warn', n_jobs = None, verbose = 0, fit_params = None,
pre_dispatch = '2* n_jobs', error_score = 'raise - deprecating')
```

　　参数：

　　estimator：需要使用交叉验证的算法。

　　X：输入样本数据。

　　y：样本标签。

　　groups：将数据集分割为训练集/测试集时，使用的样本的组标签。

　　cv：交叉验证折数或可迭代的次数。

n_jobs：同时工作的 CPU 个数(-1 代表全部)。

verbose：详细程度。

fit_params：传递给估计器(验证算法)的拟合方法的参数。

pre_dispatch：控制程序执行是并行调度的作业数量。减少这个数量,对于避免在 CPU 发送更多作业时 CPU 内存消耗的扩大是有用的。

5.9　使用交叉验证集来获取房价预测模型的最优参数

在 Jupyter Notebook 中导入 RandomForestRegressor 包,设置 max_features 列表,使用循环和交叉验证求解最优参数,代码如下:

```
1.  from sklearn.ensemble import RandomForestRegressor
2.  max_features = [.1,.3,.5,.7,.9,.99]
3.  test_scores = []
4.  for max_feat in max_features:
5.      clf = RandomForestRegressor(n_estimators = 200,max_features = max_
    feat)
6.      test_score = np.sqrt(-cross_val_score(clf,X_train,y_train,cv =
    5,scoring = 'neg_mean_squared_error'))
7.      test_scores.append(np.mean(test_score))
8.  sns.lineplot(max_features,test_scores)
```

代码运行后,可以看出当 max_features = 0.3 时,test_scores 达到了 0.137 的最优值,如图 5-12 所示。

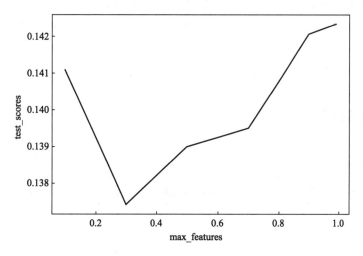

图 5-12　求解最优参数

单元6

使用逻辑回归预测共享单车用量

公共交通工具的"最后一公里"是城市居民出行采用公共交通出行的主要障碍,也是建设绿色城市、低碳城市过程中面临的主要挑战。共享单车(自行车)企业通过在校园、地铁站点、公交站点、居民区、商业区、公共服务区等提供服务,完成交通行业最后一块"拼图",带动居民使用其他公共交通工具的热情。与其他公共交通方式产生协同效应。共享单车是一种分时租赁模式,也是一种新型、绿色环保的共享经济。

在其迅猛发展,为市民提供交通方便的同时,由于交通的潮汐性以及车辆调度的不及时,使其也面临着高峰时段的站点之间存在车辆供需不平衡等问题,因此需要准确、高效地对站点之间的共享单车数量进行预测,通过建立预测模型为车辆的调度进行精确的预测,对站点之间的共享单车需求量和状态进行预测。共享单车站点的需求量不仅和时间、空间存在动态联系,同时和天气、节日、温度、风力等复杂气象有关,它的模型实际上是一个非线性的时间序列问题。本单元通过对以往的共享单车数据进行分析,建立数据模型对站点的共享单车用量进行预测,从而达到对共享单车的精确调度。

6.1 数据加载与显示

1. 导入数据加载的包

在 Jupyter Notebook 中导入 numpy、pandas、matplotlib 三个包用来加载数据,同时对数据进行分析,代码如下:

```
1.  #导入需要的包
2.  import pandas as pd
3.  #import seaborn as sns
4.  import matplotlib.pyplot as plt
5.  import numpy as np
```

2. 打开 csv 数据文件导入数据

使用 pandas 的 read_csv 打开 csv 文件导入数据,返回一个 DataFrame 对象。

逗号分隔值(comma-separated values,CSV)又称字符分隔值(因为分隔字符的也可以不是逗号)。其文件以纯文本形式存储表格数据(数字和文本)。纯文本意味着该文件是一个字符序列,不含必须像二进制数字那样被解读的数据。CSV 文件由任意数目的记录组成,记录间以某种换行符分隔;每条记录由字段组成,字段间的分隔符是其他字符或字符串,最常见的是逗号或制表符。通常,所有记录都有完全相同的字段序列。通常都是纯文本文件。建议使用 WORDPAD 或是记事本来开启,或者先另存为新文档后用 EXCEL 开启,代码如下:

```
1.  #导入数据
2.  data = pd.read_csv('kaggle_bike_competition_train.csv', header = 0,
    skip_blank_lines = True)
3.  #显示数据
4.  data.head()
```

注意:当 skip_blank_lines = True 时,这个参数忽略注释行和空行。所以,header = 0 表示第一行是数据而不是文件的第一行。

代码运行后,显示共享单车预测数据,如图 6-1 所示。

	datetime	season	holiday	workingday	weather	temp	atemp	humidity	windspeed	casual	registered	count
0	2011/1/1 0:00	1	0	0	1	9.84	14.395	81	0.0	3	13	16
1	2011/1/1 1:00	1	0	0	1	9.02	13.635	80	0.0	8	32	40
2	2011/1/1 2:00	1	0	0	1	9.02	13.635	80	0.0	5	27	32
3	2011/1/1 3:00	1	0	0	1	9.84	14.395	75	0.0	3	10	13
4	2011/1/1 4:00	1	0	0	1	9.84	14.395	75	0.0	0	1	1

图 6-1 数据的基本信息[①]

3. 显示数据、查看数据基本信息

使用 data.info()、data.isnull()、data.describe()函数显示数据基本信息,包括行数、列数、列索引、列非空值个数、列类型、内存占用等,如图 6-2 所示。

---

①本单元数据集为华盛顿地区共享单车的数据集。

```
<class 'pandas.core.frame.DataFrame'>
RangeIndex: 10886 entries, 0 to 10885
Data columns (total 12 columns):
 #   Column      Non-Null Count   Dtype
---  ------      --------------   -----
 0   datetime    10886 non-null   object
 1   season      10886 non-null   int64
 2   holiday     10886 non-null   int64
 3   workingday  10886 non-null   int64
 4   weather     10886 non-null   int64
 5   temp        10886 non-null   float64
 6   atemp       10886 non-null   float64
 7   humidity    10886 non-null   int64
 8   windspeed   10886 non-null   float64
 9   casual      10886 non-null   int64
 10  registered  10886 non-null   int64
 11  count       10886 non-null   int64
dtypes: float64(3), int64(8), object(1)
memory usage: 1020.7+ KB
```

图 6-2　数据的描述

从 info( ) 函数执行结果中可以看出，共享单车的数据类型为 int64 和 float64 类型。但是，有一个数据比较特殊，datetime 是 object 类型，这种数据类型在后续需要对它进行处理才能使用。

describe( ) 函数直接给出样本数据的一些基本的统计量，包括均值，标准差、最大值、最小值、分位数等。函数运行结果如图 6-3 所示。

	season	holiday	workingday	weather	temp	atemp	humidity	windspeed	casual	registered
count	10886.000000	10886.000000	10886.000000	10886.000000	10886.00000	10886.000000	10886.000000	10886.000000	10886.000000	10886.000000
mean	2.506614	0.028569	0.680875	1.418427	20.23086	23.655084	61.886460	12.799395	36.021955	155.552177
std	1.116174	0.166599	0.466159	0.633839	7.79159	8.474601	19.245033	8.164537	49.960477	151.039033
min	1.000000	0.000000	0.000000	1.000000	0.82000	0.760000	0.000000	0.000000	0.000000	0.000000
25%	2.000000	0.000000	0.000000	1.000000	13.94000	16.665000	47.000000	7.001500	4.000000	36.000000
50%	3.000000	0.000000	1.000000	1.000000	20.50000	24.240000	62.000000	12.998000	17.000000	118.000000
75%	4.000000	0.000000	1.000000	2.000000	26.24000	31.060000	77.000000	16.997900	49.000000	222.000000
max	4.000000	1.000000	1.000000	4.000000	41.00000	45.455000	100.000000	56.996900	367.000000	886.000000

图 6-3　共享单车的数据统计信息

isnull( ) 函数，用来判断数据中是否有缺失值。常用的方式有 isnull( ). any( ) 和 isnull( ). sum( )，代码如下：

```
1.  data.isnull().values.any()
2.  data.isnull().sum()
```

代码运行后，可以看出共享单车数据集没有缺失的数据，如图 6-4 所示。

4. 数据字段的意义

使用 data. info( ) 可以输出共享单车的数据，这部分数据是已经经过数据清洗和数据转化后的数据。数据对应字段的含义如下：

datatime：时间和日期。

season：季节。1：春季；2：夏季；3：秋季；4：冬季。

holiday：是否为假期。1：假期；0：非假期。

workingday：是否工作日。1：工作日/0 非工作日；

weather：天气类别。1：晴天；2：多云；3：小雨或小雪；4：恶劣天气。

temp：实际温度。

atemp：体感温度。

humidity：相对湿度/%。

windspeed：风速。以 km/h 为单位。

casual：随机用户，没有注册的用户。

registered：注册用户。

count：总用户数量。

```
#检测是否有null值www.baidu
data.isnull().sum()
```

executed in 12ms, finished 14:53:41 2021-02-20

```
datetime      0
season        0
holiday       0
workingday    0
weather       0
temp          0
atemp         0
humidity      0
windspeed     0
casual        0
registered    0
count         0
dtype: int64
```

图 6-4　空数据的统计

6.2　季节性数据的可视化分析

season 在数据中表示季节。根据基本的常识判断，季节性数据对单车的用量还是有一定影响的。如何按照季节对单车用量数据进行分析和统计需要做如下处理：

首先将季节中重复的数据合并。使用 unique( ) 函数，去掉 season 中重复的数据值，返回 season 中唯一的值，返回值是 list 格式。返回值中有四个值 1，2，3，4，类型是 int64 类型，使用 unique( ) 统计四个不同的季节。

使用 values_counts( ) 统计每个季节有多少条记录，汇总得到四个季节的租车总次数。代码如下：

```
data.season.value_counts()
```

使用 groupby( ) 函数对 season 数据进行分组运算，然后对季节的 counts 租车数进行汇总统计，最后得到每个季节的租车总数。

可视化呈现数据。使用 plt 的 bar( ) 函数绘制柱状图，直观地显示每个季节的租车数量，代码如下：

```
#sns.barplot(x='season',y='count',data=data)
#s_list=data.season.unique()
s_list=['1','2','3','4']
s_c=data.groupby('season')['count'].sum()
plt.bar(range(len(s_list)),s_c,color='rgb',tick_label=s_list)
plt.show()
```

代码运行后可以看出，租车数量为夏季 > 春季 > 秋季 > 冬季，结果如图 6-5 所示。

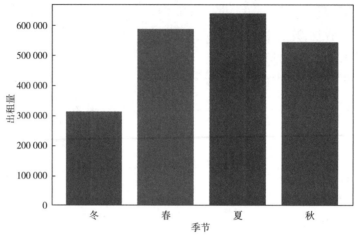

图 6-5　四个季节分组计算得到出租量的数据

6.3　天气数据、假期数据的可视化分析

天气数据的处理与季节数据的处理类似，都是先进行分组，然后进行统计计算，代码如下：

```
1.  s_list = data.season.unique()
2.  s_c = data.groupby('weather')['count'].sum()
3.  plt.bar(range(len(s_list)), s_c, color = 'rgb', tick_label = s_list)
4.  plt.show()
```

代码运行后可以看出，共享单车的出租量为 1 晴天 > 2 多云 > 3 小雨雪 > 4 恶劣天气，如图 6-6 所示。

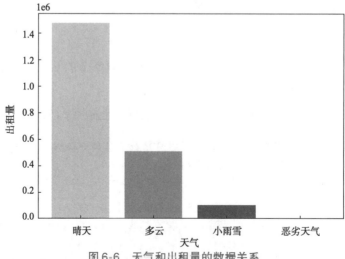

图 6-6　天气和山租量的数据关系

假期特征('holiday')分为'holiday'和'noholiday'两个类型,工作日特征('workingday')分为'workingday'和'noworkingday'两个类型,这两个特征也需要做可视化输出对比,代码如下:

```
1.  #工作日和假日的信息对比
2.  d_list = ['holiday','noholiday','workingday','noworkingday']
3.  d_s = data.groupby('holiday')['count'].sum()
4.  ar = np.array(d_s)
5.  print(ar)
6.  ##print(ar.shape)
7.  d_2 = data.groupby('workingday')['count'].sum()
8.  d_s = [data.groupby('holiday')['count'].sum()] + [data.groupby('workingday')['count'].sum()]
9.  print(d_s)
10. print(np.array(d_s).shape)
11. #plt.bar(range(len(d_list)),d_s,color = 'rgb',tick_label = d_list)
12. #plt.show()
```

6.4　时间数据的处理

从上面的分析中可以看出,datetime 是一个重要的特征,但是它的类型是 object,需要对它进行切分,把 datetime 域切成日期和时间两部分,代码如下:

```
1.  #处理时间字段
2.  temp = pd.DatetimeIndex(data['datetime'])
3.  data['date'] = temp.date
4.  data['time'] = temp.time
5.  data.head()
```

代码运行后,可以看到新生成了"date"和"time"两个特征,结果如图 6-7 所示。

	datetime	season	holiday	workingday	weather	temp	atemp	humidity	windspeed	casual	registered	count	date	time
0	2011/1/1 0:00	1	0	0	1	9.84	14.395	81	0.0	3	13	16	2011-01-01	00:00:00
1	2011/1/1 1:00	1	0	0	1	9.02	13.635	80	0.0	8	32	40	2011-01-01	01:00:00
2	2011/1/1 2:00	1	0	0	1	9.02	13.635	80	0.0	5	27	32	2011-01-01	02:00:00
3	2011/1/1 3:00	1	0	0	1	9.84	14.395	75	0.0	3	10	13	2011-01-01	03:00:00
4	2011/1/1 4:00	1	0	0	1	9.84	14.395	75	0.0	0	1	1	2011-01-01	04:00:00

图 6-7　切分结果

细化时间,把时间细化到小时,把小时字段拿出来作为更简洁的特征,代码如下:

```
1.   #设定 hour 这个字段
2.   data['hour'] = pd.to_datetime(data.time,format = "% H:% M:% S")
3.   data['hour'] = pd.Index(data['hour']).hour
4.   data
```

代码运行后,新生成了"hour"特征,结果如图6-8所示。

	datetime	season	holiday	workingday	weather	temp	atemp	humidity	windspeed	casual	registered	count	date	time	hour
0	2011/1/1 0:00	1	0	0	1	9.84	14.395	81	0.0000	3	13	16	2011-01-01	00:00:00	0
1	2011/1/1 1:00	1	0	0	1	9.02	13.635	80	0.0000	8	32	40	2011-01-01	01:00:00	1
2	2011/1/1 2:00	1	0	0	1	9.02	13.635	80	0.0000	5	27	32	2011-01-01	02:00:00	2
3	2011/1/1 3:00	1	0	0	1	9.84	14.395	75	0.0000	3	10	13	2011-01-01	03:00:00	3
4	2011/1/1 4:00	1	0	0	1	9.84	14.395	75	0.0000	0	1	1	2011-01-01	04:00:00	4
5	2011/1/1 5:00	1	0	0	2	9.84	12.880	75	6.0032	0	1	1	2011-01-01	05:00:00	5
6	2011/1/1 6:00	1	0	0	1	9.02	13.635	80	0.0000	2	0	2	2011-01-01	06:00:00	6
7	2011/1/1 7:00	1	0	0	1	8.20	12.880	86	0.0000	1	2	3	2011-01-01	07:00:00	7
8	2011/1/1 8:00	1	0	0	1	9.84	14.395	75	0.0000	1	7	8	2011-01-01	08:00:00	8
9	2011/1/1 9:00	1	0	0	1	13.12	17.425	76	0.0000	8	6	14	2011-01-01	09:00:00	9
10	2011/1/1 10:00	1	0	0	1	15.58	19.695	76	16.9979	12	24	36	2011-01-01	10:00:00	10
11	2011/1/1 11:00	1	0	0	1	14.76	16.665	81	19.0012	26	30	56	2011-01-01	11:00:00	11
12	2011/1/1 12:00	1	0	0	1	17.22	21.210	77	19.0012	29	55	84	2011-01-01	12:00:00	12

图6-8 细化时间后的结果

最后从数据中,把原始的时间字段等去掉,代码如下:

```
1.   # remove old data features
2.   dataRel = data.drop(['datetime', 'count','date','time','dayofweek
     '], axis =1)
3.   dataRel.head()
```

代码运行后,可以得到时间字段细化后的结果,如图6-9所示。

	season	holiday	workingday	weather	temp	atemp	humidity	windspeed	casual	registered	hour	dataDays	Saturday	Sunday
0	1	0	0	1	9.84	14.395	81	0.0	3	13	0	0.0	1	0
1	1	0	0	1	9.02	13.635	80	0.0	8	32	1	0.0	1	0
2	1	0	0	1	9.02	13.635	80	0.0	5	27	2	0.0	1	0
3	1	0	0	1	9.84	14.395	75	0.0	3	10	3	0.0	1	0
4	1	0	0	1	9.84	14.395	75	0.0	0	1	4	0.0	1	0

图6-9 时间字段细化后的结果

最后,按照小时统计共享单车出租量,并可视化,代码如下:

```
1.   #时间小时
2.   sns.pointplot(data['hour'],data['count'])
3.   plt.title('小时对共享单车数量的影响')
4.   plt.show()
```

代码运行后,生成一天每个时间段与共享单车用量的折线图。横坐 hour 表示"小时",纵坐标 count 表示"共享单车出租量",可以看出,出租量在 8 点左右和 17 点左右出现两个高峰,结果如图 6-10 所示。

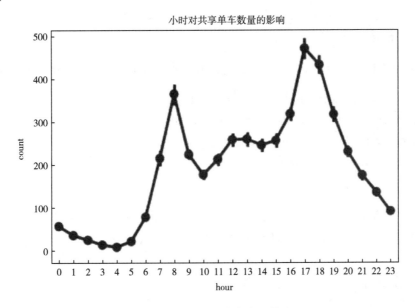

图 6-10　小时与共享单车出租量关系

6.5　特征化向量

1. 分离连续特征和离散特征

需要使用 sklearn 来建模。对于 pandas 的 dataframe 有方法/函数可以直接转成 Python 中的 dict。另外,在这里需要对离散特征和连续特征进行区分,把连续属性放入名称为 X_dictCon 的 dict 中,代码如下:

```
1.  from sklearn.feature_extraction import DictVectorizer
2.  #把连续的属性值放入一个 dict 中
3.  featureConCols = ['temp','atemp','humidity','windspeed','dataDays
    ','hour']
4.  dataFeatureCon = dataRel[featureConCols]
5.  dataFeatureCon = dataFeatureCon.fillna('NA')
6.  X_dictCon = dataFeatureCon.T.to_dict().values()
```

把离散属性放入名称为 X_dictCat 的 dict 中,代码如下:

```
1.  featureCatCols = ['season','holiday','workingday','weather','Satur-
    day','Sunday']
2.  dataFeatureCat = dataRel[featureCatCols]
3.  dataFeatureCat = dataFeatureCat.fillna('NA')
4.  X_dictCat = dataFeatureCat.T.to_dict().values()
```

使用 DictVectorizer 的 fit_transform( ) 把 X_dictCat 和 X_dictCon 做向量化特征处理,代码如下:

```
1.  vec = DictVectorizer(sparse = False)
2.  X_vec_cat = vec.fit_transform(X_dictCat)
3.  X_vec_con = vec.fit_transform(X_dictCon)
4.  dataFeatureCon.head()
```

2. 标准化连续特征值

接下来需要对连续特征值做标准化处理,让连续特征值处理过后的均值为 0,方差为 1。这样的数据放到模型里,对模型训练的收敛和模型的准确度都有好处,代码如下:

```
1.  from sklearn import preprocessing
2.  #标准化连续的数值
3.  scaler = preprocessing.StandardScaler().fit(X_vec_con)
4.  X_vec_con = scaler.transform(X_vec_con)
5.  X_vec_con
```

3. 类别特征编码

使用 one-hot 编码,对类别特征做处理。代码如下:

```
1.  from sklearn import preprocessing
2.  enc = preprocessing.OneHotEncoder()
3.  enc.fit(X_vec_cat)
4.  X_vec_cat = enc.transform(X_vec_cat).toarray()
5.  X_vec_cat
```

特征拼接,把离散的和连续的特征都组合在一起,最后的特征,前 6 列是标准化过后的连续特征,后面是编码后的离散特征,代码如下:

```
1.  import numpy as np
2.  X_vec = np.concatenate((X_vec_con,X_vec_cat),axis = 1)
3.  X_vec
```

标签也需要处理。对标签进行浮点数处理,最后得到结果的浮点数值,代码如下:

```
1.  #对 Y 向量化
2.  Y_vec_reg = dataRel['registered'].values.astype(float)
3.  Y_vec_cas = dataRel['casual'].values.astype(float)
4.  Y_vec_reg
```

程序运行后,得到 float 化后的 registered 值和 casual 值。

6.6　数据预处理

首先将训练集和测试集进行合并,使用 append 将 train_std、test 两个字段进行合并,代码如下:

```
1.  train_std_y = np.log(train_std.pop('count'))
2.  #将 train_std、test 合并,便于修改
3.  full = train_std.append( test , ignore_index = True )
4.  print ('合并后的数据集:',full.shape)
```

然后从合并后的特征集合中选定特征 year, month, weekday, hour, season, holiday, weather, temp, humidity, windspeed,代码如下:

```
1.   full['datetime'] = full['datetime'].astype(str)
2.   full['datetime'] = full['datetime'].apply(lambda x:datetime.strp-
     time(x,'%Y-%m-%d %H:%M:%S'))
3.   # 提取年、月、小时、星期日数,并以列的形式添加到数据中
4.   full['year'] = full['datetime'].apply(lambda x: x.year)
5.   full['month'] = full['datetime'].apply(lambda x: x.month)
6.   full['hour'] = full['datetime'].apply(lambda x: x.hour)
7.   full['weekday'] = full['datetime'].apply(lambda x: x.strftime('%a'))
8.   #选择需要的特征
9.   columns = ['year','month','weekday','hour','season','holiday',
10.   'weather','temp','humidity','windspeed']
11.   full_feature = full[columns]
12.   full_feature.head()
```

最后使用 get_dummies( )将多类别型数据使用 one-hot 转化成多个二分型类别,代码如下:

```
1.  #将多类别型数据使用 one-hot 转化成多个二分型类别
2.  dummies_year = pd.get_dummies(full_feature['year'],prefix = 'year')
3.  dummies_month = pd.get_dummies(full_feature['month'], prefix = 'month')
```

```
4.  dummies_weekday = pd.get_dummies(full_feature['weekday'],prefix='
    weekday')
5.  dummies_hour = pd.get_dummies(full_feature['hour'],prefix='hour')
6.  dummies_season =pd.get_dummies(full_feature['season'],prefix='season')
7.  dummies_holiday =pd.get_dummies(full_feature['holiday'],prefix='holiday')
8.  dummies_weather =pd.get_dummies(full_feature['weather'],prefix='weather')
9.  full_feature =pd.concat([full_feature,dummies_year,dummies_month,
    dummies_weekday,dummies_hour,dummies_season,dummies_holiday,
10.  dummies_weather],axis =1)
11.  #删除原有列
12.  dropFeatures = ['season','weather','year','month','weekday','hour
    ','holiday']
13.  full_feature = full_feature.drop(dropFeatures ,axis = 1)
14.  full_feature.head()
```

6.7 建立模型

设置训练集和测试集,拆分训练样本-测试样本,使用 sample 和 subtractByKey 的方法,将数据集拆分为训练集和测试集,代码如下：

```
1.  #拆分训练样本-测试样本,使用 sample 和 subtractByKey 的方法
2.  data_with_idx = data.zipWithIndex().map(lambda point_index: (point_
    index[1], point_index[0]))
3.  test = data_with_idx.sample(False, 0.2, 42)
4.  train = data_with_idx.subtractByKey(test)
5.  train_data = train.map(lambda index_point:index_point[1])
6.  test_data = test.map(lambda index_point:index_point[1])
7.  train_size = train_data.count()
8.  test_size = test_data.count()
9.  print ("Training data size: % d" % train_size)
10.  print ("Test data size: % d" % test_size)
11.  print ("Total data size: % d " % num_data)
```

使用线性回归模型预测共享单车用量,由于城市、时间、气温、天气状况、风速五个变量对使用量影响较大,所以以此作为特征变量拟合模型,代码如下：

```
1.  #定义评价模型的度量
2.  def evaluate(train, test, iterations, step, regParam, regType,intercept):
3.      model = LinearRegressionWithSGD.train(train, iterations, step,
    regParam = regParam, regType = regType, intercept = intercept)
4.      tp = test.map(lambda p: (p.label, model.predict(p.features)))
5.      rmsle = np.sqrt(tp.map(lambda lp: squared_log_error(lp[0], lp
    [1])).mean())
6.      return rmsle
```

调节参数,选择最优解。可调节参数包括:Iterations、step、L2/L1 正则化系数、Intercept。代码
如下:

```
1.  #示例:调节参数之 Iterations 迭代次数
2.  params = [1, 5, 10, 20, 50, 100, 200, 300]
3.  metrics = [evaluate(train_data, test_data, param, 0.01, 0.0, 'l2',
    False) for param in params]
4.  print(params)
5.  print(metrics)
6.  plt.plot(params, metrics)
7.  fig = plt.gcf()
8.  plt.xlabel('log')
```

代码运行后,输出模型的 MSE 误差为 1 208.955 627 135 937 7,在训练集上的 R2 值为
−0.483 487 258 551 061 45。

逻辑回归和决策边界

在线性回归中利用样本通过有监督学习,学习由 x 到 y 的映射 f,因为 y 是连续值,所以是回归问题。如果需要预测的变量 y 是一个离散值,需要使用逻辑回归算法来解决分类问题。

现实生活中有很多逻辑回归的例子。一个在线交易网站判断一次交易是否带有欺诈性、判断一个肿瘤是良性的还是恶性的、判定是否是垃圾邮件的这些问题都是分类问题。

在这些例子中,预测的是一个二值的变量,或者为 0,或者为 1;或者是带有欺诈性的交易,或者不是;或者是一个恶性肿瘤,或者不是;或者是一封垃圾邮件,或者不是。将因变量(dependant variable)可能属于的两个类分别称为负向类(negative class)和正向类(positive class)。可以使用 0 来代表负向类,1 来代表正向类。上面的分类问题仅仅局限在两类上:0 或者 1。之后会讨论多分类问题,也就是说,变量 y 可以取多个值,例如 0,1,2,3。

7.1 机器学习分类

机器学习通常分为监督学习、无监督学习、强化学习三类,如图 7-1 所示。

1. 监督学习

在监督学习下,输入数据称为"训练数据",每组训练数据有一个明确的标识或结果,如对防垃圾邮件系统中的"垃圾邮件""非垃圾邮件";对手写数字识别中的"1","2","3"和"4"等,如图 7-2 所示。

在建立预测模型的时候,监督学习建立一个学习过程,将预测结果与"训练数据"的实际结果进行比较,不断调整预测模型,直到模型的预测结果达到一个预期的准确率。监督学习的常见应用场景有分类问题和回归问题。常见算法有逻辑回归(logistic regression)和反向传递神经网络。

半监督学习是监督学习的特例。半监督学习方式下,输入数据部分被标识,部分没有被标识,这种学习模型可以用来进行预测,但是模型首先需要学习数据的内在结构以便合理地组织数据来进行预测。应用场景包括分类和回归,算法包括一些对常用监督学习算法的延伸,这些算法首先试图对未标识数据进行建模,在此基础上再对标识的数据进行预测。如图论推理算法(graph infer-ence)或者拉普拉斯支持向量机(Laplacian SVM)等。

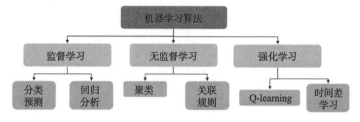

图 7-1　机器学习分类

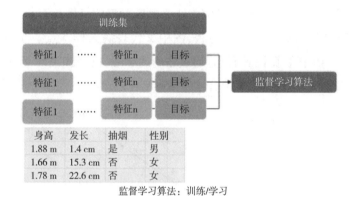

图 7-2　监督学习

2. 无监督学习

在无监督学习中,数据并不被特别标识,学习模型是为了推断出数据的一些内在结构。常见的应用场景包括关联规则的学习以及聚类等。常见算法包括 Apriori 算法以及 k-means 算法,如图 7-3 所示。

3. 强化学习

在这种学习模式下,输入数据作为对模型的反馈,不像监督学习模型那样,输入数据仅仅是作为一个检查模型对错的方式。在强化学习下,输入数据直接反馈到模型,模型必须对此立刻做出调整。常见的应用场景包括动态系统以及机器人控制等。

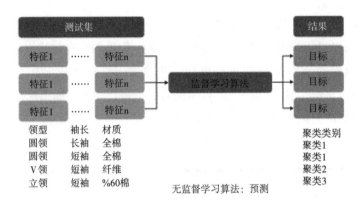

图7-3　无监督学习

7.2　逻辑回归

1. "线性回归 + 阈值"判断是否是恶性肿瘤

现在有这样一个分类任务，需要根据肿瘤大小来判断肿瘤的良性与否。训练集如图7-4所示，横轴代表肿瘤大小，纵轴代表肿瘤的良性与否，图中×代表病人样本点。注意，纵轴只有两个取值，1（代表恶性肿瘤）和0（代表良性肿瘤）。通过之前的学习，已经知道对于以上数据集使用线性回归来处理。实际上就是用一条直线来拟合这些数据。

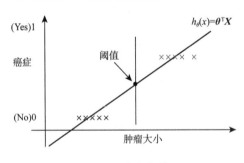

图7-4　肿瘤分类

可以先使用线性回归去拟合，然后设定一个阈值0.5，小于阈值的是0（良性肿瘤），大于阈值的是1（恶性肿瘤）。用公式表示为

$$若\ h_{\theta}(x) \geq 0.5，预测\ y = 1$$
$$若\ h_{\theta}(x) < 0.5，预测\ y = 0$$

式中，$h_{\theta}(x)$表示判别肿瘤的逻辑回归模型；y表示预测的结果。

上面的例子似乎很好地解决了恶性肿瘤的预测问题，但是这种模型有一个最大问题是对噪声很敏感（鲁棒性不够），如果增加两个训练样本，按照"线性回归 + 阈值"的思路，使用线性回归会得到一条直线，然后设置阈值为0.5，如图7-5所示。

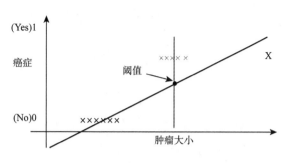

图 7-5　增加训练样本的肿瘤分类

这时会产生两个误判点,如果再使用 0.5 作为阈值来预测是否是恶性肿瘤就不合适了。

2. 使用 sigmoid 实现逻辑回归

上面的例子中,可以使用概率来作为判定恶性肿瘤的依据,概率的值在 0 到 1 之间,需要使用一个函数将 $h_\theta(x)$ 的值转换为 0 到 1,然后使用这个值作为判断是否是恶性肿瘤的依据。

可以使用 sigmoid 函数将一个($-\infty$, $+\infty$)之内的实数值变换到区间 0 ~ 1,单调增,定义域是($-\infty$, $+\infty$),值域是(0,1),sigmoid 函数为

$$y = \frac{1}{1 + e^{-x}}$$

函数图象如图 7-6 所示。

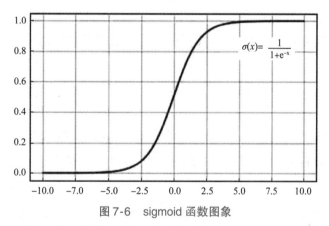

图 7-6　sigmoid 函数图象

其中的 e^x 是一个指数函数。指数函数一般表达式为 $y = a^x$ ($a > 0$ 且 $a \neq 1$),其函数图象如图 7-7 所示。

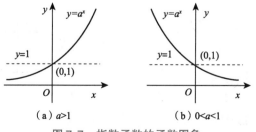

（a）$a > 1$　　　　　　（b）$0 < a < 1$

图 7-7　指数函数的函数图象

用$h_\theta(x)$替换z，就可以将线性回归的值变换到$0 \sim 1$，公式如下：

$$g(\boldsymbol{\theta}^\mathrm{T}\boldsymbol{X}) = \frac{1}{1+\mathrm{e}^{-\boldsymbol{\theta}^\mathrm{T}\boldsymbol{X}}}$$

$$g(z) = \frac{1}{1+\mathrm{e}^{-z}}$$

这个函数有一个特殊的点，就是$h_\theta(x) = 0.5$这个点，这时$\boldsymbol{\theta}^\mathrm{T}\boldsymbol{X}$的值为0。如果$\boldsymbol{\theta}^\mathrm{T}\boldsymbol{X}$的值大于0，概率就大于0.5；如果$\boldsymbol{\theta}^\mathrm{T}\boldsymbol{X}$的值小于0，概率就小于0.5。

7.3 决策边界

1. 线性判定边界

图7-8是一个典型的二分类问题，可以找到一条直线，把两类点分开。如何解决这个问题，先做一个线性回归$h_\theta(x) = \theta_0 + \theta_1 x_1 + \theta_2 x_2$，当$\theta_0 + \theta_1 x_1 + \theta_2 x_2 = 0$时，可以得到一条直线 $-3 + x_1 + x_2 = 0$，如图7-8所示。

如何使用这条直线分隔两类点？把$h_\theta(x)$外面包裹一个sigmoid函数，$h_\theta(x) = g(\theta_0 + \theta_1 x_1 + \theta_2 x_2)$，这时直线上方的点（$-3 + x_1 + x_2 > 0$）的概率都大于0.5（正样本），直线下方的点（$-3 + x_1 + x_2 < 0$）的概率都小于0.5（负样本）。

$-3 + x_1 + x_2 = 0$的这条直线就是要找的判定边界，因为是线性的，所以称为线性判定边界。

2. 非线性判定边界

实际使用经常会遇到样本线性不可分的情况，这时要区分样本点，就需要用到多项式。可以定义一个多项式$h_\theta(x) = g(\theta_0 + \theta_1 x_1 + \theta_2 x_2 + \theta_3 x_1^2 + \theta_4 x_2^2)$，如果取到一组$\theta = (-1, 0, 0, 1, 1)$，可以得到$g(x_1^2 + x_2^2 - 1)$，当$x_1^2 + x_2^2 - 1 = 0$时，就是一个圆形。在圆内，样本使用sigmoid函数后概率小于0.5，圆外点使用sigmoid函数后概率大于0.5。$x_1^2 + x_2^2 - 1 = 0$就是一个非线性判定边界，如图7-9所示。

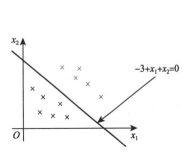

图7-8 线性判定边界

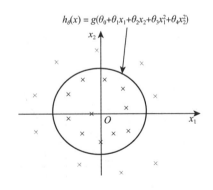

图7-9 非线性判定边界

通过上面的例子可以看出,逻辑回归就是去寻找判定边界。如果是高维的多项式,可以得到很复杂的判定边界。

图 7-10 所示横坐表示样本的一个特征,纵坐标表示样本的另一个特征,使用不同的颜色表示不同类型的样本,γ 表示正则化系数,图 7-10(a)表示使用正则化对决策边界进行了抑制,图 7-10 (b)没有使用正则化,可以实现复杂的决策边界。

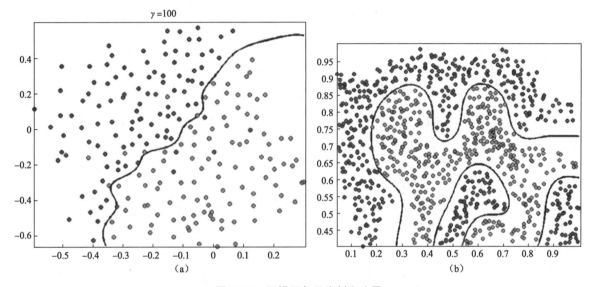

图 7-10 逻辑回归寻找判定边界

7.4 逻辑回归损失函数

逻辑回归中,对于每一组样本点可以训练出一组 θ,每组 θ 可以确定一个判定边界,需要定义一个函数来衡量这些判定边界是好还是坏。在线性回归中,使用 MSE 作为损失函数。如果逻辑回归使用线性回归的损失函数,会得到如图 7-11 所示的曲线,这条曲线是非凸函数,所以逻辑回归不能使用 MSE 作为损失函数。

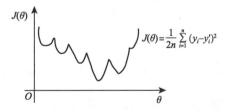

图 7-11 逻辑回归使用线性回归的损失函数

逻辑回归中,通常使用对数函数作为损失函数。对数函数的函数图象如图 7-12 所示。

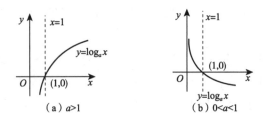

图7-12　对数函数的函数图象

逻辑回归的损失函数为

$$\mathrm{cost}(h_\theta(x),y)=\begin{cases}-\log(h_\theta(x)), & y=1\\ -\log(1-h_\theta(x)), & y=0\end{cases}$$

当 $y=1$，$h_\theta(x)$ 是一个概率，这时 $h_\theta(x)=0.01$，这种情况就是在正样本的情况下，模型预测成了负样本，这时 $-\log(h_\theta(x))$ 是一个很大的值，表示误差很大，如图7-13（a）所示。

当 $y=0$，$h_\theta(x)$ 是一个概率，这时 $h_\theta(x)=0.99$，这种情况就是在负样本的情况下，模型预测成了正样本，这时 $-\log(1-h_\theta(x))$ 是一个很大的值，表示误差很大，如图7-13（b）所示。

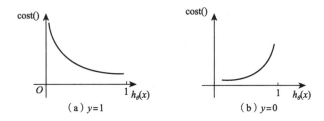

图7-13　逻辑回归的损失函数

通常将两种情况下的损失函数合并，得到逻辑回归的损失函数-互熵损失函数 $J(\theta)$，公式如下：

$$J(\theta)=\frac{1}{m}\sum_{i=1}^{m}\mathrm{cost}(h_\theta(x),y)$$

$$J(\theta)=-\frac{1}{m}\left[\sum_{i=1}^{m}y^i\log(h_\theta(x^i))+(1-y^i)\log(1-h_\theta(x^i))\right]$$

上面的式子是从第1个样本到第 m 个样本都计算一遍，还要加上正则化项L2，最后得到带有L2的逻辑回归损失函数 $J(\theta)$，公式如下：

$$J(\theta)=-\frac{1}{m}\left[\sum_{i=1}^{m}y^i\log(h_\theta(x^i))+(1-y^i)\log(1-h_\theta(x^i))\right]+\frac{\lambda}{2m}\sum_{j=1}^{n}\theta_j^2$$

继续对损失函数做梯度下降，求偏导数可以得到逻辑回归的梯度下降函数并求出梯度值 θ_j，公式如下：

$$\theta_j:=\theta_j-\alpha\sum_{i=1}^{m}(h_\theta(x^i)-y^i)x_j^i$$

$$\theta_j : = \theta_j - \alpha \frac{\partial}{\partial \theta_j} J_\theta$$

式中,$h_\theta(x^i)$表示模型预测值;y^i表示实际值;α表示学习率;θ_j:表示求出的新的梯度值。

7.5　线性决策边界代码实现

1.加载成绩数据

现在有两门课程(语文和数学)的考试成绩,x轴是语文的成绩,y轴是数学的成绩,现在需要判定学生成绩是否合格。

在 Jupyter Notebook 中导入 minimize、PolynomialFeatures 模块,代码如下:

```
1.  import pandas as pd
2.  import numpy as np
3.  import matplotlib as mpl
4.  import matplotlib.pyplot as plt
5.  from scipy.optimize import minimize
6.  from sklearn.preprocessing import PolynomialFeatures
7.  pd.set_option('display.notebook_repr_html', False)
8.  pd.set_option('display.max_columns', None)
9.  pd.set_option('display.max_rows', 150)
10. pd.set_option('display.max_seq_items', None)
11. #% config InlineBackend.figure_formats = {'pdf',}
12. % matplotlib inline
13.
14. import seaborn as sns
15. sns.set_context('notebook')
16. sns.set_style('white')
17. #显示中文
18. plt.rcParams['font.sans - serif'] = ['SimHei']
19. plt.rcParams['axes.unicode_minus'] = False
```

读取数据,定义函数 loaddata( )读取文件,定义函数 plotData( )绘制样本点。读取数据代码如下:

```
1.  def loaddata(file, delimeter):
2.      data = np.loadtxt(file, delimiter = delimeter)
3.      print('Dimensions: ',data.shape)
4.      print(data[1:6,:])
```

```
5.      return(data)
6. def plotData(data, label_x, label_y, label_pos, label_neg, axes =None):
7.     #获得正负样本的下标(即哪些是正样本,哪些是负样本)
8.     neg = data[:,2] = = 0
9.     pos = data[:,2] = = 1
10.
11.    if axes = = None:
12.        axes = plt.gca()
13.    axes.scatter(data[pos][:,0], data[pos][:,1], marker ='+', c ='k', s
    =60, linewidth =2, label =label_pos)
14.     axes.scatter(data[neg][:,0], data[neg][:,1], c ='y', s =60, la-
    bel =label_neg)
15.   axes.set_xlabel(label_x)
16.   axes.set_ylabel(label_y)
17.   axes.legend(frameon = True, fancybox = True);
18. data = loaddata('data1.txt', ',')
19. X = np.c_[np.ones((data.shape[0],1)), data[:,0:2]]
20. y = np.c_[data[:,2]]
21. plotData(data, '语文成绩', '数学成绩', '通过', '不通过')
```

代码运行后输出语文、数学成绩散点图,如图 7-14 所示。

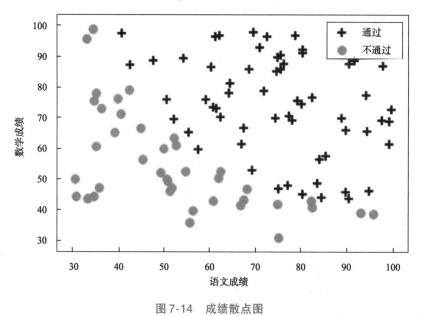

图 7-14　成绩散点图

2. 定义模型

定义逻辑回归、定义损失函数、定义梯度下降、调用损失函数和梯度下降,代码如下:

```
1.   #定义 sigmoid 函数
2.   def sigmoid(z):
3.       return(1 / (1 + np.exp(-z)))
4.   #定义损失函数
5.   def costFunction(theta, X, y):
6.       m = y.size
7.       h = sigmoid(X.dot(theta))
8.
9.       J = -1* (1/m)* (np.log(h).T.dot(y) + np.log(1 - h).T.dot(1 - y))
10.
11.      if np.isnan(J[0]):
12.          return(np.inf)
13.      return(J[0])
14.  #求解梯度
15.  def gradient(theta, X, y):
16.      m = y.size
17.      h = sigmoid(X.dot(theta.reshape(-1,1)))
18.
19.      grad = (1/m)* X.T.dot(h - y)
20.
21.      return(grad.flatten())
22.  initial_theta = np.zeros(X.shape[1])
23.  cost = costFunction(initial_theta, X, y)
24.  grad = gradient(initial_theta, X, y)
25.  print('Cost: \n', cost)
26.  print('Grad: \n', grad)
```

3. 训练模型

最小化损失函数,定义预测函数并调用,绘制决策边界,代码如下:

```
1.   # 直接调用 scipy 里面的最小化损失函数的 minimize 函数
2.   res = minimize(costFunction, initial_theta, args = (X,y), method =
     None, jac = gradient, options = {'maxiter':400})
3.   res
4.   def predict(theta, X, threshold = 0.5):
```

```
5.      p = sigmoid(X.dot(theta.T)) > = threshold
6.      return(p.astype('int'))
7.
8.  #第一门课45分，第二门课85分的同学
9.  sigmoid(np.array([1, 45, 85]).dot(res.x.T))
10.
11. p = predict(res.x, X)
12. print('Train accuracy {}% '.format(100* sum(p = = y.ravel())/p.size))
13. plt.scatter(45, 85, s =60, c ='r', marker = 'v', label = '(45, 85)')
14. #plotData(data, 'Exam 1 score', 'Exam 2 score', 'Pass', 'Failed')
15. plotData(data, '语文成绩', '数学成绩', '通过', '不通过')
16. x1_min, x1_max = X[:,1].min(), X[:,1].max(),
17. x2_min, x2_max = X[:,2].min(), X[:,2].max(),
18. xx1, xx2 = np.meshgrid(np.linspace(x1_min, x1_max), np.linspace(x2_
    min, x2_max))
19. h = sigmoid(np.c_[np.ones((xx1.ravel().shape[0],1)), xx1.ravel(),
    xx2.ravel()].dot(res.x))
20. h = h.reshape(xx1.shape)
21. plt.contour(xx1, xx2, h, [0.6], linewidths =1, colors = 'b');
```

7.6　非线性决策边界代码实现

首先使用 PolynomialFeatures 生成多个特征，然后使用正则化 Lambda = 0，Lambda = 1，绘制非线性决策边界，代码如下：

```
1.  #读取数据
2.  data2 = loaddata('data2.txt', ',')
3.  #拆分数据
4.  y = np.c_[data2[:,2]]
5.  X = data2[:,0:2]
6.  #绘制散点图
```

```
7.  plotData(data2, '语文成绩', '数学成绩', 'y = 通过', 'y = 不通过')
8.  #特征映射,代码如下:做特征映射,生成多项式特征,最高的次数为6个
9.  poly = PolynomialFeatures(6)
10. XX = poly.fit_transform(data2[:,0:2])
11. XX.shape
12. #定义损失函数,代码如下
13. def costFunctionReg(theta, reg, * args):
14.     m = y.size
15.     h = sigmoid(XX.dot(theta))
16.
17.     J = -1* (1/m)* (np.log(h).T.dot(y) + np.log(1 - h).T.dot(1 - y))
        + (reg/(2* m))* np.sum(np.square(theta[1:]))
18.
19.     if np.isnan(J[0]):
20.         return(np.inf)
21.     return(J[0])
22. #定义梯度下降函数,代码如下
23. def gradientReg(theta, reg, * args):
24.     m = y.size
25.     h = sigmoid(XX.dot(theta.reshape(-1,1)))
26.
27.     grad = (1/m)* XX.T.dot(h - y) + (reg/m)* np.r_[[[0]],theta[1:].
    reshape(-1,1)]
28.
29.     return(grad.flatten())
30. #训练模型,代码如下
31. initial_theta = np.zeros(XX.shape[1])
32. costFunctionReg(initial_theta, 1, XX, y)
33.
34. fig, axes = plt.subplots(1,3, sharey = True, figsize = (17,5))
35.
36. #决策边界,正则化系数Lambda太大太小分别会出现什么情况
```

```
37. # Lambda = 0：就是没有正则化,出现过拟合
38. # Lambda = 1：这才是正确的打开方式
39. # Lambda = 100：过度正则化,导致没拟合出决策边界
40.
41. for i, C in enumerate([0, 1, 100]):
42.     # 最优化 costFunctionReg
43.     res2 = minimize(costFunctionReg, initial_theta, args = (C, XX, y),
    method = None, jac = gradientReg, options = {'maxiter':3000})
44.
45.     # 准确率
46.     accuracy = 100* sum(predict(res2.x, XX) = = y.ravel())/y.size
47.
48.     # 对 X,y 的散列绘图
49.     plotData(data2, 'Microchip Test 1', 'Microchip Test 2', 'y =1', 'y =
    0', axes.flatten()[i])
50.
51.     # 画出决策边界
52.     x1_min, x1_max = X[:,0].min(), X[:,0].max(),
53.     x2_min, x2_max = X[:,1].min(), X[:,1].max(),
54.     xx1, xx2 = np.meshgrid(np.linspace(x1_min, x1_max), np.linspace
    (x2_min, x2_max))
55.     h = sigmoid(poly.fit_transform(np.c_[xx1.ravel(), xx2.ravel()]).
    dot(res2.x))
56.     h = h.reshape(xx1.shape)
57.     axes.flatten()[i].contour(xx1,xx2,h,[0.5],linewidths =1,colors ='g');
58.     axes.flatten()[i].set_title('Train accuracy {}% with Lambda = {}'.
format(np.round(accuracy, decimals =2), C))
```

代码运行后,输出非线性决策边界。图 7-15(a)中没有使用正则化系数,可以得到 91.53% 的训练精度;图 7-15(b)设置正则化系数为 1,可以得到 83.05% 的训练精度。从图 7-15 中可以看出,图 7-15(a)存在过拟合的现象。

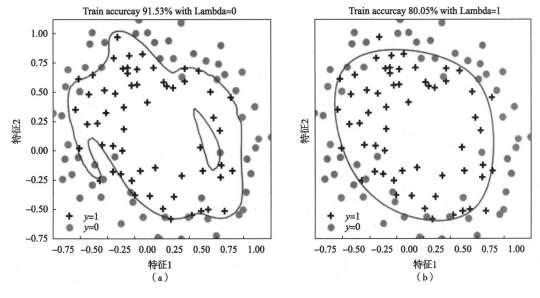

图 7-15　非线性决策边界

使用逻辑回归实现心脏病患者分类

心脏病是人类健康的头号杀手。如果可以通过提取人体相关的体测指标,通过数据挖掘方式来分析不同特征对于心脏病的影响,将对预防心脏病起到至关重要的作用。

本单元数据集包含了 303 条美国某区域的心脏病检查患者的体测数据,一共有 14 个特征。

8.1 数据加载与显示

导入数据加载的包

在 Jupyter Notebook 中导入 numpy、pandas、matplotlib 三个包,用来加载心脏病检查患者数据集,同时对数据进行分析,使用 read_csv 打开 CSV 文件,使用 info( ) 函数查看数据。代码如下:

```
1.  import numpy as np
2.  import pandas as pd
3.  import seaborn as sns
4.  import matplotlib.pyplot as plt
5.  #显示中文
6.  plt.rcParams['font.sans - serif'] = ['SimHei']
7.  plt.rcParams['axes.unicode_minus'] = False
8.  df_data = pd.read_csv("../data/heart.csv")
9.  df_data.head()
```

代码运行后,输出 heart.csv 中心脏病的数据,共有 14 个特征,303 条样本,如图 8-1 所示。心脏病数据集中字段的含义见表 8-1。

	age	sex	cp	trestbps	chol	fbs	restecg	thalach	exang	oldpeak	slope	ca	thal	target
0	63	1	3	145	233	1	0	150	0	2.3	0	0	1	1
1	37	1	2	130	250	0	1	187	0	3.5	0	0	2	1
2	41	0	1	130	204	0	0	172	0	1.4	2	0	2	1
3	56	1	1	120	236	0	1	178	0	0.8	2	0	2	1
4	57	0	0	120	354	0	1	163	1	0.6	2	0	2	1

图 8-1　心脏病数据显示结果

表 8-1　心脏病数据集中字段的含义

字段名	类型	描述
age	STRING	年龄
sex	STRING	性别,取值为 female 或 male
cp	STRING	胸部疼痛类型,痛感由重到轻依次为 typical、atypical、non-anginal 及 asymptomatic
trestbps	STRING	血压
chol	STRING	胆固醇
fbs	STRING	空腹血糖。如果血糖含量大于 120 mg/dL,则取值为 true,否则取值为 false
restecg	STRING	心电图结果是否有 T 波,由轻到重依次为 norm 和 hyp
thalach	STRING	最大心跳数
exang	STRING	是否有心绞痛。true 表示有心绞痛;false 表示没有心绞痛
oldpeak	STRING	运动相对于休息的 ST Depression,即 ST 段压值
slope	STRING	心电图 ST Segment 的倾斜度,程度取值包括 down、flat 及 up
ca	STRING	透视检查发现的主要血管数
thal	STRING	病发种类,由轻到重依次为 norm、fix 及 rev
target	STRING	是否患病。buff 表示健康,sick 表示患病

8.2　数据分析

1. 数据关系系数分析

使用 seaborn 的 heatmap( )绘制关系系数图,代码如下:

```
1.  plt.figure(figsize = (10,10))
2.  sns.heatmap(df_data.corr(),cmap = "YlGnBu",annot = True)
```

代码运行后,生成心脏病数据集特征的相关系数热力图,如图 8-2 所示。从结果中可以看出。
①心脏病和 cp(胸部疼痛类型)、thalach(最大心跳数)、exang(是否有心绞痛)、oldpeak(ST 段

压值）几个特征之间有较强的相关性，相关系数超过 0.4。

②心脏病和 age（年龄）、sex（性别）、slope（心电图 ST Segment 的倾斜度）、ca（主要血管数）、thal（病发种类）几个特征存在一定的相关性。

③心脏病和 chol（胆固醇）、fbs（空腹血糖）之间的相关性较弱。

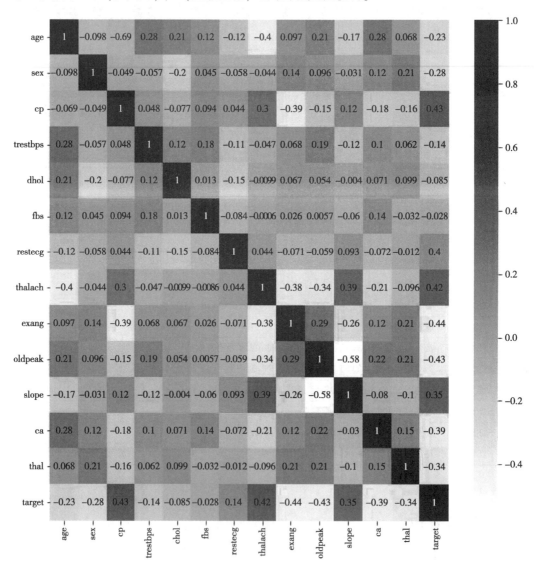

图 8-2　心脏病数据热力图

2. 绘制年龄-血压散点图

从热力图中可以看出，年龄-血压和心脏病之间存在相关性，使用 seaborn 的 scatterplot（）绘制三者之间的散点图的，代码如下：

```
1.  df_heart_target = df_data[['age','thalach','target']]
2.  markers = {'心脏病患者':'s','健康':'x'}
3.  sns.scatterplot(x = 'age',y = 'thalach',hue = 'target',style = 'target
    ',data = df_heart_target)
4.  plt.legend(title = '年龄 – 血压 – 心脏病', labels = ['心脏病患者', '健康'])
5.  plt.show()
```

代码运行后,可以看出 35 ~ 50 之间的年龄段的心脏病患者较多,这些患者中血压都偏高,所以血压是判定是否患心脏病的一个重要的特征。散点图如图 8-3 所示。

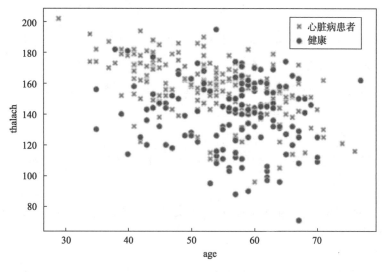

图 8-3　年龄-血压散点图

8.3　特征数据处理

1. 类型数据的处理

cp,restecg,slope,thal 特征是类型数据非连续性分类数据。采用 get_dummies( )编码方式处理非连续性分类数据,代码如下:

```
1.  df_data.cp.unique()
2.  df_data.cp.value_counts()
3.
```

```
4.  df_data.restecg.unique()
5.  df_data.restecg.value_counts()
6.
7.  df_data.slope.unique()
8.  df_data.slope.value_counts()
9.
10. df = df_data[['cp','thal','slope']]
11. df.apply(pd.value_counts)
12.
13. a = pd.get_dummies(df_data['cp'],prefix = 'cp')
14. b = pd.get_dummies(df_data['thal'],prefix = 'thal')
15. c = pd.get_dummies(df_data['slope'],prefix = 'slope')
16. df_heart_all = [df_data,a,b,c]
17. df_heart = pd.concat(df_heart_all,axis = 1)
18. df_heart = df_data.drop(columns = ['cp','thal','slope'])
19. df_heart.head()
```

代码运行后，将数据集中的 cp、thal、slope 三个类型特征继续进行 one-hot 编码，生成新的特征字段，结果如图 8-4 所示。

	age	sex	trestbps	chol	fbs	restecg	thalach	exang	oldpeak	ca	...	cp_1	cp_2	cp_3	thal_0	thal_1	thal_2	thal_3	slope_0	slope_1	slope_2
0	63	1	145	233	1	0	150	0	2.3	0	...	0	0	1	0	1	0	0	1	0	0
1	37	1	130	250	0	1	187	0	3.5	0	...	0	1	0	0	0	1	0	1	0	0
2	41	0	130	204	0	0	172	0	1.4	0	...	1	0	0	0	0	1	0	0	0	1
3	56	1	120	236	0	1	178	0	0.8	0	...	1	0	0	0	0	1	0	0	0	1
4	57	0	120	354	0	1	163	1	0.6	0	...	0	0	0	0	0	1	0	0	0	1

5 rows × 22 columns

图 8-4   cp、thal、slope 分类数据处理结果

2. 拆分数据集

先将标签取出，然后通过拆分组件将数据分为两部分，按照训练集和预测集 8∶2 的比例拆分。训练集数据用于逻辑回归二分类模型的训练，预测集数据用于训练模型的预测，代码如下：

```
1.  X = df_data.drop(['target'],axis = 1)
2.  y = df_data.target.values
3.  y = y.reshape(-1,1)
4.  X.shape,y.shape
```

```
5.
6.  from sklearn.model_selection import train_test_split
7.  from sklearn.preprocessing import MinMaxScaler
8.
9.  X_train,X_test,y_train,y_test = train_test_split(X,y,test_size = 0.2)
```

3. 归一化操作

使用 Min-Max Scaling 归一化将每个特征的数值范围变为 0 到 1 之间,可以去除量纲对结果的影响。在 sklearn 当中,使用 preprocessing. MinMaxScaler 来实这个功能,代码如下:

```
1.  scaler = MinMaxScaler()
2.  X_train = scaler.fit_transform(X_train)
3.  X_test = scaler.fit_transform(X_test)
```

8.4　逻辑回归建模

1. sigmoid 函数

逻辑回归的输出结果是判定二分类的,在实际问题中可用来解决二分类问题。逻辑回归最关键的就是理解 sigmoid 函数,又称 Logistic 函数,其表达式为

$$h(x) = \frac{1}{1 + e^{-\theta^T X}}$$

在 Logistic 函数的输出是 0 ~ 1 之间的数。把这个函数的输出值看作属于类 1 的概率,如果这条数据是类 1,就让这个函数的值接近 1;如果这条数据是类 0,就让 1 减去这个函数的值接近 1。

sigmoid 函数用来实现二分类问题,使用它对运算结果计算得到一个概率,然后和阈值进行比较,判断是否属于某一类,代码如下:

```
1.  def sigmoid(z):
2.      y_hat = 1/(1 + np.exp(-z))
3.      return y_hat
```

2. 损失函数

sigmoid 函数表示 0 ~ 1 中取 1 的概率。所以,损失函数可以定义为

当 $y = 0$ 时,$\mathrm{cost} = -\log(1 - h(x))$;

当 $y = 1$ 时,$\mathrm{cost} = -\log(h(x))$。

将损失函数与 0-1 分布的分布律对应起来,设 $p = h(x)$,损失函数就是在 0-1 分布的基础上取对数然后再取负数。损失函数的要求就是预测结果与真实结果越相近,函数值越小,所以会在前

面加上负号。当 $y=0$ 时，$1-p$ 的概率会比较大，在前面加上负号，cost 值就会很小；当 $y=1$ 时，p 的概率会比较大，在前面加上负号，cost 值就会很小。至于取对数，就是跟最大似然函数有关系，取对数不影响原本函数的单调性，而且会放大概率之间的差异，更好地区分各个样本的类别。逻辑回归的损失函数代码实现如下：

```
1.  def loss_function(X,y,w,b):
2.      y_hat = sigmoid(np.dot(X,w) +b)
3.      loss = - (y* np.log(y_hat) + (1 - y)* np.log(1 - y_hat))
4.      cost = np.sum(loss)/X.shape[0]
5.      return cost
```

3. 定义梯度下降函数

逻辑回归的梯度下降函数和线性回归一样，先微分，再把计算出来的导数乘以学习率，不断迭代更新 w 和 b。通过求偏导可以得到梯度，梯度下降函数代码如下：

```
1.      l_history = np.zeros(iter)
2.      w_history = np.zeros((iter,w.shape[0],w.shape[1]))
3.      b_history = np.zeros(iter)
4.      for i in range(iter):
5.          y_hat = sigmoid(np.dot(X,w) +b)
6.          loss = - (y* np.log(y_hat) + (1 - y)* np.log(1 - y_hat))
7.          derivative_w = np.dot(X.T,((y_hat - y)))/X.shape[0]
8.          derivative_b = np.sum(y_hat - y)/X.shape[0]
9.          w = w - lr* derivative_w
10.         b = b - lr* derivative_b
11.         l_history[i] = loss_function(X,y,w,b)
12.         print("轮次",i +1,"当前轮次训练损失:",l_history[i])
13.         w_history[i] = w
14.         b_history[i] = b
15.     return l_history,w_history,b_history
```

4. 预测分类结果

定义 predict( ) 函数预测分类的结果。这里定义了一个阈值，概率为 0.5，高于阈值的样本可以预测为 1，低于阈值的样本可以预测为 0，代码如下：

```
1.  def predict(X,w,b):
2.      z = np.dot(X,w) +b
3.      y_hat = sigmoid(z)
4.      y_pred = np.zeros((y_hat.shape[0],1))
```

```
5.        for i in range(y_hat.shape[0]):
6.            if y_hat[i,0]<0.5:
7.                y_pred[i,0]=0
8.            else:
9.                y_pred[i,0]=1
10.     return y_pred
```

5.定义模型

定义 logistic_regression( )函数训练模型,同时初始化参数开始训练,完成 500 轮次的训练,输出最终损失为 0.331 373,准确率为 87.19%,代码如下:

```
1.  def logistic_regression(X,y,w,b,lr,iter):
2.      l_history,w_history,b_history = gradient_descent(X,y,w,b,lr,
    iter)
3.      print("训练最终损失:", l_history[-1]) #打印最终损失
4.      y_pred = predict(X,w_history[-1],b_history[-1]) #进行预测
5.      traning_acc = 100 - np.mean(np.abs(y_pred - y_train))* 100 #计算
    准确率
6.      print("逻辑回归训练准确率: {:.2f}% ".format(traning_acc))  # 打印
    准确率
7.      return l_history, w_history, b_history #返回训练历史记录
8.  #初始化参数
9.  dimension = X.shape[1]
10. weight = np.full((dimension,1),0.1) #权重向量,向量一般是1D,但这里实际
    上创建了2D张量
11. bias = 0 #偏置值
12. #初始化超参数
13. alpha = 1 #学习速率
14. iterations = 500 #迭代次数
15. #用逻辑回归函数训练机器
16. loss_history, weight_history, bias_history = logistic_regression(X_
    train,y_train,weight,bias,alpha,iterations)
```

8.5　模型测试

模型训练完毕后需要使用测试集对模型进行测试,输出模型的 acc 并绘制损失曲线,代码

如下：

```
1.  y_pred = predict(X_test,weight_history[-1],bias_history[-1])
2.  testing_acc = 100 - np.mean(np.abs(y_pred - y_test)) * 100
3.  print("逻辑回归测试准确率: {:.2f}% ".format(testing_acc))
4.  loss_history_test = np.zeros(iterations) #初始化历史损失
5.  for i in range(iterations): #求训练过程中不同参数带来的测试集损失
6.      loss_history_test[i] = loss_function(X_test,y_test,
7.      weight_history[i],bias_history[i])
8.  index = np.arange(0,iterations,1)
9.  plt.plot(index, loss_history,c = 'blue',linestyle = 'solid')
10. plt.plot(index, loss_history_test,c = 'red',linestyle = 'dashed')
11. plt.legend(["Training Loss", "Test Loss"])
12. plt.xlabel("Number of Iteration")
13. plt.ylabel("Cost")
14. plt.show() # 同时显示训练集和测试集损失曲线
```

代码运行后，输出针对训练集和测试集的损失曲线，如图 8-5 所示。横坐标 Number of Iteration 表示训练迭代次数，纵坐标 Cost 表示损失。从图 8-5 中可以看出，当迭代到 80～100 次后，Training Loss（训练集的损失）进一步下降，越来越小，但是 Test Loss（测试集的损失）没有下降，反而呈现上升趋势，这是明显的过拟合现象，因此最佳迭代次数应该在 80～100 之间。

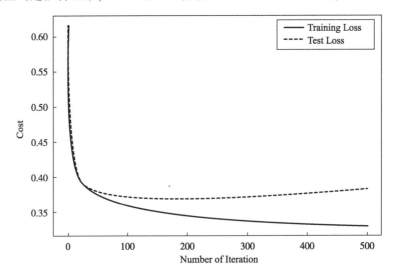

图 8-5　损失曲线

8.6 使用 sklearn 实现逻辑回归

1. 使用 sklearn 建立逻辑回归模型

在实际工作中,不会使用前面的方法来搭建模型,而是使用 sklearn 库的方法直接调用模型。下面的代码是直接调用 LogisticRegression( )函数实现逻辑回归,然后使用 score( )函数来计算准确率、训练集 score 分数,代码如下:

```
1.  from sklearn.linear_model import LogisticRegression #导入逻辑回归模型
2.  lr = LogisticRegression() # lr,就代表逻辑回归模型
3.  lr.fit(X_train,y_train) # fit,就相当于梯度下降
4.  print("sklearn 逻辑回归测试准确率{:.2f}% ".format(lr.score(X_test,y_
    test)* 100))
```

计算分类的准确度,测试集 score 分数,代码如下:

```
1.  lr.score(X_test,y_test)
2.  from sklearn.metrics import accuracy_score
3.  y_pre_lr = lr.predict(X_test)
4.  accuracy_score(y_test,y_pre_lr)
```

2. sklearn 中 LogisticRegression( ) 函数的用法

(1)基本语法

LogisticRegression(C = 1.0, class_weight = None, dual = False, fit_intercept = True, intercept_scaling = 1, max_iter = 100, multi_class = ' ovr ', n_jobs = 1, penalty = 'L2 ', random_state = None, solver = 'liblinear ', tol = 0.0001, verbose = 0, warm_start = False)

(2)主要参数说明

penalty:惩罚项,str 类型,可选参数为 L1 和 L2,默认为 L2。用于指定惩罚项中使用的规范。

dual:对偶或原始方法,bool 类型,默认为 False。对偶方法只用在求解线性多核(liblinear)的 L2 惩罚项上。当样本数量 >样本特征的时候,dual 通常设置为 False。

tol:停止求解的标准,float 类型,默认为 1e - 4。就是求解到多少的时候停止,认为已经求出最优解。

C:正则化系数 λ 的倒数,float 类型,默认为 1.0。必须是正浮点型数。像 SVM 一样,越小的数值表示越强的正则化。

fit_intercept:是否存在截距或偏差,bool 类型,默认为 True。

intercept_scaling:仅在正则化项为"liblinear",且 fit_intercept 设置为 True 时有用。float 类型,默认为 1。

class_weight:用于标示分类模型中各种类型的权重,可以是一个字典或者'balanced'字符串,默认为不输入,也就是不考虑权重,即为 None。

8.7 使用网格搜索做超参数调优

1. 初始化模型参数范围

机器学习中的参数是从训练数据中学习出的变量,如网络中的权重、偏差等。而超参数是用来确定模型的一些参数。超参数不同,模型不同。超参数一般需要根据经验确定,这些参数无法直接从数据中学习到,需要在模型训练前设置。

例如 LogisticRegression 模型中 C、penalty、class_weight 等参数;深度学习中,学习速率、迭代次数、层数、每层神经元的个数等都是超参数。

对于 LogisticRegression 模型,可以根据经验首先确定模型超参数的范围,然后通过查找搜索范围内的所有的点来确定最优值。这里选取 C、penalty、class_weight 三个超参数,代码如下:

```
1.  parm_grid = [
2.      {
3.          'C':[0.01,0.1,1,10,100],
4.          'penalty':['L2','L1',],
5.          'class_weight':['balanced',None]    #参数为'balanced'则依照输
    入样本的频率自动调整权重
6.      }
7.  ]
```

2. 使用网格搜索参数

网格搜索(grid search):一种调参手段。穷举搜索:在所有候选的参数中,通过循环遍历,尝试每一种可能性,表现最好的参数就是最终的结果。其原理就像是在数组里找最大值。为什么叫网格搜索?以有两个参数的模型为例,参数 a 有 3 种可能,参数 b 有 4 种可能,把所有可能性列出来,可以表示成一个 3×4 的表格,其中每个 cell 就是一个网格,循环过程就像是在每个网格里遍历、搜索,所以称为网格搜索。可以使用 sklearn 提供的 GrideSearchCV 函数实现网格搜索功能。

(1)GrideSearchCV 的基本语法

sklearn. model_selection. GridSearchCV(estimator, param_grid, scoring, refit, cv, n_jobs)

(2)主要参数说明

estimator:选择使用的分类器,并且传入除需要确定最佳的参数之外的其他参数。

param_grid:需要最优化的参数的取值,其值为字典或者列表。

scoring:模型评价标准。根据所选模型不同,评价准则不同。默认值为 None,表示使用 estimator 的误差估计函数。

n_jobs:训练模型时使用的进程个数,默认为 1。若值为 -1,则用所有的 CPU 进行运算;若值为 1,则不进行并行运算,这样方便调试。

refit:默认为 True,程序将会以交叉验证训练集得到的最佳参数,作为最终用于性能评估的最佳模型参数。在搜索参数结束后,用最佳参数结果再次训练一遍全部数据集。

cv:交叉验证参数,默认 None,使用三折交叉验证。

3. GrideSearchCV 的使用

使用 GrideSearchCV 的 fit 方法训练模型,然后使用. best_estimator_输出最优参数,代码如下:

```
1.  from sklearn.model_selection import GridSearchCV
2.  gride_search = GridSearchCV(lr,parm_grid,cv =10,n_jobs = -1)
3.  %%time
4.  gride_search.fit(X_train,y_train)
5.  gride_search.best_estimator_
```

代码运行后,输出当前模型的最优参数,C =0.01、class_weight = 'balanced'、penalty = 'L2'。

8.8 逻辑回归模型的衡量指标

1.混淆矩阵

混淆矩阵是机器学习中总结分类模型、预测结果的情形分析表。以矩阵形式将数据集中的记录按照真实的类别与分类模型预测的类别判断两个标准进行汇总。其中,矩阵的行表示真实值,矩阵的列表示预测值。下面先以二分类为例,看一下混淆矩阵表现形式,如图 8-6 所示,其中 a,b,c,d 表示样本数量。

混淆矩阵		预测值	
		正	负
真实值	正	a	b
	负	c	d

图 8-6 混淆矩阵

混淆矩阵中分为 TP、FN、FP、TN 四个类别,如图 8-7 所示,四个类别的含义如下所示:

TP(true positive,真正):将正类预测为正类,真实为 0,预测也为 0。

FN(false negative,假负):将正类预测为负类,真实为 0,预测为 1。

FP(false positive,假正):将负类预测为正类,真实为 1,预测为 0。

TN(true negative,真负):将负类预测为负类,真实为 1,预测也为 1。

2.混淆矩阵的评价指标

(1)正确率/准确率(accuracy)

正确率是指被正确分类的样本比例或数量。准确率的定义是预测正确的结果占总样本的百分比,如图 8-7 所示。

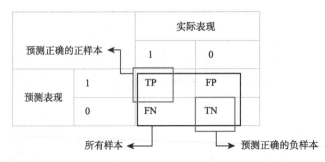

图 8-7　Accuracy 图示

其公式如下：

$$\mathrm{Accuracy} = \frac{\mathrm{TP + TN}}{\mathrm{TP + FP + TN + FN}}$$

虽然准确率可以判断总的正确率,但是在样本不平衡的情况下,并不能作为很好的指标来衡量结果。举个简单的例子,比如在一个总样本中,正样本占 90%,负样本占 10%,样本是严重不平衡的。对于这种情况,只需要将全部样本预测为正样本即可得到 90% 的高准确率,但实际这种分类是无效的。这就说明了,由于样本不平衡的问题,导致了得到的高准确率结果含有很大的水分。即如果样本不平衡,准确率就会失效。正因为如此,也就衍生出了其他两种指标:精准率和召回率。

（2）精确率（precision）

精密性（精确率/精度/查准率）是对"TP（真阳性率）"预测的评估。即预测为阳性的数据（分母 TP + FP）中,实际对了多少（分子 TP）。本例中,指准确预测出患心脏病人数的比例,表示预测为正例（TP + FP）的实例中实际为正例的比例,如图 8-8 所示。

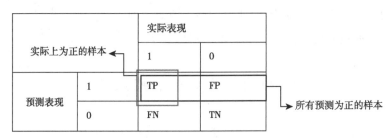

图 8-8　Precision 图示

其公式如下：

$$\mathrm{Precision} = \frac{\mathrm{TP}}{\mathrm{TP + FP}}$$

（3）召回率（recall）

召回率又称查全率,它是针对原样本而言的,它的含义是在实际为正的样本中被预测为正样

本的概率。查全率关心的是"预测出正例的保证性",即从正例中挑选出正例的问题,如图 8-9
所示。

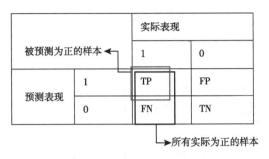

图 8-9 Recall 图示

其公式如下:

$$Recall = \frac{TP}{TP + FN}$$

TP/(TP + FN)即实际为阳性 P(分母 TP + FN),其中预测正确的比例(分子 TP)。

召回率的应用场景:以网贷违约率为例,相对好用户更关心坏用户,不能错放过任何一个坏用
户。因为如果过多地将坏用户当成好用户,这样后续可能发生的违约金额会远超过好用户偿还的
借贷利息金额,造成严重偿失。召回率越高,代表实际坏用户被预测出来的概率越高。

理想情况下,精确率和召回率两者都越高越好。然而,事实上这两者在某些情况下是矛盾的,
精确率高时,召回率低;精确率低时,召回率高。

举例:有 10 000 个人,预测某种癌症,见表 8-2。

表 8-2 假设样本

真值/预测结果	样本标签	
	0	1
0	9990	0
1	10	0

$$Accuracy = \frac{9990}{9990 + 10} = 99.9\%$$

$$Precision = 0\%$$

$$Recall = 0\%$$

这种癌症本身的发病率只有 0.1%,即使不训练模型而直接预测所有人都是健康的,这样的预
测准确率也能达到 99.9%,但是精确率和召回率为 0%,所以这个模型是个无效的模型。

(4)F1 分数

二分类算法中通常使用准确率和召回率这两个指标来评价二分类模型的分析效果。但是这
两个指标是存在冲突的,当准确率高的时候,召回率就低,反之亦然。当这两个指标发生冲突时,
很难在模型之间进行比较。比如,有如下两个模型 A、B,A 模型的召回率高于 B 模型,但是 B 模型

的准确率高于 A 模型, A 和 B 这两个模型的综合性能,哪一个更优呢? 见表 8-3。

表 8-3　A 和 B 模型性能

模型	准确率	召回率
A	80%	90%
B	90%	80%

这时会将准确率和召回率结合成一个指标,即 F1 分数。特别是在需要使用一个简单的办法对比两个分类器时。F1 分数是准确率与召回率的调和平均数。

3. 生成混淆矩阵

可以导入 sklearn 中 metrics 库来生成混淆矩阵。首先获取分类模型的衡量指标,代码如下:

```
1. gride_search.best_estimator_
2. gride_search.best_params_
3. gride_search.best_score_
4. lr = gride_search.best_estimator_
5. lr.score(X_test,y_test)
6. lr.score(X_train,y_train)
```

然后使用网络搜索后得到的 lr 预测 y_pre_lr 的值,代码如下:

```
1. from sklearn.metrics import f1_score
2. y_pre_lr = lr.predict(X_test)
3. f1_score(y_test,y_pre_lr)
```

最后导入 sklearn 的 metrics 包,计算得到混淆矩阵,代码如下:

```
1. from sklearn.metrics import classification_report
2. print(classification_report(y_test,y_pre_lr))
```

代码运行后,输出 precision、recall、f1 的值,如图 8-10 所示。

```
              precision    recall  f1-score   support

           0       0.85      0.65      0.74        26
           1       0.78      0.91      0.84        35

    accuracy                           0.80        61
   macro avg       0.82      0.78      0.79        61
weighted avg       0.81      0.80      0.80        61
```

图 8-10　模型 metrics 输出结果

4. 绘制混淆矩阵

在使用混淆矩阵时,通常需要绘制混淆矩阵的图像,这样可以更加直观地观察结果,可以定义一个函数来实现这个功能,代码如下:

```
1.  from sklearn.metrics import confusion_matrix
2.  cnf_matrix = confusion_matrix(y_test,y_pre_lr)
3.  cnf_matrix
4.
5.  def plot(matrix):
6.      f,ax = plt.subplots()
7.      print(matrix)
8.      sns.heatmap(matrix,annot = True,cmap = "Blues",ax = ax)
9.      ax.set_title("心脏病预测混淆矩阵")
10.     ax.set_xlabel("实际值 0/1")
11.     ax.set_ylabel("预测值 0/1")
12.
13. plot(cnf_matrix)
```

代码运行后,绘制了当前模型的分类结果的混淆矩阵,如图 8-11 所示。从图中可以看出,如果混淆矩阵中的非对角线元素均为 0,就会得到一个近乎完美的分类器,图中的数据含义如下:

00(TN)代表有 22 人真实没患心脏病;模型预测 22 人也没患心脏病。

01(FP)代表有 5 人没患心脏病;模型却预测 5 人患了心脏病。

10(FP)代表有 4 人患了心脏病;模型却预测 4 人没患心脏病。

12(TP)代表有 30 人患了心脏病;模型预测 30 人患了心脏病。

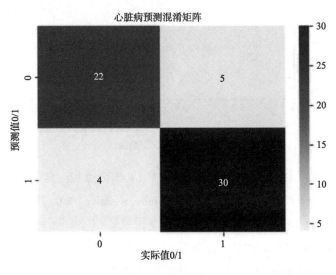

图 8-11　运行结果

8.9 **逻辑回归评价指标**

逻辑回归中,通常还需要使用 ROC 和 AUC 这两个指标来判断模型是否最优。

1. ROC

在上面的例子中,逻辑回归设置的阈值为 0.5,在实际项目中,需要根据要求设定阈值。例如在一些病毒检测实验中,要求能够识别出所有的真阳性,可以容忍一定程度的假阳性,此时阈值 0.1 是最佳选择。如果不能容忍假阳性,要求识别出来的样本必须为真阳性,此时阈值可以设置为 0.9。

在阈值的变动过程中,可以分别求出 TPR(真阳性率)和 FPR(假阳性率),将 FPR 作为横轴, TPR 作为纵轴,将所有的点连接起来就得到 ROC 曲线,如图 8-12 所示。

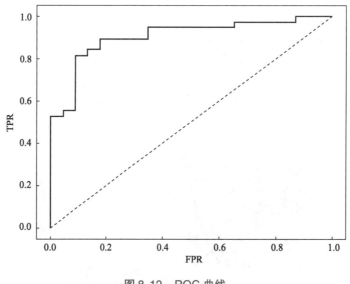

图 8-12 ROC 曲线

可以先看图 8-12 中的 (0,1)、(1,0)、(0,0)、(1,1) 四个点和对角线。

(0,1)点:即 FPR =0, TPR =1,这意味着 FN(false negative) =0,并且 FP(false positive) =0。表示分类器很完美,因为它将所有的样本都正确分类。

(1,0) 点:即 FPR =1,TPR =0,这个分类器是最差的,因为它成功避开了所有的正确答案。

(0,0) 点:即 FPR = TPR =0,FP = TP =0,此时分类器将所有的样本都预测为负样本。

(1,1) 点 :分类器将所有的样本都预测为正样本。

对角线上的点表示分类器将一半的样本预测为正样本,另外一半的样本预测为负样本。因此,ROC 曲线越接近左上角,分类器的性能越好。绘制 ROC 曲线代码如下:

```
1.  from sklearn.metrics import roc_curve
2.  fprs,tprs,thresholds = roc_curve(y_test,decision_scores)
3.
4.  def plot_roc_curve(fprs,tprs):
5.      plt.figure(figsize = (8,6),dpi =80)
6.      plt.plot(fprs,tprs)
7.      plt.plot([0,1],linestyle ='--')
8.      plt.xticks(fontsize =13)
9.      plt.yticks(fontsize =13)
10.     plt.ylabel('TP rate',fontsize =15)
11.     plt.xlabel('FP rate',fontsize =15)
12.     plt.title('ROC 曲线',fontsize =17)
13.     plt.show()
14.
15. plot_roc_curve(fprs,tprs)
```

2. AUC

AUC 值为 ROC 曲线所覆盖的区域面积。显然,AUC 越大,分类器分类效果越好。AUC 是一个数值,当仅仅看 ROC 曲线分辨不出哪个分类器的效果更好时,可用 AUC 来判断。

AUC =1,是完美分类器,采用这个预测模型时,不管设定什么阈值都能得出完美预测。绝大多数预测的场合,不存在完美分类器。

0.5 < AUC <1,优于随机猜测。若这个分类器(模型)妥善设定阈值,能有预测价值。

AUC =0.5,和随机猜测一样(例:丢铜板),模型没有预测价值。

AUC <0.5,比随机猜测还差;但只要总是反预测而行,就优于随机猜测。计算 AUC 的面积代码如下:

```
1.  # AUC - ROC 曲线的面积
2.  from sklearn.metrics import roc_auc_score
3.  roc_auc_score(y_test,decision_scores)
```

单元9

k-means、DBSCAN 聚类算法

前面学习的分类算法在训练模型时数据集都是有标签的，例如在水果的分类算法中，训练集中已经知道了某一个水果是哪一类，这种分类算法在这里就是有监督学习。聚类算法也是要确定一个物体的类别，但和之前的分类问题不同的是，这里没有事先定义好的类别，也就是数据集是没有标签的，聚类算法要自己想办法把一批样本分开，分成多个类，保证每一个类中的样本之间是相似的，而不同类的样本之间是不同的。在这里，类型被称为"簇"（cluster），聚类算法是一种典型的无监督学习算法。

常用的聚类算法有两种 *k*-means 和 DBSCAN。聚类算法的应用场景：用户画像、广告推荐、数据分割、搜索引擎的流量推荐、恶意流量识别；基于位置信息的商业推送、新闻聚类、筛选排序；图像分割、降维、识别；离群点检测；信用卡异常消费；发掘相同功能的基因片段，如图 9-1、图 9-2 所示。

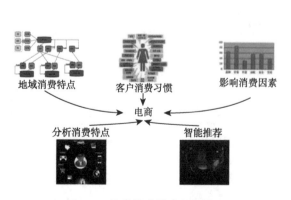

图 9-1　聚类算法的应用场景 1

（a）原始图像

（b）*k* =2时RGB通道分割结果

（c）*k* =3时RGB通道分割结果

（d）*k* =4时RGB通道分割结果

图 9-2　聚类算法的应用场景 2

9.1 ## *k*-means 算法流程

k-means 算法是一种很简单的算法。它属于无监督分类,通过按照一定的方式度量样本之间的相似度,通过迭代更新聚类中心,当聚类中心不再移动或移动差值小于阈值时,则将样本分为不同的类别。使用 *k*-means 算法实现图像分类,需要五个步骤:

① 随机选取聚类中心。

② 根据当前聚类中心,利用选定的度量方式,分类所有样本点。

③ 计算当前每一类的样本点的均值,作为下一次迭代的聚类中心。

④ 计算下一次迭代的聚类中心与当前聚类中心的差距。

⑤ 如差距小于给定迭代阈值时,迭代结束;反之,继续下一次迭代。

首先需要选择合适的"簇"值,也就是 *k* 的值,例如指定 $k=3$,如图 9-3 所示。

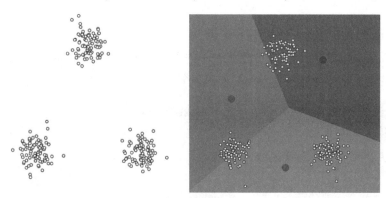

图 9-3　指定 $k=3$ 的样本簇

计算样本到"簇"质心的距离,根据"簇"的值,随机选择样本数据,或者自行指定一个数组(n_clusters,n_features),对 *k* 个簇的质心进行初始化,并计算每个样本点到每个质心的距离(欧式距离或者余弦相似度)。

按照样本到质心的距离重新划分簇,按照优化目标 $\min \sum_{i=1}^{k} \sum_{x \in C_i} \mathrm{dist}(C_i, x)^2$ 重新划分簇,公式中 $\mathrm{dist}(C_i, x)^2$ 表示样本中的点 x 到所有簇的质心的距离,min 表示从所有的距离中找到最小值,将样本划分到对应的簇中,重复上述步骤,如图 9-4 所示。

重复计算质心,更新簇,如图 9-5 所示。

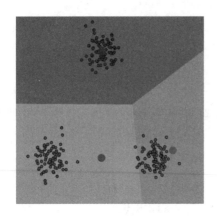 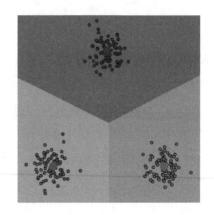

图9-4 重新划分簇　　　　　　　　　图9-5 更新簇

9.2 *k*-means 算法实现

1. 生成数据集

在 Jupyter Notebook 中导入包，设置随机种子的数量。定义五个点 blok_centers，作为聚类算法的五个簇的质心，定义和簇对应的方差 std，代码如下：

```
1.  from sklearn.datasets import make_blobs
2.  #生成簇的中心点,共五个簇
3.  blok_centers = np.array(
4.      [[0.2,2.3],
5.       [-1.5,2.3],
6.       [-2.8,1.8],
7.       [-2.8,2.8],
8.       [-2.8,1.3]])
9.  #定义方差,和中心点之间的距离
10. blok_std = np.array([0.4,0.3,0.1,0.1,0.1])
```

2. 聚类中的簇和质心

通常用 k 表示聚类最终要得到簇的个数。代码中使用 n_clusters 指定簇的个数，需要提前指定。

质心是簇的中心，一个簇所有样本点的中心，通常取向量各维度的平均值。

3. make_blobsk( ) 函数

使用 make_blobsk( ) 生成 X, y 坐标，这里需要指定四个参数，n_samples = 2000 生成样本点的数量，centers = blok_centers 指定簇的质心，cluster_std = blok_std 指定生成样本点方差，random_state = 7 指定随机种子的数量，然后使用 shape 检测 X, y 的形状，代码如下：

```
1.  X,y =make_blobsk(n_samples =2000,centers =blok_centers,
2.                cluster_std =blok_std,random_state =7)
3.  X.shape,y.shape
```

代码运行后,输出结果:((2000,2),(2000,))。

4. 生成数据集的可视化

使用 plot 的 scatter 方法显示数据的原始分布,代码如下:

```
1.  def plot_clusters(X,y =None):
2.      plt.figure(figsize =(8,4))
3.      plt.scatter(X[:,0],X[:,1],c =y,s =1)
4.      plt.xlabel("x1")
5.      plt.ylabel("x2")
6.      plt.show()
7.  plot_clusters(X)
```

代码运行结果如图 9-6 所示。

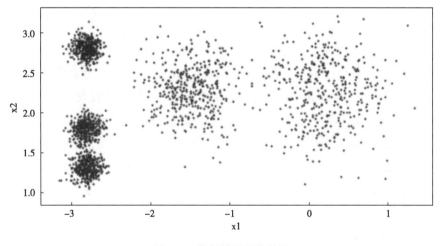

图 9-6 数据的可视化结果

5. 建立 *k*-means 模型,并预测结果

(1)建立 *k*-means 模型

使用 sklearn 库实现 *k*-means 算法,然后使用 fit_predict 对 X 中的样本进行聚类操作,得到训练好的模型 *k*-means,并返回预测的聚类结果 y_pred,代码如下:

```
1.  from sklearn.cluster import KMeans
2.  k =5
3.  kmeans =KMeans(n_clusters =k,random_state =42)
4.  y_pred =kmeans.fit_predict(X)
```

```
5.　y_pred,y_pred.shape
```

代码运行后输出：(array([4, 1, 0, ..., 3, 0, 1]), (2000,))，其中，第一个输出是聚类的结果，第二个输出是 y_pre 的 shape。

（2）输出聚类的标签和簇的质心

使用 labels_ 输出簇的编号，同时使用 cluster_centers_ 输出簇的质心，代码如下：

```
kmeans.labels_,kmeans.cluster_centers_
```

代码运行后输出结果如图 9-7 所示。输出的内容包括每个样本所属的标签（类别）和五个簇的质心坐标。

（3）k-means 函数

可以使用 k-means 函数建立 k-means 聚类模型。Kmeans 函数有多个参数，它的语法如下：

```
(array([4, 1, 0, ..., 3, 0, 1]),
array([[ 0.20876306,  2.25551336],
       [-2.80389616,  1.80117999],
       [-1.46679593,  2.28585348],
       [-2.79290307,  2.79641063],
       [-2.80037642,  1.30082566]]))
```

图 9-7　聚类的标签和簇的质心

```
KMeans (n_clusters =8,init ='k -means + +',n_init =10 , max_iter = 300 , tol =
0.0001 , verbose = 0 , random_state = None , copy_x = True , algorithm = 'auto' )
```

参数说明：

n_clusters：簇的数量 int 类型值，要形成的簇数以及要生成的质心数，默认为 8。

init：初始化方法，可自行指定一个数组（n_clusters, n_features），对簇进行初始化，默认值 default = 'kmeans' 以智能方式为 k-means 聚类选择初始聚类中心以加速收敛。

n_init：int 类型值，使用不同的质心种子运行的 k-means 算法的次数，默认为 10。

max_iter：int 类型值，最大迭代次数，默认为 300，是 k-means 算法单次运行的最大迭代次数。

random_state：int 类型值，随机种子的数量，确定质心初始化的随机数生成。

（4）k-means 方法函数

fit(X)：训练 k-means 模型，返回训练好的 k-means 模型。

fit_predict(X)：计算聚类中心并返回每个样本的聚类索引。

fit_transform(X)：计算样本点到簇质心的距离，返回一个 ndarray(n_samples,)，n_samples 为样本数量。

6.预测实际值

随机生成样本点 X_new，使用 k-means 预测结果，代码如下：

```
1.  X_new =np.array([[0,2],[3,2],[ -3,3],[ -3,2.5]])
2.  kmeans.predict(X_new)
```

返回结果：array([0, 0, 3, 3])。

结果是一个 ndarray(4,1)，值表示的是每个点属于哪个簇。使用 transform() 函数输出样本到簇质心的距离，代码如下：

```
kmeans.transform(X_new)
```

输出结果如图 9-8 所示。

```
array([[0.32995317, 2.81093633, 1.49439034, 2.9042344 , 2.88633901],
       [2.80290755, 5.80730058, 4.4759332 , 5.84739223, 5.84236351],
       [3.29399768, 1.21475352, 1.69136631, 0.29040966, 1.71086031],
       [3.21806371, 0.72581411, 1.54808703, 0.36159148, 1.21567622]])
```

图 9-8 样本到簇质心的距离

输出结果是一个 ndarray(n_samples，n_clusters)，每个样本点输出四个值，表示这个样本点距离四个簇质心的距离，可以找到四个值最小的，判定这个样本点属于哪个簇。

7. 对预测结果做可视化展示

定义函数实现数据的可视化。展示所有数据，代码如下：

```
1.  #展示所有数据
2.  def plot_data(X):
3.      plt.plot(X[:,0],X[:,1],'k.',markersize=2)
```

展示簇的中心点，代码如下：

```
1.  #展示簇的中心点
2.  def plot_centroids(centroids,crircle_color='w',cross_color='k'):
3.      plt.scatter(centroids[:,0],centroids[:,1],marker='o',s=30,
    linewidths=8,color=crircle_color,zorder=10,alpha=0.9)
4.      plt.scatter(centroids[:,0],centroids[:,1],marker='x',s=50,
    linewidths=8,color=crircle_color,zorder=11,alpha=1)
```

绘制决策边界，代码如下：

```
1.  #绘制决策边界
2.  def plot_decision_boundaries(clusterer,X,resolution=1000,show_
    centroids=True,show_xlabels=True,show_ylabels=True):
3.      #获取坐标棋盘
4.      mins=X.min(axis=0)-0.1
5.      maxs=X.max(axis=0)+0.1
6.      xx,yy=np.meshgrid(np.linspace(mins[0],maxs[0],resolution),
7.                        np.linspace(mins[1],maxs[1],resolution))
8.      Z=clusterer.predict(np.c_[xx.ravel(),yy.ravel()])
```

```
9.      Z = Z.reshape(xx.shape)
10.     print(xx.shape,yy.shape,Z.shape)
11.     #绘制等高线
12.     plt.contourf(Z,extent = (mins[0],maxs[0],mins[1],maxs[1]),cmap
        = 'Pastel2')
13.     plt.contour(Z,extent = (mins[0],maxs[0],mins[1],maxs[1]),line-
        widths = 1,colors = 'k')
14.
15.     plot_data(X)
16.     plot_centroids(clusterer.cluster_centers_)
17.     plt.xlabel("x1")
18.     plt.ylabel("x2")
19.     plt.tick_params(labelbottom = 'off')
20.     plt.tick_params(labelleft = 'off')
```

调用函数,代码如下:

```
1.  plt.figure(figsize = (15,5))
2.  plot_decision_boundaries(kmeans,X)
3.  plt.show()
```

代码运行后,可以看出,数据被划分为五个簇,结果如图9-9所示。

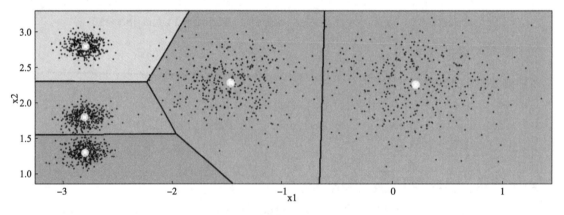

图9-9　绘制决策边界可视化

上面的代码只是显示了五次更新迭代后的最终结果,可以编写代码将五次"簇"更新的过程显示出来,代码如下:

```
1.  plt.figure(figsize = (15,15))
2.  #第一个图原始数据的初始化点和质心
3.  plt.subplot(421)
4.  plot_data(X)
5.  plot_centroids(kmeans_iter1.cluster_centers_, crircle_color = 'r',
    cross_color = 'k')
6.  plt.title('原始的簇质心')
7.
8.  #第一次按照质心划分簇
9.  plt.subplot(422)
10. plot_decision_boundaries(kmeans_iter1,X,show_xlabels = False,show_
    ylabels = False)
11. plt.title('第一次划分簇')
12.
13. #按照划分的簇,重新计算质心
14. plt.subplot(423)
15. plot_decision_boundaries(kmeans_iter1,X,show_xlabels = False,show_
    ylabels = False)
16. plot_centroids(kmeans_iter2.cluster_centers_, crircle_color = 'r',
    cross_color = 'k')
17. plt.title('第一次更新的簇质心')
18. plt.subplot(424)
19. plot_decision_boundaries(kmeans_iter2,X,show_xlabels = False,show_
    ylabels = False)
20. plt.title('更新簇后,第二次划分簇')
21.
22. #按照划分的簇,第二次更新簇质心
23. plt.subplot(425)
24. plot_decision_boundaries(kmeans_iter2,X,show_xlabels = False,show_
    ylabels = False)
25. plot_centroids(kmeans_iter3.cluster_centers_, crircle_color = 'r',
    cross_color = 'k')
26. plt.title('第二次更新质心')
27.
28. #第三次划分簇
29. plt.subplot(426)
30. plot_decision_boundaries(kmeans_iter3,X,show_xlabels = False,show_
    ylabels = False)
```

```
31. plt.title('第三次划分簇')
32.
33. #按照划分的簇,重新计算质心,
34. plt.subplot(427)
35. plot_decision_boundaries(kmeans_iter4,X,show_xlabels = False,show_
    ylabels = False)
36. plot_centroids(kmeans_iter5.cluster_centers_,crircle_color = 'r',
    cross_color = 'k')
37. plt.title('第四次更新质心')
38.
39. #第五次划分簇
40. plt.subplot(428)
41. plot_decision_boundaries(kmeans_iter5,X,show_xlabels = False,show_
    ylabels = False)
42. plt.title('第五次划分簇')
43.
44.
45. plt.show()
```

代码运行后,输出每次划分簇后的结果,并将结果可视化。从结果中可以看出,k-means 算法五次更新簇的质心和重新划分"簇"的过程,如图9-10所示。

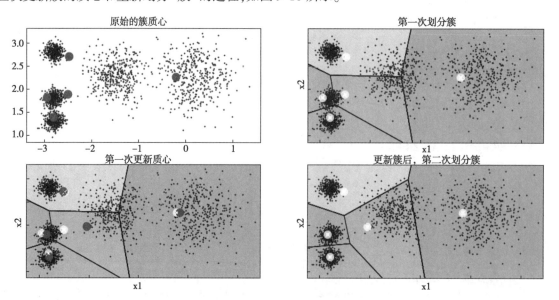

图 9-10　簇划分的结果

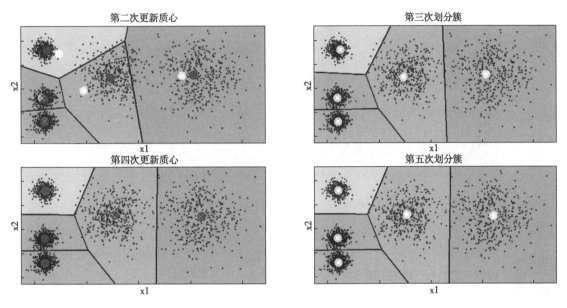

图 9-10 簇划分的结果(续)

9.3 ***k*-means 算法的不稳定性**

使用 *k*-means 算法时,相同的算法,不同的随机种子数量产生的结果不同。这是因为它高度依赖于第一个质心(随机选择)的位置。该算法可以找到一个局部最小值,后面几次的算法都会局限在一定的范围内。如下面代码所示,两次运行 *k*-means 算法,因为 random_state 设置不同,导致产生不同的结果,代码如下:

```
1.  c1 = KMeans(n_clusters = 5,init = 'random',n_init = 1,random_state = 10)
2.  c2 = KMeans(n_clusters = 5,init = 'random',n_init = 1,random_state = 20)
3.  def  plot_clusterer_comparsion(c1,c2,X):
4.      c1.fit(X)
5.      c2.fit(X)
6.
7.      plt.figure(figsize = (15,5))
8.      plt.subplot(121)
9.      plot_decision_boundaries(c1,X)
10.     plt.subplot(122)
11.     plot_decision_boundaries(c2,X)
12.
13. plot_clusterer_comparsion(c1,c2,X)
```

程序运行后的效果如图 9-11 所示。从图中可以看出,簇的初始化位置会影响簇的划分结果。

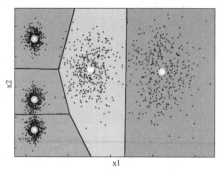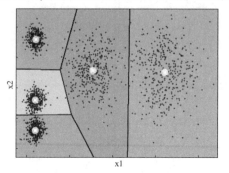

图 9-11 k-means 算法运行结果

9.4   DBSCAN 聚类算法

1. DBSCAN 聚类算法简介

k-means 算法简单明了,但是也有缺点,例如每次需要先确定 k 值,同时 k 值会直接影响聚类的结果。聚类算法受初始值影响较大,因为需要计算每个样本点到所有"簇"质心的距离,算法复杂度与样本规模成线性关系,并且很难发现任意形状的簇。

聚类算法还有另外一种,即 DBSCAN 聚类算法简称 DBSCAN 算法,它基于密度聚类算法。所谓密度聚类算法就是根据样本的紧密程度来进行聚类。

2. DBSCAN 算法基本概念

DBSCAN 算法包含很多基本术语,如 r 邻域、核心对象等,具体含义如下:

r 邻域:给定对象半径为 r 内的区域称为该对象的 r 邻域。如图 9-12(a)所示,p 对象在半径 r 区域内构成的圆就是该对象的 r 邻域。

核心对象:如果给定对象 r 邻域内的样本点数大于或等于 MinPoints(简称为 MinPts),则称该对象为核心对象。如图 9-12(b)所示,设置 MinPoints 的点为 2,那么在对象 p 的 e 邻域内有四个点,大于 MinPoints,那么 p 对象就是核心对象。

ε-邻域的距离阈值:设定的半径为 r。

直接密度可达:如果样本点 q 在 p 的 r 邻域内,并且 p 为核心对象,那么对象 p-q 直接密度可达,如图 9-12(c)所示。

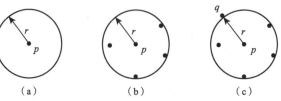

图 9-12   DBSCAN 算法基本概念

3. 密度可达与密度相连

密度可达:若有一个点的序列 q_0、q_1,…,q_k,对任意 $q_0 \sim q_k$ 是直接密度可达的,则称从 q_0 到 q_k 密度可达,这实际上是直接密度可达的"传播",如图 9-13 所示,q-p 直接密度可达,m-q 直接密度可达,那么 m-p 密度可达。

密度相连:若从某核心点 p 出发,点 q 和点 k 都是密度可达的,则称点 q 和点 k 是密度相连的,如图 9-14 所示,q-o 是密度可达,p-o 是密度可达,q-p 是密度相连。在 DBSCAN 算法中,某些样本可以看成一个类(又称簇),即最大的密度相连的样本集合。

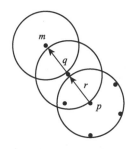

图 9-13　密度可达

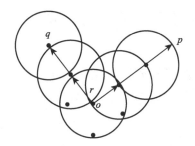

图 9-14　最大的密度相连

如图 9-15 所示,如果设置 MinPts = 5,箭头上的点都是核心对象,因为其 ε-邻域至少有 5 个样本。单箭头连线之外的样本是非核心对象。所有核心对象密度直达的样本在以单箭头连线起点为核心对象为中心的圆内,如果不在圆内,则不能密度直达。图中用箭头线连起来的核心对象组成了密度可达的样本序列。在这些密度可达的样本序列的 ε-邻域内所有的样本相互都是密度相连的。

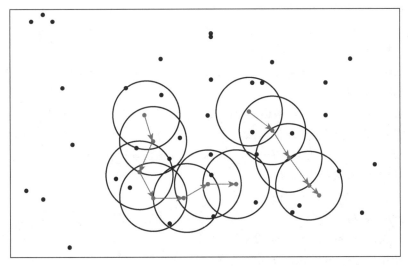

图 9-15　密度相连

4. 边界点与噪声点

边界点属于某一个类的非核心点，不能发展下线了，如图9-16中 B、C 点就是边界点。

噪声点是不属于任何一个类簇的点，从任何一个核心点出发都是密度不可达的，如图9-16中的 N 点。

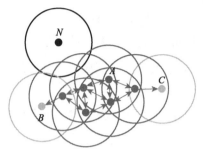

图9-16 边界点

9.5 DBSCAN 聚类算法实现

DBSCAN 算法需要设置三个参数，即 D 输入数据集、参数 ε 指定邻域半径、MinPts 密度阈值（邻域内最少的样本点数）。首先设置参数 ε，可以根据 K 距离来设定。找突变点，然后设定 K 距离，给定数据集 $P = \{p(i); i = 0, 1, \cdots, n\}$，计算点 $p(i)$ 到集合 D 的子集 S 中所有点之间的距离，距离按照从小到大的顺序排序。最后求出 MinPts，一般先取小一些，需要多次尝试，具体代码如下：

```
1.   np.unique(dbscan.labels_)
2.   # 调整不同的 eps 和 min_samples,可以得到不同的结果
3.   dbscan2 = DBSCAN(eps = 0.2, min_samples = 5)
4.   dbscan2.fit(X)
5.   # 定义函数
6.   def plot_dbscan(dbscan, X, size, show_xlabels = True, show_ylabels = True):
7.       #定义分类的集合1000 个
8.       core_mask = np.zeros_like(dbscan.labels_, dtype = bool)
9.       #设定核心点为 True
10.      core_mask[dbscan.core_sample_indices_] = = True
11.      #找到离群点
12.      anomalies_maske = dbscan.labels_ = = -1
13.      #不是离群点,也不是核心点的点
14.      non_core_mask = ~ (core_mask |anomalies_maske)
15.      #得到核心点的坐标
16.      cores = dbscan.components_
17.      #得到离群点
18.      anomalies = X[anomalies_maske]
19.      non_cores = X[non_core_mask]
20.      #绘制特征点
21.      plt.scatter(cores[:,0], cores[:,1],
```

```
22.     c =dbscan.labels_[core_mask],marker ='o',s =size,cmap ='Paired')
23.     #绘制
24.     plt.scatter(cores[:,0],cores[:,1],marker ='* ',s =20,c =dbscan.
labels_[core_mask])
25.     plt.scatter(anomalies[:,0],anomalies[:,1],c ='r',marker ="x",s =100)
26.     plt.scatter(non_cores[:,0],non_cores[:,1],c =dbscan.labels_
[non_core_mask],marker =".")
27.     plt.title("DBSCAN 聚类算法 eps ={:.2f}, min_samples ={}".format
(dbscan.eps, dbscan.min_samples), fontsize =14)
28. #数据结果可视化
29. plt.figure(figsize =(10,5))
30. plt.subplot(121)
31. plot_dbscan(dbscan,X,size =100)
32.
33. plt.subplot(122)
34. plot_dbscan(dbscan2,X,size =600)
35. plt.show()
```

代码运行后,可以看出同样的数据样本,如果 eps 也就是 ε-邻域的距离阈值设置不同,会得到不同的结果。图 9-17(a)中设置 eps 为 0.08,DBSCAN 算法将数据分为了四个"簇",图 9-17(b)中设置 eps 为 0.20,DBSCAN 算法将数据分为两个"簇"。

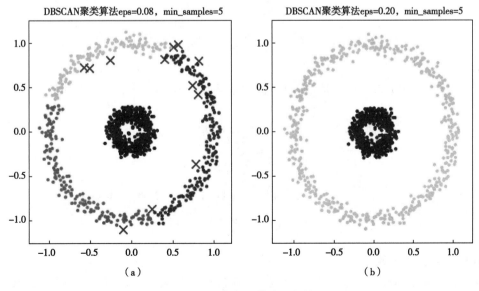

图 9-17 DBSCAN 算法运行结果

9.6 使用 *k*-means 算法完成图像分割

图像分割是利用图像的灰度、颜色、纹理、形状等特征，把图像分成若干个互不重叠的区域，并使这些特征在同一区域内呈现相似性；在不同的区域之间存在明显的差异性。然后就可以将分割的图像中具有独特性质的区域提取出来用于不同的研究。

无论是灰度图还是 RGB 彩色图，实际上都是存有灰度值的矩阵，所以，图像的数据格式决定了在图像分割方向上，使用 *k*-means 聚类算法是十分容易也是十分具体的。

1. 在 Jupyter Notebook 中导入包并读取图片

代码如下：

```
1.  #ladybug.png
2.  from matplotlib.image import imread
3.  #image = imread('ladybug.png')
4.  image = imread('street.png')
5.  #image = imread('street.jpg')
6.  image.shape
```

2. 定义 *k*-means 算法

指定了 8 个簇，随机种子数量为 42。代码如下：

```
kmeans = KMeans(n_clusters = 8,random_state =42).fit(X)
```

输出簇质心的信息。代码如下：

```
1.  kmeans = KMeans(n_clusters = 8,random_state =42).fit(X)
2.  kmeans.cluster_centers_
```

代码运行后，输出了 8 个簇的质心信息，运行结果如图 9-18 所示。

```
array([[0.11479929, 0.131998  , 0.07966766, 1.        ],
       [0.8284333 , 0.8657306 , 0.8369371 , 1.        ],
       [0.34023207, 0.36337155, 0.3184119 , 1.        ],
       [0.21859105, 0.23419523, 0.21826884, 1.        ],
       [0.11070266, 0.6104674 , 0.5286546 , 1.        ],
       [0.5076874 , 0.51638305, 0.46688914, 1.        ],
       [0.5988326 , 0.6959467 , 0.65107864, 1.        ],
       [0.9320028 , 0.95879555, 0.95750713, 1.        ]], dtype=float32)
```

图 9-18 8 个簇的质心信息

3. 计算聚类后的图像

使用 *k*-means 得到簇的数量（10,8,6,4,2），通过当前的聚类中心，得到聚类后的图像，代码如下：

```
1.  segmented_img = kmeans.cluster_centers_[kmeans.labels_].reshape
    (311,500,4)
2.  segmented_imgs = []
3.  n_colors = (10,8,6,4,2)
4.  for n_cluster in n_colors:
5.      kmeans = KMeans(n_clusters = n_cluster,random_state =42).fit(X)
6.      segmented_img = kmeans.cluster_centers_[kmeans.labels_]
7.      segmented_imgs.append(segmented_img.reshape(image.shape))
```

4. 使用 matplotlib 显示聚类后的图片

代码如下：

```
1.  plt.figure(figsize = (20,10))
2.  plt.subplot(231)
3.  plt.imshow(image)
4.  plt.title('Original image')
5.
6.  for idx,n_clusters in enumerate(n_colors):
7.      plt.subplot(232 + idx)
8.      plt.imshow(segmented_imgs[idx])
9.      plt.title('{}colors'.format(n_clusters))
```

代码运行后，可以看出使用 10,8,6,4,2 作为簇的数量对图像聚类分割，通过设置不同的簇可以将图像中的物体进行分割。这种方法通常用在图像的预处理中。例如当 $K=2$ 时，就可以将公路的边界和中线提取出来，然后再做后续的处理，输出结果如图 9-19 所示。

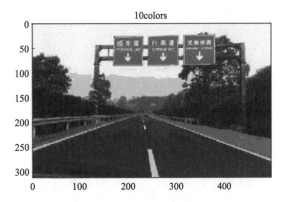

图 9-19 图像聚类运行结果

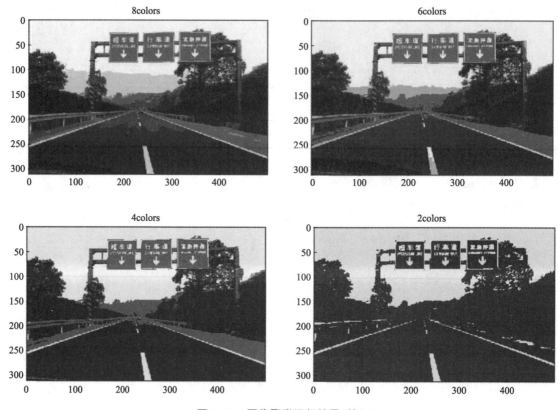

图 9-19　图像聚类运行结果（续）

单元10

使用决策树完成鸢尾花分类问题

鸢尾花分类是一个典型的分类问题,植物学家收集了鸢尾花的一些测量数据,包括花瓣的长度和宽度以及花萼的长度和宽度,测量的单位是 cm。这些花之前已经被植物学专家鉴定为属于 Setosa、Versicolor 或 Virginica 三个品种之一,如图 10-1 所示。可以根据测量的数据区判定每朵鸢尾花所属的品种。

图 10-1　鸢尾花的三个品种

现在目标是构建一个决策树模型,从这些已知品种的鸢尾花测量数据中进行学习,从而能够预测新鸢尾花的品种。因为有已知的鸢尾花的测量数据并且这些数据是带有标签的,所以这是一个监督学习问题。在这个问题中,要在多个选项中预测其中一个(鸢尾花的品种)。这是一个分类(classification)问题。输出(鸢尾花的品种)称为类别(class)。数据集中的每朵鸢尾花都属于三个类别之一,所以这是一个三分类问题。

10.1 决策树

1. 决策树概述

决策树算法起源于 E. B. 亨特等人于 1966 年发表的论文"Experiments in Induction"，但真正让决策树成为机器学习主流算法的还是 Quinlan（昆兰），他在 2011 年获得了数据挖掘领域最高奖 KDD 创新奖。昆兰在 1979 年提出了 ID3 算法，掀起了决策树研究的高潮。现在最常用的决策树算法 C4.5 是昆兰在 1993 年提出的。（关于为什么叫 C4.5，还有个轶事：因为昆兰提出 ID3 后，掀起了决策树研究的高潮，然后 ID4、ID5 等名字就被占用了，因此昆兰只好让讲自己对 ID3 的改进叫做 C4.0，C4.5 是 C4.0 的改进）。现在商业应用新版本是 C5.0link。

2. 决策树基本概念

决策树就是一棵树。一棵决策树包含一个根节点、若干个内部结点和若干个叶节点；叶节点对应于决策结果，其他每个节点则对应于一个属性测试；每个节点包含的样本集合根据属性测试的结果被划分到子节点中；根节点包含样本全集，从根节点到每个叶子节点的路径对应了一个判定测试序列。可以按照样本的特征进行决策，如图 10-2 所示为一个样本按照特征根据决策树进行类别决策。

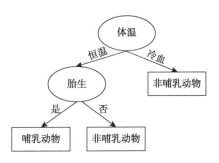

名字	体温	胎生	类别
小喵咪	恒温	是	?

图 10-2　决策树的基本结构

10.2 信息熵

决策树学习的关键在于如何选择最优的划分属性。所谓最优划分属性，对于二分类而言，就是尽量使划分的样本属于同一类别，即"纯度"最高的属性。那么如何来度量特征的纯度，这时候就要用到信息熵（information entropy）。先来看看信息熵的定义：假如当前样本集 D 中第 k 类样本所占的比例为 $p_k(k=1,2,3,4,\cdots,|y|)$，$y$ 为类别总数，则样本集的信息熵为

$$\mathrm{Ent}(D) = -\sum_{k=1}^{|y|} p_k\log_2 p_k$$

$\mathrm{Ent}(D)$ 的值越小，则 D 的纯度越高，这个公式决定了信息增益的一个缺点，即信息增益对可取值数目多的特征有偏好（即该属性能取的值越多，信息增益越偏向这个属性）。因为特征可取的值

越多,会导致"纯度"越大,即 Ent(D) 会很小。如果一个特征的离散个数与样本数相等,那么 Ent(D) 的值会为 0。

下面举例计算图 10-3 所示数据集的信息熵。

特征（属性）

编号	色泽	根蒂	敲声	纹理	脐部	触感	好瓜	label（标签）
1	青绿	蜷缩	浊响	清晰	凹陷	硬滑	是	
2	乌黑	蜷缩	沉闷	清晰	凹陷	硬滑	是	
3	乌黑	蜷缩	浊响	清晰	凹陷	硬滑	是	
4	青绿	蜷缩	沉闷	清晰	凹陷	硬滑	是	
5	浅白	蜷缩	浊响	清晰	凹陷	硬滑	是	
6	青绿	稍蜷	浊响	清晰	稍凹	软黏	是	
7	乌黑	稍蜷	浊响	稍糊	稍凹	软黏	是	
8	乌黑	稍蜷	浊响	清晰	稍凹	硬滑	是	
9	乌黑	稍蜷	沉闷	稍糊	稍凹	硬滑	否	
10	青绿	硬挺	清脆	清晰	平坦	软黏	否	
11	浅白	硬挺	清脆	模糊	平坦	硬滑	否	
12	浅白	蜷缩	浊响	模糊	平坦	软黏	否	
13	青绿	稍蜷	浊响	稍糊	凹陷	硬滑	否	
14	浅白	稍蜷	沉闷	稍糊	凹陷	硬滑	否	
15	乌黑	稍蜷	浊响	清晰	稍凹	软黏	否	
16	浅白	蜷缩	浊响	模糊	平坦	硬滑	否	
17	青绿	蜷缩	沉闷	稍糊	稍凹	硬滑	否	

图 10-3　西瓜数据集

该数据集包含 17 个样本,类别为二元的。正例(类别为 1 的样本)占的比例为 $p_1 = \dfrac{8}{17}$,反例(类别为 0 的样本)占的比例为 $p_2 = \dfrac{9}{17}$。根据信息熵的公式能够计算出数据集 D 的信息熵为

$$\text{Ent}(D) = -\left(\frac{8}{17}\log_2\frac{8}{17} + \frac{9}{17}\log_2\frac{9}{17}\right) = 0.998$$

10.3　信息增益

假定离散属性 a 有 V 个可能的取值 $\{a^1, a^2, a^3, \cdots, a^v\}$,如果使用特征 a 来对数据集 D 进行划分,则会产生 V 个分支节点,其中第 v 个节点包含了数据集 D 中所有在特征 a 上取值为 a^v 的样本总数,记为 D^v。因此,可以根据上面信息熵的公式计算出信息熵,再考虑到不同的分支节点所包含的样本数量不同,给分支节点赋予权重 $\dfrac{|D^v|}{|D|}$,即样本数越多的分支节点的影响越大,因此,能够计算出特征 a 对样本集 D 进行划分所获得的"信息增益",即

$$\text{Gain}(D, a) = \text{Ent}(D) - \sum_{v=1}^{V} \frac{|D^v|}{|D|}\text{Ent}(D^v)$$

一般而言,信息增益越大,则表示使用某个特征对数据集划分所获得的"纯度提升"越大。所

以,信息增益可用于决策树划分属性的选择,其实就是选择信息增益最大的属性。ID3 算法就是采用信息增益来划分属性的。

10.4 计算信息增益

1. 计算色泽的信息增益

现在以图 10-3 所示的西瓜数据集为例来计算信息增益。从数据集中能够看出特征集为{色泽、根蒂、敲声、纹理、脐部、触感}。下面来计算每个特征的信息增益。先看"色泽",它有三个值:{青绿、乌黑、浅白},若使用"色泽"对数据集 D 进行划分,则可得到三个子集,分别为 D^1(色泽 = 青绿)、D^2(色泽 = 乌黑)、(D^3色泽 = 浅白),如图 10-4 所示。

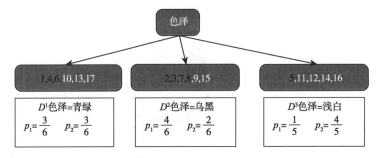

图 10-4 使用"色泽"对数据集 D 进行划分

D^1 包含六个样本{1,4,6,10,13,17},其中正例占的比例为 $p_1 = \dfrac{3}{6}$,反例占的比例为 $p_2 = 3$。D^2 包含六个样本{2,3,7,8,9,15},其中正例占的比例为 $p_1 = \dfrac{4}{6}$,反例占的比例为 $p_2 = \dfrac{2}{6}$。D^3 包含了五个样本{5,11,12,14,16},其中正例占的比例为 $p_1 = \dfrac{1}{5}$,反例占的比例为 $p_2 = \dfrac{4}{5}$。因此,可以计算出用"色泽"划分之后所获得的三个分支节点的信息熵为

$$\text{Ent}(D^1) = -\left(\frac{3}{6}\log_2\frac{3}{6} + \frac{3}{6}\log_2\frac{3}{6}\right) = 1.00$$

$$\text{Ent}(D^2) = -\left(\frac{4}{6}\log_2\frac{4}{6} + \frac{2}{6}\log_2\frac{2}{6}\right) = 0.918$$

$$\text{Ent}(D^3) = -\left(\frac{1}{5}\log_2\frac{1}{5} + \frac{4}{5}\log_2\frac{4}{5}\right) = 0.722$$

计算特征"色泽"、"根蒂"和"纹理"的信息增益:

$$\text{Gain}(D,色泽) = \text{Ent}(D) - \sum_{v=1}^{3}\frac{|D^v|}{|D|}\text{Ent}(D^v)$$

$$= 0.998 - \left(\frac{6}{17} \times 1.000 + \frac{6}{17} \times 0.918 + \frac{5}{17} \times 0.772\right) = 0.109$$

2. 所有特征的信息增益

同理计算特征的信息增益,确定决策树的根节点:

$$\text{Gain}(D,根蒂)=0.143$$
$$\text{Gain}(D,敲声)=0.141$$
$$\text{Gain}(D,纹理)=0.381$$
$$\text{Gain}(D,脐部)=0.289$$
$$\text{Gain}(D,触感)=0.006$$

比较发现,特征"纹理"的信息增益最大,于是它被选为决策树的根节点,因此可得图 10-5。

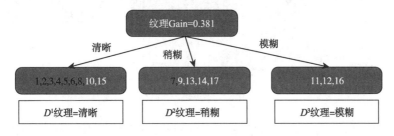

图 10-5　"纹理"的信息增益

3. 计算信息增益确定第二层的特征节点

继续对图 10-5 中每个分支进行划分,图 10-5 中第一个分支节点{"纹理 = 清晰"}为例,对这个节点进行划分。设该节点的样本集 $D^1\{1,2,3,4,5,6,8,10,15\}$,共 9 个样本,可用特征集合为{色泽,根蒂,敲声,脐部,触感},因此基于 D^1 能够计算出各个特征的信息增益:

$$\text{Gain}(D^1,根蒂)=0.458$$
$$\text{Gain}(D^1,敲声)=0.331$$
$$\text{Gain}(D^1,色泽)=0.043$$
$$\text{Gain}(D^1,脐部)=0.458$$
$$\text{Gain}(D,触感)=0.458$$

比较发现,"根蒂"、"脐部"和"触感"这三个属性均取得了最大的信息增益,可以随机选择其中之一作为划分属性,这里选择"根蒂"。因此可得图 10-6。

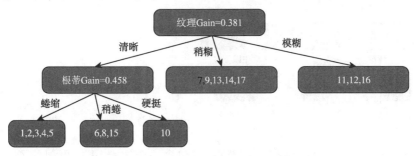

图 10-6　按"根蒂"划分

4. 计算信息增益确定第三层的特征节点

继续对图 10-6 中的每个分支节点递归地进行划分，图 10-6 中的节点{"根蒂 = 蜷缩"}为例，设该节点的样本集为{1,2,3,4,5}，共五个样本，但这五个样本的 label 均为"好瓜"，因此当前节点包含的样本全部属于同一类别无须划分，将当前节点标记为 C 类（在这个例子中为"好瓜"）叶节点，递归返回。因此图 10-6 变为图 10-7。

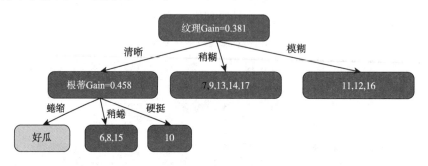

图 10-7　按"根蒂"划分

接下来对图 10-7 中节点{"根蒂 = 稍蜷"}进行划分，该点的样本集为 D^1{6,8,15}，共有三个样本。可用特征集为{色泽,敲声,脐部,触感}，同样可以计算出各个特征的信息增益，计算过程如下：

$$\text{Gain}(D^1,触感) = 0.251$$
$$\text{Gain}(D^1,敲声) = 0$$
$$\text{Gain}(D^1,脐部) = 0$$
$$\text{Gain}(D^1,色泽) = 0.251$$

从计算结果可知，"色泽"和"触感"两个属性均取得了最大的信息增益，选择一个属性进行划分。这里选择"色泽"进行划分，可得到图 10-8。

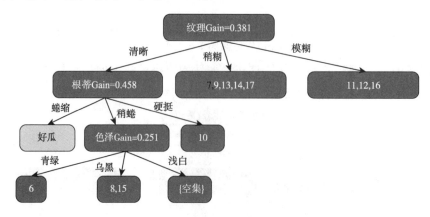

图 10-8　按"色泽"划分

继续递归地进行，看"色泽 = 青绿"这个节点，只包含一个样本，无须再划分了，直接把当前节点标记为叶节点，类别为当前样本的类别，即好瓜。递归返回。然后对递归地对"色泽 = 乌黑"这

个节点进行划分,就不再赘述了。等到递归的深度处理完"色泽＝乌黑"分支后,返回来处理"色泽
＝浅白"这个节点,因为当前节点包含的样本集为空集,不能划分,对应的处理措施为:将其设置为
叶节点,类别设置为其父节点(根蒂＝稍蜷)所含样本最多的类别,"根蒂＝稍蜷"包含{6,8,15}三
个样本,6,8 为正样本,15 为负样本,因此"色泽＝浅白"节点的类别为正(好瓜)。最终,得到的决
策树如图 10-9 所示。

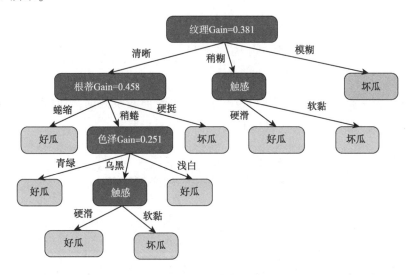

图 10-9　西瓜分类决策树

10.5　使用决策树完成鸢尾花分类的代码实现

1. 导入数据加载的包并打开数据集

这里用到了鸢尾花(iris)数据集,这是机器学习和统计学中一个经典的数据集。它包含在 Scikit-learn 的 datasets 模型中。可以调用 load_iris( )函数来加载数据,也可以下载数据集到本地然后
打开,代码如下:

```
1.  import numpy as np
2.  import os
3.  % matplotlib inline
4.  import matplotlib
5.  import matplotlib.pyplot as plt
6.  plt.rcParams['axes.labelsize'] = 14
```

```
7.plt.rcParams['xtick.labelsize'] = 12
8.plt.rcParams['ytick.labelsize'] = 12
9.#显示中文
10.plt.rcParams['font.sans - serif'] = ['SimHei']
11.plt.rcParams['axes.unicode_minus'] = False
12.import warnings
13.warnings.filterwarnings('ignore')
```

2. 加载数据

使用 load_iris 方法打开数据集，使用 head( ) 函数查看前五条数据，同时可以使用 info( ) 函数或者 head( ) 函数查看数据的信息，代码如下：

```
1.  from sklearn.datasets import load_iris
2.  from sklearn.tree import DecisionTreeClassifier
3.  iris = load_iris()
4.  X = iris.data[:,2:]
5.  y = iris.target
```

每一行数据代表一个花的信息，有四个特征一个标签，花萼长度'sepal length(cm) '，花萼宽度 'sepal width(cm) '，花瓣长度'petal length(cm) '，花瓣宽度'petal width(cm) '。Species 字段是花的种类，即标签。

3. 定义决策树

这里定义了一个决策树，设置 max_depth 的值为 2，然后使用 fit 方法进行训练。从代码执行效果中可以看到生成的决策树 tree_clf 的信息，代码如下：

```
1.  tree_clf = DecisionTreeClassifier(max_depth = 2)
2.  tree_clf.fit(X,y)
```

代码运行后，输出定义的决策树的基本参数。使用的 gini 系数作为决策树的生成依据，最大深度 max_depth 为 2，最小叶子节点数为 2，结果如图 10-10 所示。

```
DecisionTreeClassifier(ccp_alpha=0.0, class_weight=None, criterion='gini',
                       max_depth=2, max_features=None, max_leaf_nodes=None,
                       min_impurity_decrease=0.0, min_impurity_split=None,
                       min_samples_leaf=1, min_samples_split=2,
                       min_weight_fraction_leaf=0.0, presort='deprecated',
                       random_state=None, splitter='best')
```

图 10-10　DecisionTreeClassifier 运行结果

4. 决策树的可视化

可以使用 graphviz 包中的 dot 命令行工具将此 dot 文件转换为各种格式，如 PDF 或 png。下面代码将 dot 文件转换为 png 文件：

```
1.  from sklearn.tree import export_graphviz
2.  #修改可视的函数的参数
3.  export_graphviz(
4.      tree_clf,
5.      out_file = 'iris_tree1.dot',
6.      #传入特征值
7.      feature_names = iris.feature_names[2:],
8.      #传入标签的名字
9.      class_names = iris.target_names,
10.     rounded = True,
11.     filled = True
12. )
```

代码运行后，生成一个 dot 文件，需要在 Jupyter Notebook 中执行 \$ dot-Tpng iris_tree1. dot-o iris _tree1. png 命令将 dot 文件转换为 png 文件，然后再显示，代码如下：

```
1.  from IPython.display import Image
2.  Image(filename = 'iris_tree1.png',width = 400,height = 400)
```

代码运行后，使用图形化的方式将决策树的生成过程进行展示，同时将每次决策过程中 gini 系数计算的结果、集合分类的情况都做了展示，如图 10-11 所示。

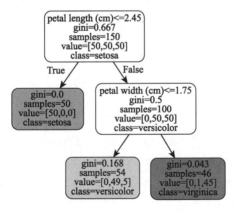

图 10-11　运行结果

10.6　鸢尾花分类效果的可视化

分类完毕可以绘制决策树的决策边界。通过决策边界可以更加直观地展示分类的结果，代码如下：

```
1.  from matplotlib.colors import ListedColormap
2.
3.  def plot_decision_boundary(clf, X, y, axes=[0, 7.5, 0, 3], iris=True,
    legend=False, plot_training=True):
4.      x1s = np.linspace(axes[0], axes[1], 100)
5.      x2s = np.linspace(axes[2], axes[3], 100)
6.      x1, x2 = np.meshgrid(x1s, x2s)
7.      X_new = np.c_[x1.ravel(), x2.ravel()]
8.      y_pred = clf.predict(X_new).reshape(x1.shape)
9.      custom_cmap = ListedColormap(['#fafab0','#9898ff','#a0faa0'])
10.     plt.contourf(x1, x2, y_pred, alpha=0.3, cmap=custom_cmap)
11.     if not iris:
12.         custom_cmap2 = ListedColormap(['#7d7d58','#4c4c7f','#507d50'])
13.         plt.contour(x1, x2, y_pred, cmap=custom_cmap2, alpha=0.8)
14.     if plot_training:
15.         plt.plot(X[:, 0][y==0], X[:, 1][y==0], "yo", label="Iris-
    Setosa")
16.         plt.plot(X[:, 0][y==1], X[:, 1][y==1], "bs", label="Iris-
    Versicolor")
17.         plt.plot(X[:, 0][y==2], X[:, 1][y==2], "g^", label="Iris-
    Virginica")
18.         plt.axis(axes)
19.     if iris:
20.         plt.xlabel("Petal length", fontsize=14)
21.         plt.ylabel("Petal width", fontsize=14)
22.     else:
23.         plt.xlabel(r"$x_1$", fontsize=18)
24.         plt.ylabel(r"$x_2$", fontsize=18, rotation=0)
25.     if legend:
26.         plt.legend(loc="lower right", fontsize=14)
27.
28. plt.figure(figsize=(8, 4))
29. plot_decision_boundary(tree_clf, X, y)
30. plt.plot([2.45, 2.45], [0, 3], "k-", linewidth=2)
31. plt.plot([2.45, 7.5], [1.75, 1.75], "k--", linewidth=2)
32. plt.plot([4.95, 4.95], [0, 1.75], "k:", linewidth=2)
33. plt.plot([4.85, 4.85], [1.75, 3], "k:", linewidth=2)
34. plt.text(1.40, 1.0, "Depth=0", fontsize=15)
```

```
35. plt.text(3.2, 1.80, "Depth =1", fontsize =13)
36. plt.text(4.05, 0.5, "(Depth =2)", fontsize =11)
37. plt.title('Decision Tree decision boundaries')
38.
39. plt.show()
```

代码运行后,可以看出整个决策树生成的过程,如图 10-12 所示。横轴表示花瓣长度,纵轴表示花瓣宽度。图 10-12 中的横线按照花瓣的宽度进行分类,竖线按照花瓣的长度进行分类,其中黑色实线表示第一次分类,粗虚线表示第二次分类,细虚线表示第三次分类,通过三次分类,将鸢尾花数据样本分为了三类。

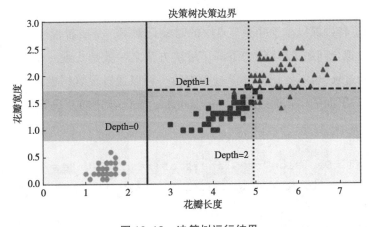

图 10-12　决策树运行结果

10.7 鸢尾花分类的概率估计

虽然决策树可以做到对样本分类,但在算法使用中也会遇到这种情况,模型对样本 X 进行预测,样本 X 属于 A 类的概率为 51%,属于 B 类的概率为 49%,这时虽然算法给出结论是样本 X 属于 A 类,但是其对这次预测的结果把握性并不是很大。对于现实中某些宁可不做也不要出现错误的场景来说,这次预测并不是所需要的。所以,做算法时需要模型输出样本属于每个类别的概率。

本例中输入数据为:花瓣长 5 cm,宽 1.5 cm 的花。相应的叶节点是深度为 2 的左节点,因此决策树应输出以下概率:

```
tree_clf.predict_proba([[5,1.5]])
```

再使用 predict( ) 函数：

```
tree_clf.predict([[5,1.5]])
```

代码的输出表示花瓣长 5 cm，宽 1.5 cm 的花是三类鸢尾花的概率分别为：

```
Iris-Setosa 为 0% (0/54);
Iris-Versicolor 为 90.7% (49/54);
Iris-Virginica 为 9.3% (5/54).
```

10.8　鸢尾花分类的过拟合

当决策树的深度特别深以至于叶子节点中的对象只剩下一个或者很少，导致决策树的模型过于复杂，容易造成过拟合问题，泛化能力下降。解决方法之一就是找到一个点（深度）让决策树停止分裂，不要让树长过长，也不要让树分得过于细致。可以通过设置树的最大深度和叶子的最小尺寸来防止过拟合。下面的代码通过设置 min_samples_leaf = 4 来防止决策树出现过拟合现象，代码如下：

```
1.  tree_clf1 = DecisionTreeClassifier(random_state = 42)
2.  tree_clf2 = DecisionTreeClassifier(min_samples_leaf = 4,random_state = 42)
3.  tree_clf1.fit(X,y)
4.  tree_clf2.fit(X,y)
```

绘制决策边界输出，代码如下：

```
1.  plt.figure(figsize = (12,4))
2.  plt.subplot(121)
3.  plot_decision_boundary(tree_clf1,X,y,axes = [-1.5,2.5,-1,1.5])
4.
5.  plt.subplot(122)
6.  plot_decision_boundary(tree_clf2,X,y,axes = [-1.5,2.5,-1,1.5])
7.  plt.show()
```

代码运行后，将两个数据测试的决策边界可视化呈现。可以看到图 10-13（a）是决策树的过拟合状态；图 10-13（b）通过设置 min_samples_leaf = 4，抑制了过拟合现象的发生。

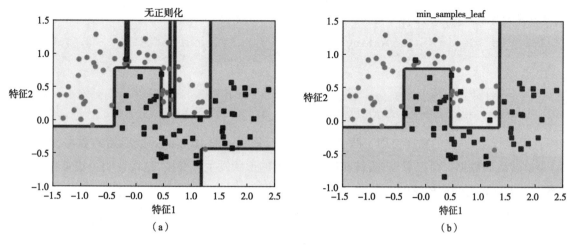

图 10-13　不同决策树运行结果

10.9　决策树对数据的敏感性

决策树对数据具有敏感性。最明显的一点就是关于决策边界的划分。决策树的决策边界是非常规整的,都是直线,垂直于横轴或者垂直于纵轴,这就导致了绘制出的决策边界很可能不是真实情况的决策边界。而且决策树对数据旋转十分敏感,对训练细节也很敏感。下面的代码演示了数据旋转前和数据旋转后使用决策树对数据进行分类得到完全不同的结果,具体如下:

```
1.  np.random.seed(6)
2.  Xs = np.random.rand(100,2) - 0.5
3.  ys = (Xs[:,0] > 0).astype(np.float32) * 2
4.
5.  angle = np.pi/4
6.  rotation_matrix = np.array([[np.cos(angle), - np.sin(angle)],[np.
    sin(angle),np.cos(angle)]])
7.  Xsr = Xs.dot(rotation_matrix)
8.
9.  tree_clf_s = DecisionTreeClassifier(random_state = 42)
10. tree_clf_sr = DecisionTreeClassifier(random_state = 42)
11. tree_clf_s.fit(Xs,ys)
12. tree_clf_sr.fit(Xsr,ys)
13.
14. plt.figure(figsize = (12,4))
```

```
15. plt.subplot(121)
16. plot_decision_boundary(tree_clf_s,Xs,ys,axes = [-1,1, -0.6,1])
17.
18. plt.subplot(122)
19. plot_decision_boundary(tree_clf_sr,Xsr,ys,axes = [-1,1, -0.6,1])
```

代码运行后,同样的数据使用了两个决策树进行训练。图 10-14(a)是原始的数据,图 10-14 (b)是旋转后的数据。可以看出,旋转前后决策边界的形状发生了变换,这就证明了决策树的决策边界对数据旋转是高度敏感的。

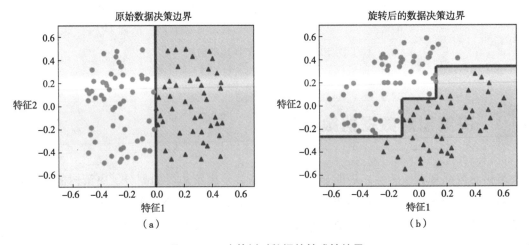

图 10-14　决策树对数据的敏感性结果

单元11

使用支持向量机完成鸢尾花分类

支持向量机(support vector machine, SVM)是一种用于回归和分类的监督学习算法,但更常用于分类。SVM 在各种设置下均表现出色,通常被认为是最好的"开箱即用"分类器。本单元使用 SVM 完成鸢尾花的分类任务。

11.1 支持向量机原理

支持向量机是一种二分类模型,它的基本模型是定义在特征空间上的间隔最大的线性分类器。间隔最大使它有别于感知机。SVM 还包括核技巧,这使它成为实质上的非线性分类器。SVM 的学习策略就是间隔最大化,可形式化为一个求解凸二次规划的问题,也等价于正则化的合页损失函数的最小化问题。SVM 的学习算法就是求解凸二次规划的最优化算法。

SVM 学习的基本想法是求解能够正确划分训练数据集并且几何间隔最大的分离超平面。如图 11-1 所示,$w \cdot x + b = 0$ 即为分离超平面,对于线性可分的数据集来说,这样的超平面有无穷多个(即感知机),但是几何间隔最大的分离超平面却是唯一的。

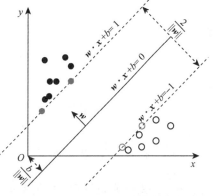

图 11-1 支持向量机原理图

11.2 线性可分支持向量机

1. 线性可分支持向量机基本概念

对于线性可分的数据集，学习的目标是在特征空间中找到一个分离超平面，能将实例分到不同的类。分离超平面将特征空间划分为两部分：一部分是正类，另一部分是负类。分离超平面的法向量指向的一侧为正类，另一侧为负类，并且要求这个分离超平面距离最近的两个点的距离之和最大，这个分离超平面被称为间隔最大分离超平面。线性可分支持向量机的数学模型为

$$f(x) = \mathrm{sgn}(wx + b)$$

其中，$wx + b = 0$ 使用向量表示可以写成 $w \cdot x + b = 0$（w、x 分别表示权重 w 和特征 x 的向量），这就是间隔最大分离超平面。所需要求得的模型就是这个间隔最大分离超平面，如图 11-2 所示。其中，negative objects 表示负样本对象，positive objects 表示正样本对象，$w \cdot x + b = \pm 1$ 表示上下两个超平面的方程，这两个超平面上的点就是支持向量。

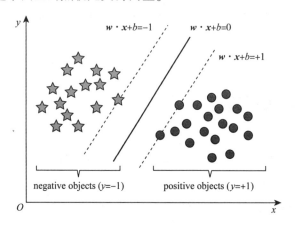

图 11-2 间隔最大的分离超平面

2. 代码实现

线性支持向量机有两种：SVC（支持向量分类）用于分类问题；SVR（支持向量回归）用于回归问题。下例使用鸢尾花数据集，它是基于鸢尾花的花萼长度和宽度进行分类的。只用其中二维特征，能够方便可视化，代码如下：

```
1.  from sklearn.svm import SVC
2.  from sklearn import datasets
3.
4.  iris = datasets.load_iris()
5.  X = iris['data'][:,(2,3)]
```

```
6.   y = iris['target']
7.
8.   setosa_or_versicolor = (y==0)|(y==1)
9.   X = X[setosa_or_versicolor]
10.  y = y[setosa_or_versicolor]
11.  #kernel="linear"(线性核函数)给了线性的决策边界。两类之间的分离边界是直线
12.  svm_clf = SVC(kernel='linear',C=float('inf'))
13.  svm_clf.fit(X,y)
```

首先生成一个 SVC 向量机,核函数设置为线性核函数,分类边界是直线,然后将 SVC 分类的效果可视化,代码如下:

```
1.   # 一般的模型
2.   x0 = np.linspace(0, 5.5, 200)
3.   pred_1 = 5* x0 - 20
4.   pred_2 = x0 - 1.8
5.   pred_3 = 0.1 * x0 + 0.5
6.   def plot_svc_decision_boundary(svm_clf, xmin, xmax,sv=True):
7.       w = svm_clf.coef_[0]
8.       b = svm_clf.intercept_[0]
9.       print (w)
10.      x0 = np.linspace(xmin, xmax, 200)
11.      decision_boundary = - w[0]/w[1] * x0 - b/w[1]
12.      margin = 1/w[1]
13.      gutter_up = decision_boundary + margin
14.      gutter_down = decision_boundary - margin
15.      if sv:
16.          svs = svm_clf.support_vectors_
17.          plt.scatter(svs[:,0],svs[:,1],s=180,facecolors='#FFAAAA')
18.      plt.plot(x0,decision_boundary,'k-',linewidth=2)
19.      plt.plot(x0,gutter_up,'k--',linewidth=2)
20.      plt.plot(x0,gutter_down,'k--',linewidth=2)
21.  plt.figure(figsize=(14,4))
22.  plt.subplot(121)
23.  plt.plot(X[:,0][y==1],X[:,1][y==1],'bs')
24.  plt.plot(X[:,0][y==0],X[:,1][y==0],'ys')
25.  plt.plot(x0,pred_1,'g--',linewidth=2)
26.  plt.plot(x0,pred_2,'m-',linewidth=2)
```

```
27. plt.plot(x0,pred_3,'r-',linewidth=2)
28. plt.axis([0,5.5,0,2])
29.
30. plt.subplot(122)
31. plot_svc_decision_boundary(svm_clf, 0, 5.5)
32. plt.plot(X[:,0][y==1],X[:,1][y==1],'bs')
33. plt.plot(X[:,0][y==0],X[:,1][y==0],'ys')
34. plt.axis([0,5.5,0,2])
```

代码运行后，结果如图 11-3 所示。从可视化的结果中可以看到，两类数据已经完全分开，并且标注出了边界点，如图 11-3 所示。

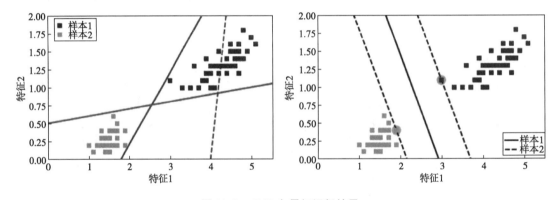

图 11-3　SVC 向量机运行结果

3. 软间隔支持向量机

支持向量机的作用是在分类完全正确的前提下，尽可能使两类数据点越远越好。如图 11-4 所示鸢尾花分类中，横轴为花瓣长度（petal length），纵轴为花瓣宽度（petal width），图 11-4（a）因为离群点和异常点（outlier）带来的影响，使得决策边界不好[图 11-4（b）的决策边界出现了偏离]。如何解决该问题？可使用软间隔，实质就是调整一下决策边界的划分条件，允许出现错误过拟合。

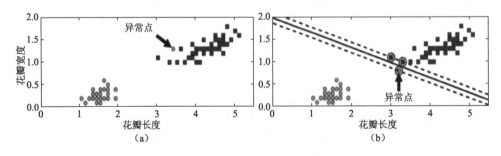

图 11-4　软间隔示例

下面的例子中增加了松弛因子 C，当 C 趋近于很大时，意味着分类严格不能有错误；当 C 趋近

于很小时,意味着可以有更大的错误容忍,代码如下:

```
1.  iris = datasets.load_iris()
2.  X = iris["data"][:,(2,3)] # petal length, petal width
3.  y = (iris["target"] == 2).astype(np.float64) # Iris-Viginica
4.  svm_clf = Pipeline((
5.      ('std',StandardScaler()),
6.      ('linear_svc',LinearSVC(C=1))
7.  ))
8.  svm_clf.fit(X,y)
```

设置 C=1,输出了 LinearSVC 模型的结构,代码运行结果如图 11-5 所示。

```
Pipeline(memory=None,
         steps=[('std',
                 StandardScaler(copy=True, with_mean=True, with_std=True)),
                ('linear_svc',
                 LinearSVC(C=1, class_weight=None, dual=True,
                           fit_intercept=True, intercept_scaling=1,
                           loss='squared_hinge', max_iter=1000,
                           multi_class='ovr', penalty='l2', random_state=None,
                           tol=0.0001, verbose=0))],
         verbose=False)
```

图 11-5　LinearSVC 运行结果

可以通过调节 C 的值来实现不同的分类效果,代码如下:

```
1.  scaler = StandardScaler()
2.  svm_clf1 = LinearSVC(C=1,random_state=42)
3.  svm_clf2 = LinearSVC(C=100,random_state=42)
4.
5.  scaled_svm_clf1 = Pipeline((
6.      ('std',scaler),
7.      ('linear_svc',svm_clf1)
8.  ))
9.
10. scaled_svm_clf2 = Pipeline((
11.     ('std',scaler),
12.     ('linear_svc',svm_clf2)
13. ))
14. scaled_svm_clf1.fit(X,y)
15. scaled_svm_clf2.fit(X,y)
16.
17. b1 = svm_clf1.decision_function([-scaler.mean_ / scaler.scale_])
18. b2 = svm_clf2.decision_function([-scaler.mean_ / scaler.scale_])
19. w1 = svm_clf1.coef_[0] / scaler.scale_
20. w2 = svm_clf2.coef_[0] / scaler.scale_
```

```
21. svm_clf1.intercept_ = np.array([b1])
22. svm_clf2.intercept_ = np.array([b2])
23. svm_clf1.coef_ = np.array([w1])
24. svm_clf2.coef_ = np.array([w2])
```

将结果可视化,代码如下:

```
1. plt.figure(figsize = (14,4.2))
2. plt.subplot(121)
3. plt.plot(X[:,0][y = =1], X[:,1][y = =1], "g^", label = "Iris -Virginica")
4. plt.plot(X[:,0][y = =0], X[:,1][y = =0], "bs", label = "Iris -Versicolor")
5. plot_svc_decision_boundary(svm_clf1, 4, 6, sv = False)
6. plt.xlabel("petal length", fontsize = 14)
7. plt.ylabel("petal width", fontsize = 14)
8. plt.legend(loc = "upper left", fontsize = 14)
9. plt.title("$C = {}$".format(svm_clf1.C), fontsize = 16)
10. plt.axis([4, 6, 0.8, 2.8])
11.
12. plt.subplot(122)
13. plt.plot(X[:,0][y = =1], X[:,1][y = =1], "g^")
14. plt.plot(X[:,0][y = =0], X[:,1][y = =0], "bs")
15. plot_svc_decision_boundary(svm_clf2, 4, 6, sv = False)
16. plt.xlabel("petal length", fontsize = 14)
17. plt.title("$C = {}$".format(svm_clf2.C), fontsize = 16)
18. plt.axis([4, 6, 0.8, 2.8])
```

代码运行后,结果如图 11-6 所示,从可视化结果可以看到,图 11-6(b)使用较高的 C 值,分类器会减少误分类,但最终会有较小间隔。图 11-6(a)使用较低的 C 值,间隔要大得多,但很多实例最终会出现在间隔之内。

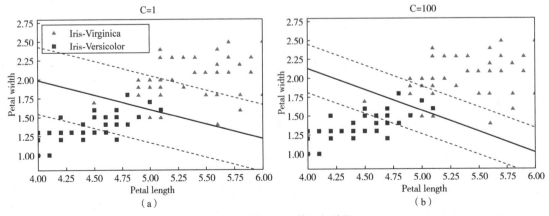

图 11-6　不同 C 值的运行结果

11.3 非线性支持向量机

1. 非线性支持向量机基本概念

对于一组既非线性可分,也非近似线性可分的数据集,称为线性不可分,如图 11-7 所示。

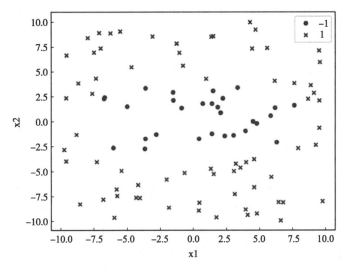

图 11-7　线性可分、近似线性可分和线性不可分数据集

图 11-7 中分别用圆和叉来代表不同的标签分类,很明显该数据集无法通过一条直线正确分类,但是该数据集可以被一条椭圆曲线正确分类,如图 11-8 所示。

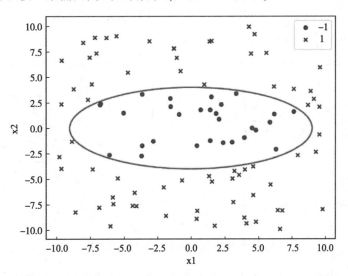

图 11-8　椭圆曲线正确分类

但是想要求解图 11-8 中的椭圆曲线这种非线性分类的问题，相对来说难度较大。既然线性分类问题相对容易求解，那么该数据集能否通过一定的非线性变化转换为一个线性分类的数据呢？

将数据集进行非线性转换，对特征做平方（$Z = x \times x$），这时可以看到转换后的数据可以通过一条直线正确分类，原来椭圆曲线变换成了图 11-9 中的直线，原来非线性分类问题变换成了线性分类问题。

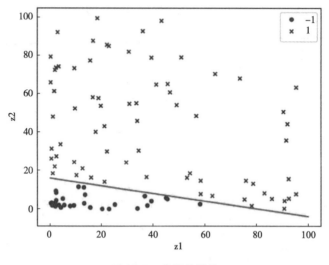

图 11-9　非线性转换

上例中，将原空间称为输入空间 X，映射后的新空间称为特征空间 H。可以使用核函数进行变换。常用的核函数有线性核函数、高斯核函数、多项式核函数。通常是使用核函数将低维数据转化为高维数据。

如图 11-10 所示，可以使用核函数将一个 $\mathbf{R}^2$ 空间中的线映射到 $\mathbf{R}^3$ 空间中的面，这时在 $\mathbf{R}^3$ 空间中就可以使用平面把两类数据分开。

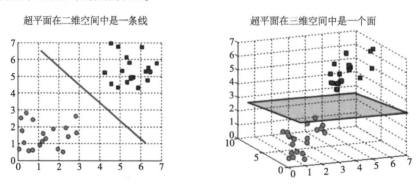

图 11-10　低维数据转换为高维数据

将一个二维数据映射到三维，可以使用高斯核函数，然后使用超平面进行分隔。

在癌症肿瘤分类中，癌症（cancer）数据集和良性（normal）数据集分布如图 11-11 所示，可以使用和函数(kernel)将数据集映射到三维空间,这时就可以找到决策平面(decision surface),将癌症和良性数据分开。

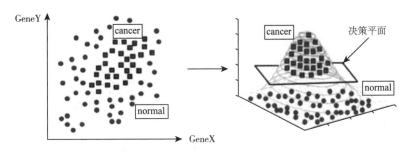

图 11-11　二维数据映射到三维

2. 代码实现

使用 make_moons( )函数生成一个非线性的数据,代码如下:

```
1.  from sklearn.datasets import make_moons
2.  X, y = make_moons(n_samples =100, noise =0.15, random_state =42)
3.
4.  def plot_dataset(X, y, axes):
5.      plt.plot(X[:, 0][y = =0], X[:, 1][y = =0], "bs")
6.      plt.plot(X[:, 0][y = =1], X[:, 1][y = =1], "g^")
7.      plt.axis(axes)
8.      plt.grid(True, which = 'both')
9.      plt.xlabel(r"$x_1$", fontsize =20)
10.     plt.ylabel(r"$x_2$", fontsize =20, rotation =0)
11.
12. plot_dataset(X, y, [-1.5, 2.5, -1, 1.5])
13. plt.show()
```

代码运行后,生成了两个月牙形的数据集,两个数据集相互交错,如图 11-12 所示。

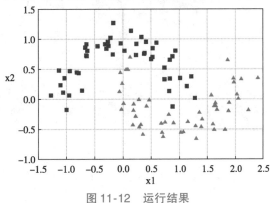

图 11-12　运行结果

定义一个非线性支持向量机，代码如下：

```
1.  from sklearn.datasets import make_moons
2.  from sklearn.pipeline import Pipeline
3.  from sklearn.preprocessing import PolynomialFeatures
4.
5.  polynomial_svm_clf = Pipeline((
6.                  ("poly_features",PolynomialFeatures(degree=3)),))
7.                  ("scaler",StandardScaler()),
8.                  ("svm_clf",LinearSVC(C=10,loss="hinge"))
9.
10. polynomial_svm_clf.fit(X,y)
```

这里使用了管道Pipeline，最终生成一个LinearSVC，可以将分类的结果可视化，代码如下：

```
1.  def plot_predictions(clf,axes):
2.      x0s = np.linspace(axes[0],axes[1],100)
3.      x1s = np.linspace(axes[2],axes[3],100)
4.      x0,x1 = np.meshgrid(x0s,x1s)
5.      X = np.c_[x0.ravel(),x1.ravel()]
6.      y_pred = clf.predict(X).reshape(x0.shape)
7.      plt.contourf(x0,x1,y_pred,cmap=plt.cm.brg,alpha=0.2)
8.
9.  plot_predictions(polynomial_svm_clf,[-1.5,2.5,-1,1.5])
10. plot_dataset(X,y,[-1.5,2.5,-1,1.5])
```

代码运行后，可以看到支持向量机能够很好地将两类数据点分隔开，可视化结果如图11-13所示。

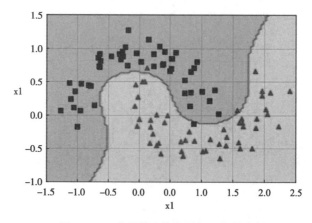

图11-13　非线性支持向量机运行结果

11.4　线性可分、线性不可分、非线性支持向量机的区别

　　数据线性可分那么肯定存在唯一的超平面将样本完全分开,并满足间隔最大化,此时分类器就是线性可分支持向量机。但是现实情况中完全线性可分情况很少,如图 11-14 所示,由于两个样本的存在,实际上根本不可能存在一个超平面(二维中为直线)将数据完全分开。由于只是少量样本导致的线性不可分(完全或者大多数线性不可分时,需要非线性支持向量机),可以将数据集近似看成线性可分,实际上仍然存在无穷超平面可以切分数据集,从中选取保证间隔尽量大的同时误分类个数尽量小的超平面即可。这就是所谓的基于软间隔最大化的线性不可分支持向量机,但是当数据完全线性不可分时,根本不存在这样的超平面,只能用超曲面划分,如图 11-15 所示。但是非线性问题往往不好求解,需要变换成线性问题,可以将原空间映射到新的空间,使数据线性可分。这就需要用到核函数,利用非线性支持向量机进行计算。

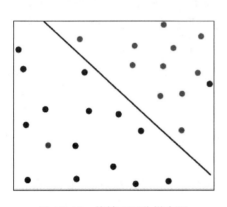

图 11-14　线性不可分样本图

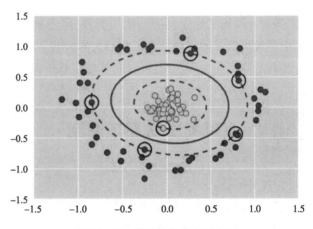

图 11-15　非线性数据集样本图

单元12

使用贝叶斯算法完成垃圾邮件分类

随着互联网的迅速发展和应用普及,电子邮件的广泛应用给生产和生活带来了相当大的便利,但是垃圾邮件的出现带来了相当大的烦恼。针对垃圾邮件问题本单元以贝叶斯算法为理论基础,将理论应用于工程实际,设计和实现了基于贝叶斯算法的垃圾邮件过滤系统。

12.1 贝叶斯算法原理介绍

1. 贝叶斯公式

贝叶斯定理是关于随机事件 A 和 B 的条件概率(或边缘概率)的定理。其中 $P(A|B)$ 是在 B 发生的情况下 A 发生的可能性,又称贝叶斯公式,表示如下:

$$P(A|B) = \frac{P(B|A)P(A)}{P(B)}$$

由联合概率公式推导而来:

$$P(A,B) = P(B|A)P(A) = P(A|B)P(B)$$

式中,$P(B)$ 称为先验概率;$P(B|A)$ 称为后验概率;$P(B,A)$ 称为联合概率。

2. 贝叶斯算法

贝叶斯算法,又称朴素贝叶斯算法。朴素:特征条件独立。贝叶斯:基于贝叶斯定理。属于监督学习的生成模型,实现简单,没有迭代,并有坚实的数学理论(即贝叶斯定理)作为支撑。在大量样本下会有较好的表现,不适用于输入向量的特征条件有关联的场景。

朴素贝叶斯算法是基于贝叶斯定理和特征条件独立假设的分类方法,它通过特征计算分类的概率,选取概率大的情况,是基于概率论的一种机器学习分类(监督学习)方法,被广泛应用于情感分类领域。

12.2　贝叶斯算法常用的三个模型

1. 多项式模型

重复的词语视为出现多次,如下所示:

$P(($ "代开","发票","增值税","发票","正规","发票" $)|S) = P($ "代开" $|S)P($ "发票" $|S)$ $P($ "增值税" $|S)P($ "发票" $|S)P($ "正规" $|S)P($ "发票" $|S) = P($ "代开" $|S)P^3($ "发票" $|S)P($ "增值税" $|S)P($ "正规" $|S)$

统计与判断时,都要关注重复次数。

2. 伯努利模型

将重复的词语都视为其只出现一次,如下所示:

$P(($ "代开","发票","增值税","发票","正规","发票" $)|S) = P($ "代开" $|S)P($ "发票" $|S)$ $P($ "增值税" $|S)P($ "正规" $|S)$

这种方法更加简化与方便。当然它丢失了词频的信息,因此效果会差一些。

3. 混合模型

在计算句子概率时,不考虑重复词语出现的次数,但是在计算词语的概率 $P($ "词语" $|S)$ 时,考虑重复词语的出现次数。

贝叶斯算法在数据较少的情况下仍然有效,可以处理多类别问题。但是对于输入数据的准备方式较为敏感。适用于数据类型是标称型数据的应用场景中。

12.3　垃圾邮件分类

垃圾邮件曾经是一个令人头痛的问题,长期困扰着邮件运营商和用户。据统计,用户收到的电子邮件中多数是垃圾邮件。传统的垃圾邮件过滤方法主要有关键词法和校验码法等。关键词法的过滤依据是特定的词语,如垃圾邮件的关键词:"发票"、"贷款"、"利率"、"中奖"、"办证"、"抽奖"、"号码"、"钱"、"款"和"幸运"等等。但这种方法效果很不理想,而且容易规避。一是正常邮件中也可能有这些关键词,非常容易误判。

直到提出了使用"贝叶斯"算法才使得垃圾邮件的分类达到一个较好的效果,而且随着邮件数目越来越多,贝叶斯分类的效果会越来越好。采用的分类方法是通过多个词来判断是否为垃圾邮件,但这个概率难以估计,通过贝叶斯公式,可以转化为求垃圾邮件中这些词出现的概率。

1. 读取文本数据

在 Jupyter Notebook 中读取文本文件,获取数据集,代码如下:

```
1.  import numpy as np
2.  import pandas as pd
3.  import seaborn as sns
4.  import matplotlib.pyplot as plt
5.  import seaborn as sns
6.  #显示中文
7.  plt.rcParams['font.sans - serif'] = ['SimHei']
8.  plt.rcParams['axes.unicode_minus'] = False
9.  df = pd.read_table("../input/SMSSpamCollection", sep = '\t', names =
    ['label', 'sms_message'])
10. df.head()
```

代码运行结果如图 12-1 所示，数据集中的列目前没有命名，可以看出有两列。第一列有两个值："ham"表示邮件不是垃圾邮件；"spam"表示邮件是垃圾邮件。第二列是被分类的信息的文本内容。

	label	sms_message
0	ham	Go until jurong point, crazy.. Available only ...
1	ham	Ok lar... Joking wif u oni...
2	spam	Free entry in 2 a wkly comp to win FA Cup fina...
3	ham	U dun say so early hor... U c already then say...
4	ham	Nah I don't think he goes to usf, he lives aro...

图 12-1　代码运行结果

2. 准备数据

数据预处理中已经大概了解了数据集的结构，现在将标签转换为二元变量，0 表示"ham"（非垃圾邮件），1 表示"spam"（垃圾邮件），这样比较方便计算。为何要执行这一步？原因在于 Scikit-learn 处理输入的方式。Scikit-learn 只处理数字值，因此如果标签值保留为字符串，Scikit-learn 会自己进行转换（更确切地说，字符串标签将转型为未知浮点值）。

如果标签保留为字符串，模型依然能够做出预测，但是稍后计算效果指标（例如，计算精确率和召回率分值）时可能会遇到问题。因此，为了避免稍后出现意外的陷阱，最好将分类值转换为整数，再传入模型中。

说明：使用映射方法将"标签"列中的值转换为数字值，如下所示：{ 'ham':0, 'spam':1} 这样会将"ham"值映射为 0，将"spam"值映射为 1。此外，为了知道正在处理的数据集的大小，使用"shape"输出行数和列数，代码如下：

```
1.  print(df.shape)
2.  df['label'] = df.label.map({'ham':0, 'spam':1})
3.
4.  df['label'].head()
```

3. 数据转换

Bag of Words（BoW）的数据集中有大量文本数据（5572 行数据）。大多数机器学习算法都要求输入的是数字数据，而邮件通常都是文本。下面介绍 Bag of Words（BoW）的概念。它用来表示要处理的问题具有"大量单词"或很多文本数据。BoW 的基本概念是拿出一段文本,计算该文本中单词的出现频率。**注意**:BoW 平等地对待每个单词,单词的出现顺序并不重要。

可以将文档集合转换成矩阵,每个文档是一行,每个单词(令牌)是一列,对应的(行,列)值是每个单词(令牌)在此文档中出现的频率。

假设有四个如下所示的文档:

```
['Hello, how are you! ',
'Win money, win from home.',
'Call me now',
'Hello, Call you tomorrow? ']
```

目标是将这组文本转换为频率分布矩阵,如图 12-2 所示。

	are	call	from	hello	home	how	me	money	now	tomorrow	win	you
0	1	0	0	1	0	1	0	0	0	0	0	1
1	0	0	1	0	1	0	0	1	0	0	2	0
2	0	1	0	0	0	0	1	0	1	0	0	0
3	0	1	0	1	0	0	0	0	0	1	0	1

图 12-2　频率分布矩阵

从图 12-2 中可以看出,文档在行中进行了编号,每个单词是一个列名称,相应的值是该单词在文档中出现的频率。接下来需要将文档进行转换。要处理这一步,需要使用 Count Vectorizer 方法,该方法的作用如下:

① 它会令牌化字符串(将字符串划分为单个单词)并为每个令牌设定一个整型 ID。

② 它会计算每个令牌的出现次数。

4. 生成数据集

通过在 Scikit-learn 中使用 train_test_split 方法,将数据集拆分为训练集和测试集。使用以下变量拆分数据,代码如下:

```
1.  from sklearn.model_selection import train_test_split
2.  X_train, X_test, y_train, y_test = train_test_split(
3.                                    df['sms_message'],df['label'],
4.                                        random_state=1)
5.  print('Number of rows in the total set: {}'.format(df.shape[0]))
6.  print('Number of rows in the training set: {}'.format(X_train.shape[0]))
7.  print('Number of rows in the test set: {}'.format(X_test.shape[0]))
```

接下来需要对数据集应用 Bag of Words 流程。已经拆分了数据,下一个目标是将数据转换为期望的矩阵格式。为此,像之前一样使用 CountVectorizer()。需要完成两步:

首先,需要对 CountVectorizer()拟合训练数据 (X_train) 并返回矩阵。

其次，需要转换测试数据（X_test）以返回矩阵。

X_train 是数据集中 'sms_message' 列的训练数据，将使用此数据训练模型，代码如下：

```
1.  from sklearn.feature_extraction.text import CountVectorizer
2.  count_vector = CountVectorizer()
3.
4.  #拟合训练数据,然后返回矩阵
5.  training_data = count_vector.fit_transform(X_train)
6.
7.  #转换测试数据并返回矩阵
8.  testing_data = count_vector.transform(X_test)
```

Scikit-learn 具有多个朴素贝叶斯实现方法，这样就不用从头进行计算，使用 Scikit-learn 的 sklearn. naive_bayes 方法对数据集做出预测。

这个分类器适合分类离散特征（例如单词计数文本分类）。另一方面，高斯朴素贝叶斯算法更适合连续数据，因为它假设输入数据是高斯（正态）分布，代码如下：

```
1.  from sklearn.naive_bayes import MultinomialNB
2.  naive_bayes = MultinomialNB()
3.  naive_bayes.fit(training_data, y_train)
4.  predictions = naive_bayes.predict(testing_data)
```

评估模型中已经对测试集进行了预测，下一个目标是评估模型的效果。可以采用各种衡量指标：

① 准确率衡量的是分类器做出正确预测的概率，即正确预测的数量与预测总数（测试数据点的数量）之比。

② 精确率指的是分类为垃圾邮件的邮件实际上是垃圾邮件的概率，即真正例（分类为垃圾邮件并且实际上是垃圾邮件的）与所有正例（所有分类为垃圾邮件，无论是否分类正确）之比，换句话说，就是以下公式的比值结果：

$$[TP/(TP+FP)]$$

③ 召回率（敏感性）表示实际上为垃圾邮件并且被分类为垃圾邮件的邮件所占比例，即真正例（分类为垃圾邮件并且实际上是垃圾邮件的）与所有为垃圾邮件之比，换句话说，就是以下公式的比值结果：

$$[TP/(TP+FN)]$$

④ F1 分数。对于偏态分类分布问题（数据集就属于偏态分类），例如，如果有 100 封邮件，只有 2 封是垃圾邮件，剩下的 98 封不是，则准确率本身并不是很好的指标。将 90 封邮件分类为非垃圾邮件（包括 2 封垃圾邮件，但是将其分类为非垃圾邮件，因此它们属于假负例），并将 10 封邮件分类为垃圾邮件（所有 10 封都是假正例），依然会获得比较高的准确率分数。对于此类情形，精确

率和召回率非常实用。可以通过这两个指标获得 F1 分数,即精确率和召回率分数的加权平均值。该分数的范围是 0 到 1,1 表示最佳潜在 F1 分数。

　　使用所有四个指标可以确保模型达到一个较好的效果。这四个指标的范围都在 0 到 1 之间,分数尽量接近 1 可以很好地表示模型的效果如何,代码如下:

```
1.  from sklearn.metrics import accuracy_score, precision_score, recall_
    score, f1_score
2.  print ('Accuracy score: ', format (accuracy_score (y_test, predic-
    tions)))
3.  print ('Precision score: ', format (precision_score (y_test, predic-
    tions)))
4.   print ('Recall score: ', format(recall_score(y_test, predictions)))
5.  print ('F1 score: ', format (f1_score (y_test, predictions)))
```

代码运行后,结果如图 12-3 所示。输出模型的 Accuracy、Precision、Recall、F1 的分数。

```
Accuracy score:  0.9885139985642498
Precision score:  0.9720670391061452
Recall score:  0.9405405405405406
F1 score:  0.9560439560439562
```

图 12-3　模型 Accuracy、Precision、Recall、F1 的分数

　　和其他分类算法相比,朴素贝叶斯算法具有的一大主要优势是能够处理大量特征。在垃圾邮件分类中,有数千个不同的单词,每个单词都被当作一个特征。此外,即使存在不相关的特征也有很好的效果,不容易受到这种特征的影响。另一个主要优势是相对比较简单。朴素贝叶斯算法可以直接使用,很少需要调整参数,除非通常分布数据已知的情况需要调整。它很少会过拟合数据。另一个重要优势是相对于它能处理的数据量来说,训练和预测速度很快。总之,朴素贝叶斯算法是非常实用的算法。

单元13

使用词向量Word2Vec
算法自动生成诗词

日常生活中使用的自然语言不能够直接被计算机所理解,当需要对这些自然语言进行处理时,就需要使用特定的手段对其进行分析或预处理。使用 one-hot 编码形式对文字进行处理可以得到词向量,但是,由于对文字进行唯一编号进行分析的方式存在数据稀疏的问题,Word2Vec 算法能够解决这一问题,可用来映射每个词到一个向量,表示词对词之间的关系。

13.1　Word2Vec 算法概述

Word to Vector(Word2Vec),即由词到向量。Word2Vec 使用一层神经网络将 one-hot 形式的词向量映射到分布式形式的词向量。Word2Vec 是谷歌于 2013 年提出的一种词嵌入模型,它实际上是一种浅层的神经网络模型,只有两种网络结构,分别是 CBOW 和 Skip-gram。其中,CBOW 的目标是根据上下文出现的词语来预测当前词的生成概率,而 Skip-gram 是根据当前词来预测上下文中各词的生成概率。

13.2　Word2Vec 算法的实现

在处理自然语言时,通常将词语或者字做向量化,可以使用 one-hot 编码将词语转化为向量,例如"我爱北京天安门",分词后对其进行 one-hot 编码,结果可得到:

"我":$[1,0,0,0]$

"爱":$[0,1,0,0]$

"北京":[0,0,1,0]

"天安门":[0,0,0,1]

但是,如果有 n 个词语,任何一个词的编码只有一个 1, n-1 位为 0,这会导致数据非常稀疏(0 特别多,1 很少),存储开销也很大。

于是,分布式表示被提出来了。什么是分布式表示? 它的思路是通过训练,将每个词都映射到一个较短的词向量上来。这个较短的词向量维度是多大呢? 一般需要在训练时自己来指定。

图 13-1 展示了四个不同的单词,可以用一个可变化的维度长度表示(图中只画出了前四维)。

king	queen	woman	priness
0.99	0.99	0.02	0.98
0.99	0.05	0.01	0.02
0.05	0.93	0.999	0.94
0.7	0.6	0.5	0.1
…	…	…	…

当使用词嵌入模型后,词之间可以存在一些关系,例如 king 的词向量减去 man 的词向量,再加上 woman 的词向量会等于 queen 的词向量。

图 13-1　可变化的维度长度表示四个不同的单词

如果使用热力图,可以直观地看到词向量的相似度,如图 13-2 所示。

由"king－man ＋ woman"生成的向量并不完全等同于"queen",但"queen"是在此集合中最接近它的单词,如图 13-3 所示。

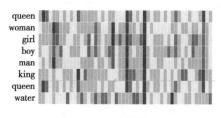

图 13-2　词向量热力图

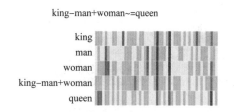

图 13-3　king-man ＋ woman 向量相似度

13.3　Word2Vec 模型

自然语言处理最典型的例子,就是智能手机输入法中的下一单词预测功能。这是被数十亿人每天使用上百次的功能。可以认为该模型接收到两个字(我打)并推荐了一组单词("电话" 就是其中最有可能被选用的一个),如图 13-4 所示。

Word2Vec 模型其实就是简单化的神经网络,输入为特征(feature),输出为预测值(predict),如图 13-5 所示。它对所有它知道的单词(模型的词库,可能有几千到几百万个单词)按可能性打分,输入法程序会选出其中分数最高的推荐给用户。

图 13-4　输入法智能输入

自然语言处理模型的输出就是模型所知单词的概率评分。通常把概率按百分比表示，但是实际上，40%这样的分数在输出向量组时表示为0.4，如图13-6所示。

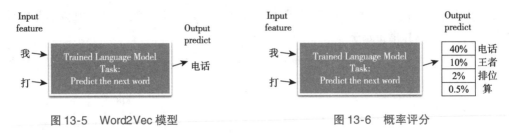

图 13-5 Word2Vec 模型　　　　　　　图 13-6 概率评分

Word2Vec 模型是一个简化的神经网络，包括输入层（input）、经过训练的语言模型（trained language model）和带 softmax() 函数的输出层（output），如图13-7所示。

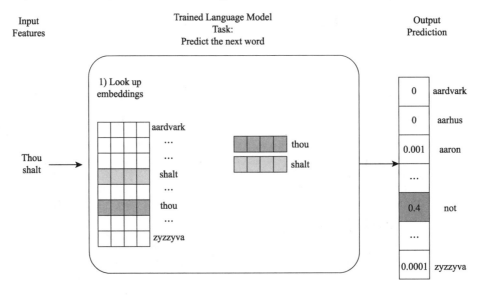

图 13-7 Word2Vec 模型结构

对于句子"I like deep learning and NLP"，基于这个句子，可以构建一个大小为6的词汇表。假设使用300个特征去表示一个单词。记上面的权重矩阵为 $w(6,300)$，用独热码 wt 表示矩阵（300，1）。$wt \times w$ 两个矩阵相乘，隐藏层神经网络输出的是一个 6×1 维矩阵，如图13-8所示。

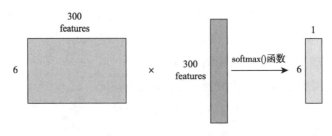

图 13-8 隐藏层神经网络输出

13.4　生成 Word2Vec 数据

先是获取大量文本数据,然后建立一个可以沿文本滑动的窗(例如一个窗里包含三个单词),利用这样滑动的窗就能为训练模型生成大量样本数据,如图 13-9 所示。

图 13-9　建立文本滑动的窗

前两个单词单做特征,第三个单词单做标签;生成了数据集中的第一个样本,如图 13-10 所示。

图 13-10　数据集中第一个样本

窗口滑动到下一个位置并生成数据集中第二个样本,如图 13-11 所示。

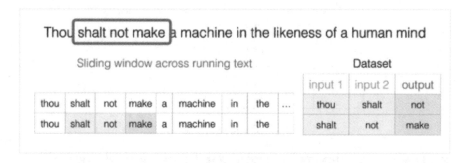

图 13-11　数据集中第二个样本

所有数据集上全部滑动后,得到一个较大的数据集,如图 13-12 所示。

Thou shalt not make | a machine in | the likeness of a human mind

Sliding window across running text

thou	shalt	not	make	a	machine	in	the	...
thou	shalt	not	make	a	machine	in	the	
thou	shalt	not	make	a	machine	in	the	
thou	shalt	not	make	a	machine	in	the	
thou	shalt	not	make	a	machine	in	the	

Dataset

input 1	input 2	output
thou	shalt	not
shalt	not	make
not	make	a
make	a	machine
a	machine	in

图 13-12　较大的数据集

13.5　训练 Word2Vec 模型

如果一个语料库稍微大一些，可能的结果简直太多了，计算起来十分耗时，这时可以输入两个单词，看它们是不是前后对应的输入和输出，也就相当于一个二分类任务。但是，此时训练集构建出来的标签全为 1，无法进行较好的训练，这时可以加入一些负样本（negative example），如图 13-13 所示，input word 表示输入的词语，output word 表示输出的词语，target 表示结果用 0 和 1 表示正负样本。

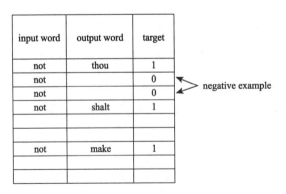

input word	output word	target
not	thou	1
not		0
not		0
not	shalt	1
not	make	1

negative example

图 13-13　待更新模型

通过神经网络反向传播来计算更新，此时不仅更新权重参数矩阵，也会更新输入数据，最终训练得到模型，模型输出使用 sigmoid 计算得到一个概率，计算损失（error），根据损失更新模型，如图 13-14 所示。

Word2Vec 优点：由于 Word2Vec 会考虑上下文，与之前的词向量方法相比，效果更好；比之前的词向量方法维度更少，所以速度更快；通用性很强，可以用在各种自然语言处理任务中。

input word	output word	target	input · output	sigmoid()	Error
not ⬜⬜⬜⬜	thou ⬛⬛⬛⬛	1	0.2	0.55	0.45
not ⬜⬜⬜⬜	aaron ⬛⬛⬛⬛	0	-1.11	0.25	-0.25
not ⬜⬜⬜⬜	taco ⬛⬛⬛⬛	0	0.74	0.68	-0.68

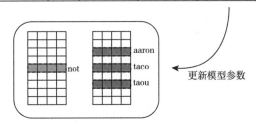

图 13-14　最终训练得到的模型

Word2Vec 缺点：由于词和向量是一对一的关系，所以多义词的问题无法解决；此外，Word2Vec 是一种静态的方式，虽然通用性强，但是无法针对特定任务做动态优化。

13.6 Word2Vec 算法的代码实现

1. 定义配置文件

在 Jupyter Notebook 中，设置 Word2Vec 的配置文件，设置滑动窗口大小、词向量维度等，代码如下：

```
1.  class CONF:
2.      path = '诗词.txt'
3.      window = 16      #滑窗大小
4.      min_count = 60  #过滤低频字
5.      size = 125       #词向量维度
6.      topn = 14        #生成诗词的开放度
7.      model_path = 'Word2Vec'
```

2. 定义模型

定义词向量模型，完成模型初始化功能，代码如下：

```
1.  Model:
2.      class def __init__(self, window, topn, model):
3.          self.window = window
4.          self.topn = topn
5.          self.model = model  #词向量模型
6.          self.chr_dict = model.wv.index_to_key  #字典
7.
```

```python
8.     @ classmethod
9.     def initialize(cls, config):
10.        """模型初始化"""
11.        if exists(config.model_path):
12.            # 模型读取
13.            model = Word2Vec.load(config.model_path)
14.        else:
15.            # 语料读取
16.            with open(config.path, encoding = 'utf-8') as f:
17.                ls_of_ls_of_c = [list(line.strip()) for line in f]
18.            # 模型训练和保存
19.            #window 表示当前词与预测词在一个句子中的最大距离是多少
20.            #min_count,可以对字典做截断。词频少于 min_count 次数的单词会
被丢弃掉，默认值为 5
21.            model = Word2Vec(ls_of_ls_of_c, window = config.window,
min_count = config.min_count)
22.            model.save(config.model_path)
23.        return cls(config.window, config.topn, model)
24.
25.    def poem_generator(self, title, form):
26.        """诗词生成"""
27.        def clean(lst): return [t[0] for t in lst if t[0] not in [',', '。']]
28.        # 标题补全
29.        if len(title) < 4:
30.            if not title:
31.                title += choice(self.chr_dict)
32.            for _ in range(4 - len(title)):
33.                #similar_by_key 计算相似度
34.                similar_chr = self.model.wv.similar_by_key(title[-
1], self.topn // 2)
35.                similar_chr = clean(similar_chr)
36.                #choice 返回一个列表项,随机选取一个数据并返回随机项
37.                char = choice([c for c in similar_chr if c not in title])
38.                title += char
```

```
39.          # 文本生成
40.          poem = list(title)
41.          for i in range(form[0]):
42.              for _ in range(form[1]):
43.                  predict_chr = self.model.predict_output_word(
44.                      poem[-self.window:], max(self.topn, len(poem) + 1))
45.                  predict_chr = clean(predict_chr)
46.                  char = choice([c for c in predict_chr if c not in poem
     [len(title):]])
47.                  poem.append(char)
48.              poem.append(',' if i % 2 == 0 else '。')
49.          length = form[0] * (form[1] + 1)
50.          return '《%s》' % ''.join(poem[:-length]) + '\n' + ''.join
     (poem[-length:])
```

3. 训练模型

载入文本数据，训练并保存模型。调用模型，实现自动书写诗词的功能，代码如下：

```
1.  def main(config=CONF):
2.      """主函数"""
3.      form = {'五言绝句': (4, 5), '七言绝句': (4, 7), '对联': (2, 9)}
4.      m = Model.initialize(config)
5.      while True:
6.          title = input('输入标题:').strip()
7.          poem = m.poem_generator(title, form['五言绝句'])
8.          print('\033[031m%s\033[0m' % poem)   # red
9.          poem = m.poem_generator(title, form['七言绝句'])
10.         print('\033[033m%s\033[0m' % poem)   # yellow
11.         poem = m.poem_generator(title, form['对联'])
12.         print('\033[036m%s\033[0m\n' % poem)  # purple
```

代码运行后，需要输入标题，例如输入：花田，程序会自动生成五言绝句、七言绝句和对联。具体如下：

输入标题:花田

五言绝句:桑零柘亩禾,种畦稻园杏。田蔬苗蚕农,黍菜麦村耕。

七言绝句:稻畦村麦蔬耕锄,田种禾亩苗桑农。柘稼粟丰蚕熟菜,枣药添肥炊烹童。

对联:梅村麦园篱亩杏禾桑,柘梨田畦耕蚕枣种熟。

Zhang Ming　Sun Xiaoli

Artificial Intelligence Technology and Applications (Bilingual)

中国铁道出版社有限公司
CHINA RAILWAY PUBLISHING HOUSE CO., LTD.

Introduction

As a Chinese-English textbook, this book first introduces linear regression to understand the basic knowledge of machine learning, such as dataset, normalization, loss function and gradient descent function. On this basis, through several typical cases, it has detailed the implementation process and related knowledge of the machine learning algorithms such as logistic regression, decision tree, clustering, support vector machine, etc. Moreover, it has also integrated the national vocational skill standard "artificial intelligence trainer" and "big data application development" 1 + X certificate and other relevant contents, so it has a strong significance in practical guidance.

This book can be used as teaching materials for AI, big data and other related professional and bilingual courses for higher vocational education, and also as a tutorial for application development technicians in the field of machine learning.

Preface

As one of the most popular fields in computer science, artificial intelligence (AI) has been applied widely in a variety of fields, including machine learning, natural language processing, image recognition and speech recognition, etc. For most industries, AI application, machine learning and big data analysis have become a very important part, so it will be a great advantage to master AI technology. To better cope with future challenges, both the management of enterprises and R&D personnel in the industry need to master AI technology.

Standing from the perspective of students, this book uses typical AI machine learning cases as carriers and the implementation process of project cases as the main thread to integrate the theoretical knowledge of AI machine learning into the project implementation process and gradually introduce the necessary basic knowledge of AI machine learning and some excellent tools, so as to help students have the relevant skills for using the sklearn library to implement machine learning tasks, be able to independently write programs to complete related projects and be competent for positions related to artificial intelligence machine learning.

This book consists of 13 units, which are divided into three parts: Unit 1-Unit 5 introduce linear regression, including simple linear regression, multiple linear regression, polynomial regression and ridge regression, etc.; Unit 6-Unit 8 introduce logistic regression, including logistic regression and decision boundary; Unit 9-Unit 13 introduce classic and commonly used machine learning methods, including clustering algorithms, decision trees, support vector machines, Bayes algorithm and Word2Vec, etc.

This book can be used as high vocational teaching materials for AI, big data and other related professional and bilingual courses, and also as a tutorial for application development technicians in the field of machine learning.

During the process of learning, readers must personally practice the case coded of this book. The online course resources accompanying this book can be available by searching for the "Artificial Intelligence Application Technology" course on the "Smart Vocational Education" website, or by joining the "Artificial Intelligence Application Technology" course in the MOOC College of Smart Vocational Education, and starting up in-depth learning through the teaching videos provided on the platform.

Despite the greatest efforts of the editor, there may still be some shortcomings in this book. We welcome valuable opinions from experts and readers from all walks of life for further improvement on this book.

If you need to apply for course resources while reading this book, or if you find any problem or have any disagreement, you can contact us via e-mail. Editor's e-mail: 45655600 @ qq. com.

Zhang Ming Sun Xiaoli
April, 2023

Contents

Prediction on the Revenue of Advertising Investment via Simple Linear Regression Model

A certain company has calculated its recent investment in wechat, weibo, TV, and other advertising media, and now needs to predict how much it can earn from its unit investment in advertisement.

1.1 Introduction to Simple Linear Regression

Simple linear regression, also known as univariate linear regression, means that the regression model only contains one independent variable; otherwise, it is called multiple linear regression. The simple linear regression has such a model as follows:

$$y = ax + b$$

Where, y is the dependent variable; x is the independent variable; b is the constant term, namely the intercept of the regression line on the vertical axis; a is the regression coefficient and the slope of the regression line.

Usually it is expressed as $y = wx + b$, where w represents weight and b represents the bias.

The company needs to predict w and b based on its given revenue y and advertising investment x, and ultimately obtain a one-dimensional linear equation, through which it can be predicted how much the enterprise can earn from the advertising funds invested, as shown in Fig. 1-1. The horizontal axis represents the investment of the enterprise, and the vertical axis represents the

expected return of the enterprise. The straight line goes through the points of the univariate linear equation obtained through training.

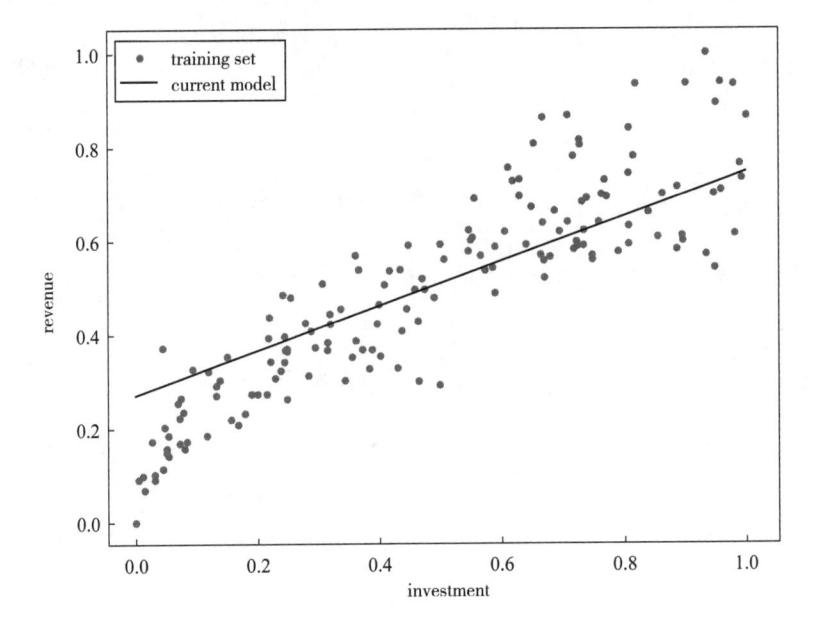

Fig. 1-1   Corporate investment and expected revenue

1.2   Dataset Reading

1. Import the corresponding package

Create a new Jupyter Notebook file and import pandas, matplotlib and numpy packages. Import these packages to complete data processing and visualization display. If you need to display Chinese on the icon, you also need to set the Chinese display format for matplotlib. The code is as follows:

```
1.   import pandas as pd
2.   import matplotlib.pyplot as plt
3.   import numpy as np
4.   #Display Chinese
5.   plt.rcParams['font.sans-serif'] = ['SimHei']
6.   plt.rcParams['axes.unicode_minus'] = False
```

2. Load data

You can use the read_csv method of pandas to open the dataset and use the head ( ) function to check the first five pieces of data and also the data information. The code is as follows:

	wechat	weibo	others	sales
0	304.4	93.6	294.4	9.7
1	1011.9	34.4	398.4	16.7
2	1091.1	32.8	295.2	17.3
3	85.5	173.6	403.2	7.0
4	1047.0	302.4	553.6	22.1

```
1.  df_data =pd.read_csv("advertising.csv")
2.  df_data.head()
```

Fig. 1-2   Code running effect

After running the code, output the first five pieces of data for reading the csv file, as shown in Fig. 1-2.

1.3 Data Correlation Analysis

In datasets with many sample attributes, some relationship exists between the features (x) and labels (y) of the sample. Some features have strong correlation with labels, while others are weak. Moreover, the correlation between features and labels can be displayed through drawing, covariance, correlation coefficient and information entropy. Usually, the heatmap ( ) function of the seaborne package is used to draw a heatmap between features and labels subject to the following code:

```
1.  import seaborn as sns
2.  sns.heatmap(df_data.corr(),cmap = 'YlGnBu',annot = True)
3.  plt.show()
```

Run the code and you can see the effect as shown in Fig. 1-3. Through visualization, you can easily see the correlation between features. According to Fig. 1-3, it is easy to see the "impact factor" between features. The darker the color of the area where the horizontal and vertical coordinates intersect, the deeper the relationship between them. The numbers marked on the area block also display such a feature.

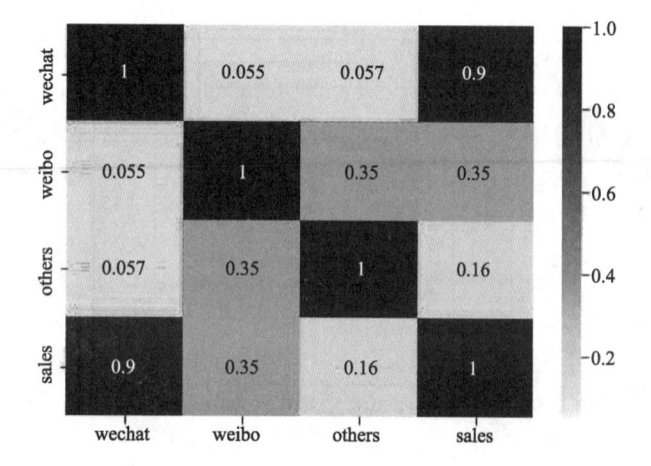

Fig. 1-3   Heatmap between features and labels

From Fig. 1-3, it can be seen that a strong correlation exists between the "wechat" investment and the corporate revenue, with a correlation coefficient of 0.9. Taking bivariate as an example, there are three types of relationships between variable x and variable y: positive linear correlation, negative linear correlation, and nonlinear correlation, as shown in Fig. 1-4. The horizontal axis represents x while the vertical axis represents y.

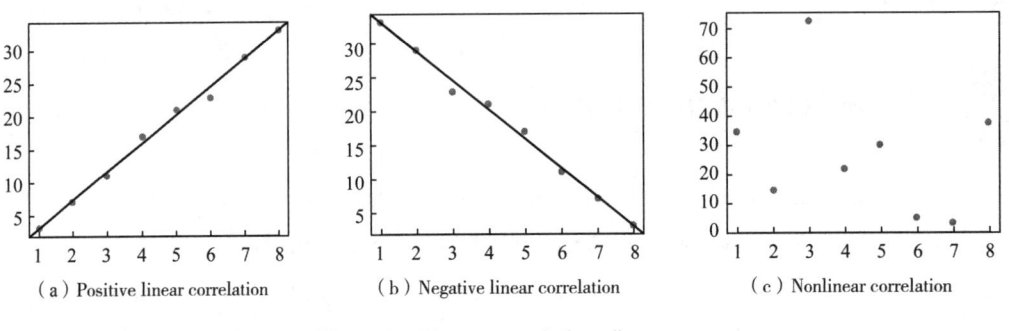

(a) Positive linear correlation　　(b) Negative linear correlation　　(c) Nonlinear correlation

Fig. 1-4　Linear correlation diagram

At the same time, use seaborn to draw a pairplot diagram and it can also be analyzed that the "wechat" investment basically shows a linear relationship with the corporate revenue. The code is as follows:

```
1.  sns.pairplot(df_data,x_vars = ['wechat','weibo','others'],
2.                  y_vars = 'sales',
3.                  height = 4,aspect = 1)
4.  plt.show()
```

Fig. 1-5 shows the code running results, which shows a linear relationship between wechat investment and revenue.

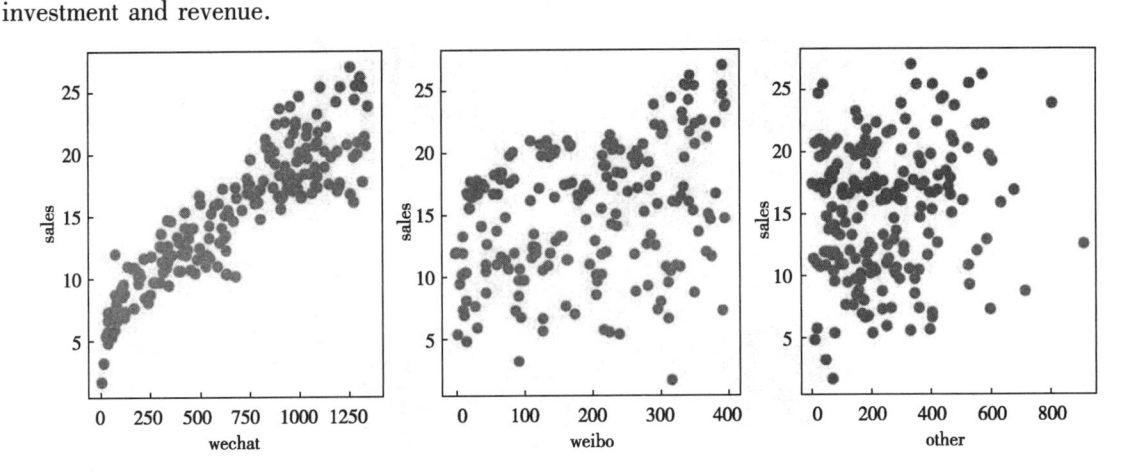

Fig. 1-5　Relationship between advertising revenue and wechat, weibo, and other investments

1.4   Data Processing

1. Select data

The first step in linear model construction is to select data. By analyzing the data correlation, you can extract the "wechat" data from the dataset as X and "sales" as y subject to the code as follows:

```
1.  X_data = df_data['wechat']
2.  y_data = df_data['sales']
3.  print(type(X_data))
4.  print(type(y_data))
```

Run the code and it will output X_data subject to the type < class 'pandas. core. series. Series ' > and the y_data subject to the type < class ' pandas. core. series. Series ' >, indicating that the extracted X and y are series objects.

The linear model should use tensors as input, converting series type data to tensors, and using the array () function to convert the type from series to ndarry. The code is as follows:

```
1.  # Tensor transformation
2.  X_data = np.array(X_data)
3.  y_data = np.array(y_data)
4.  print(type(X_data))
5.  print(X_data.shape)
6.  print(X_data.ndim)
7.  X = X_data.reshape(len(X_data),1)
8.  print(type(X))
9.  print(X.shape)
10. print(X.ndim)
```

Run the code and it will output X_data subject to the type of 'numpy. ndarray' and the X_data shape of (200,), and the value of X. ndim is 2.

2. Split the dataset into train and test sets

Model training requires dividing the dataset into train and test sets and using the sklearn model _selection() function to split the X_data and y_data according to 80% of the training set and 20% of the test set to a training set and a test set.

```
1.  from sklearn.model_selectionimporttrain_test_split
2.  X_train,X_test,y_train,y_test =train_test_split(X_data,y_data,
    test_size =0.2,random_state =0)
```

Simultaneously, draw a scatterplot based on the train and test set data to display the distribution of features and labels. The code is as follows:

```
1.  sns.scatterplot(X_data,y_data)
2.  plt.xlabel("weibo")
3.  plt.ylabel("sales")
4.  plt.show()
```

Run the code and see the relationship between weibo and sales in the training set as shown in Fig. 1-6, indicating a linear relationship between the two data.

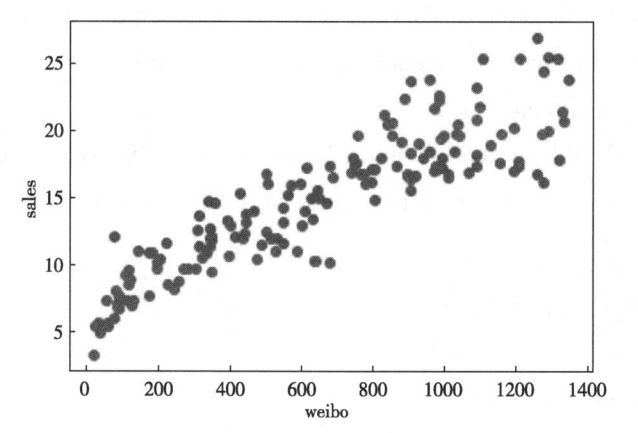

Fig. 1-6   Relationship between weibo and sales (training set)

Use the sns scatterplot method to draw a weibo and revenue scatterplot in the test set, with the code as follows:

```
1.  sns.scatterplot(X_test,y_test)
2.  plt.xlabel("weibo")
3.  plt.ylabel("sales")
4.  plt.show()
```

Run the program and you can see the relationship between weibo and revenue in the test set, as shown in Fig. 1-7. From the two diagrams, it can be seen that the data distribution is the same in the test and the training sets, but the difference is that the training set contains more data.

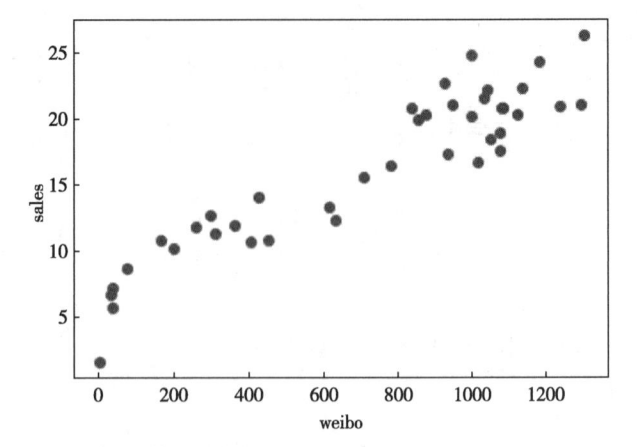

Fig. 1-7 Relationship between weibo and sales (test set)

1.5 Data Normalization

Why do we often normalize data in machine learning? The reason is that normalization accelerates the gradient descent to find the optimal solution and may improve accuracy. Here, linear normalization is used subject to such an equation as follows:

$$X = \frac{x - \min(x)}{\max(x) - \min(x)}$$

Where, X represents the normalized data; x represents the data in the dataset; min (x) represents the minimum value in the dataset; max (x) represents the maximum value in the dataset.

This normalization method is more suitable for situations with concentrated numerical values. However, this method has such a drawback as follows: if max and min are unstable, it is easy to make the normalization results unstable, which also makes the subsequent use effect unstable. In practical use, empirical constant values can be used instead of max and min.

Here, a function scalar ( ) is defined to implement normalization operations. Then the function is called to normalize the X_train, X_test, y_train and y_test while drawing a scatterplot, subject to the code as follows:

```
1.  def scalar(train,test):
2.      min = train.min(axis = 0)
3.      max = train.max(axis = 0)
4.      gap = max - min
5.      train - = min
6.      train/ = gap
```

```
7.          test - =min
8.          test/ =gap
9.          return train,test
10.
11. X_data,X_test = scalar(X_train,X_test)
12. y_train,y_test = scalar(y_train,y_test)
13.
14. sns.scatterplot(X_train,y_train)
15. plt.show()
```

Run the code and it will output the normalized relationship between wechat and revenue as shown in Fig. 1-8. By comparing the data maps after normalization and before normalization, it can be seen that the range of wechat data values has changed from 0 to 1400 through normalization to 0-1; the range of income (sales) data values ranges from 0 to 25 through normalization to 0-1. The range of values for wechat investment and sales in the two figures has changed, but the distribution between the two has not changed, as shown in Fig. 1-8.

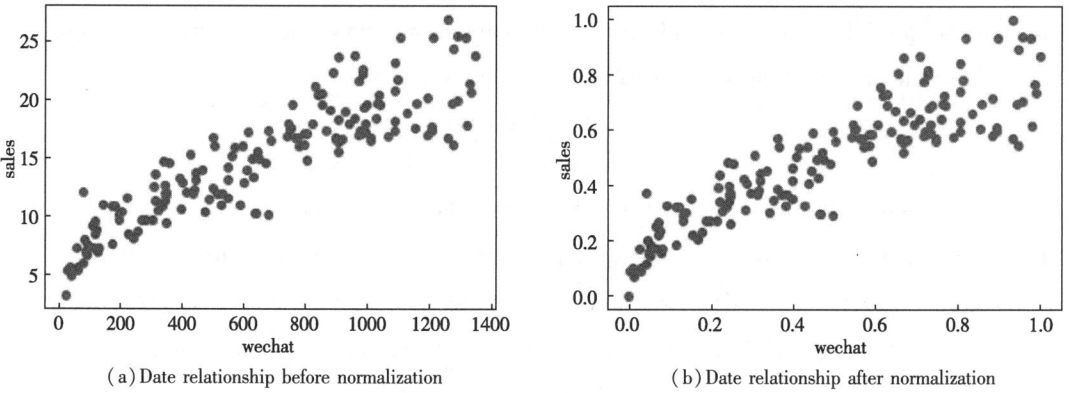

(a) Date relationship before normalization   (b) Date relationship after normalization

Fig. 1-8   Comparison of wechat investment and income before and after normalization

1.6   Define Loss Function

In machine learning, the objective function is the core guidance of the whole model optimization learning. When the objective function needs to be minimized, the objective function is also called loss function or cost function. For supervised learning tasks, the goal is usually to make the predicted value as close to the label as possible, that is, to minimize the objective function. The selection of loss function mainly depends on the requirements of learning tasks. The two most commonly used loss function MSE and cross entropy are used for regression and classification tasks respectively. The

mean square error (MSE) loss function is subject to such a formula as follows:

$$\text{MSE} = \frac{1}{2m} \sum_{i=1}^{n} (y_i - y_i')^2$$

Where, y_i' represents the predicted value; y_i represents the actual value.

The distance is expressed as $y_i - y_i'$ between the actual and predicted values, and the sum of all the distances between the actual and predicted values is the overall loss of the model. In order to ensure that the distance will not produce positive and negative offset, a square term needs to be added, and then divided by $2m$ (m is the number of samples) to calculate the average value, and then the mean square error loss function can be obtained.

After derivation, MSE is not significant in computational complexity, so it is most commonly used. The code can be easily implemented and you can write your own code to implement it. The code is as follows:

```
1.   def loss_function(X,y,weight,bias):
2.       y_hat =weight* X +bias* #y =a* x +b
3.       loss = y_hat - y
4.       cost =np.sum(loss* * 2)/(2* len(X))
5.       return cost
```

Firstly, initialize the weight w and bias value b, and then output the MSE value. The code is as follows:

```
1.   print ("When the weight is 5 and the bias value is 3, the loss function
     is", loss_function (X_train, y_train, weight =5, bias =3))
2.   print ("When the weight is 100 and the bias value is 1, the loss function
     is", loss_function (X_train, y_train, weight =5, bias =3))
```

After running the code, when the weight is 5 and the bias value is 3, the loss value is 12. 796 390 970 780 058; when the weight is 100 and the bias value is 1, the loss value is 12. 796 390 970 780 058.

Then use the scatterplot ( ) function to draw an image of the predicted and actual values of the model, with the following code:

```
1.   sns.scatterplot(X_train,y_train)
2.   line_x =np.linspace(X_train.min(),X_train.max())
3.   line_y = [ -5* xx +3forxxinline_x]
4.   sns.scatterplot(line_x,line_y)
5.   plt.show()
```

After running the code, the results are shown in Fig. 1-9. From the output results, it can be

seen that when the model uses initialization parameters, there is a significant error between the predicted and actual values of the model.

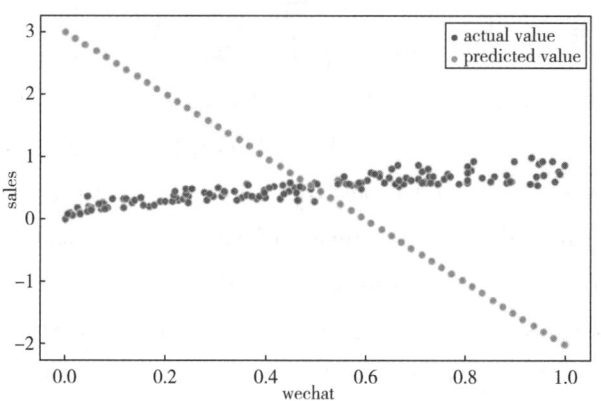

Fig. 1-9   Computational results from model initialization parameters

1.7   Define Gradient Descent Function

In the previous section, the loss function is defined. By taking x and y as known quantities, the loss function can be converted into the form of $l\ (w,b)$. In this function, w and b are unknown values. The MSE function can be expressed with the following formula:

$$l(w,b) = \frac{1}{2m}\sum_{i=1}^{m}\left[(wx + b) - y_i'\right]^2$$

Gradient descent essentially requires the values of w and b to be determined when $l\ (w,b)$ has the minimum value. $l\ (w,b)$ can be regarded as a quadratic function, where w and b are variables.

You can first associate the function $y = x^2$, as shown in Fig. 1-10. See Fig. 1-11 for the graph of the $l(w,b)$ function. The value of $l(w,b)$ can be regarded as the value of the z-axis.

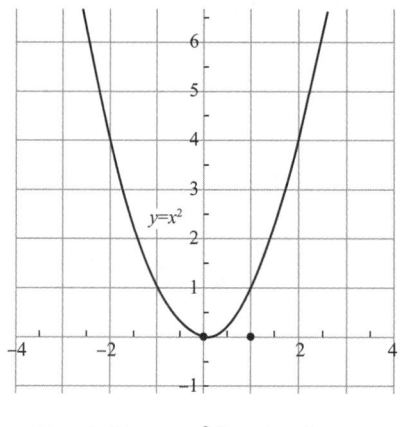

Fig. 1-10 $y = x^2$ function image

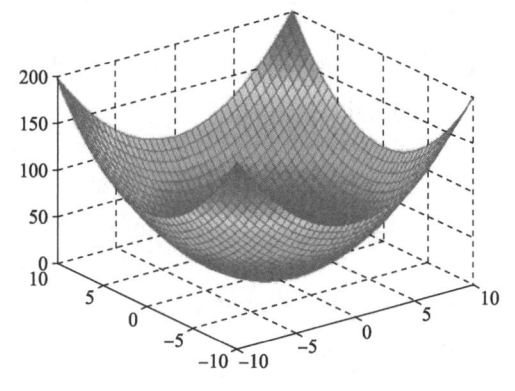

Fig. 1-11 $l(w,b)$ function image

Gradient descent requires using partial derivative to calculate the gradient of $l(w,b)$, and determine the value of the next (w,b) according to the gradient and learning rate (step size). The specific formula is as follows:

$$\frac{\partial l}{\partial b} = \frac{1}{n} \sum_{i=1}^{m} [(wx'_i + b) - y'_i]$$

$$= \frac{1}{m} \sum_{i=1}^{n} (\text{loss})$$

Where, loss represents the loss of a single sample, $\text{loss} = (wx'_i + b) - y'_i$.

$$\frac{\partial l}{\partial w} = \frac{1}{m} \sum_{i=1}^{n} [(wx'_i + b) - y'_i] x'_i$$

Use a matrix to transform the above equation and it can be obtained that $[\boldsymbol{X} = (x_0, \cdots, x_n), \boldsymbol{y} = (y^1, \cdots, y^m)]$

$$\nabla J(w) = \frac{1}{n} \boldsymbol{X}^{\mathrm{T}}((\boldsymbol{X}w + b) - \boldsymbol{y})$$

$$\nabla J(w) = \frac{1}{n} \boldsymbol{X}^{\mathrm{T}}(\text{loss})$$

Where, $\boldsymbol{X}$ represents the set of features in the dataset; $\boldsymbol{y}$ represents the set of labels in the dataset.

Use the code to implement gradient descent function, introduce such parameters as X training set, y label, w weight, b bias value, alpha learning rate/step size, iter iterations, and use l_ history, w_history and b_history list to maintain the loss, weight and bias values after individual iteration, and return the results subject to such codes as follows:

```
1.  def gradient_descent(X,y,w,b,lr,iter):
2.      l_history = np.zeros(iter)
3.      w_history = np.zeros(iter)
4.      b_history = np.zeros(iter)
5.      for i in range(iter):
6.      y_hat = w* X + b
7.      loss = y_hat - y
8.      derivative_w = X.T.dot(loss)/len(X)
9.      derivative_b = sum(loss)/len(X)
10.        w = w - derivative_w
11.        b = b - derivative_b
12.        l_history[i] = loss_function(X,y,w,b)
13.        w_history[i] = w
14.      b_history[i] = b
15.      returnl_history,w_history,b_history
```

1.8   Training Model

The model training first needs initialization of such four parameters as iter iterations, alpha learning rate/step size, weight and bias value, and working out an initial loss value.

```
1.   # Initialization parameters
2.   iter =100
3.   alpha =0.5
4.   weight = -5
5.   bias =3
6.   print ("When the weight is -5 and the bias value is 3, the Loss
     function is", loss_function (X_train, y_train, weight, bias))
```

Run the code, and when the output weight is −5 and the bias value is 3, the loss function is 1. 343 795 534 906 634.

Then call the gradient descent function to start training. Obtain the loss_history, weight_history and bias_history values through iteration, with the code as follows:

```
1.   # Start training
2.   loss_history,weight_history,bias_history =gradient_descent
     (X_train,y_train,weight, bias, alpha, iterations)
```

After the code runs, print out loss_ history; the set iter = 100, so it will output 100 loss results, as shown in Fig. 1-12.

```
array([1.16870167, 1.01830004, 0.88743937, 0.77347921, 0.67423079,
       0.5877945 , 0.5125164 , 0.44695606, 0.38985899, 0.34013266,
       0.29682557, 0.25910905, 0.22626142, 0.19765412, 0.17273979,
       0.15104168, 0.13214461, 0.11568699, 0.10135391, 0.0888711 ,
       0.07799972, 0.06853174, 0.06028599, 0.0531047 , 0.04685045,
       0.04140357, 0.03665984, 0.03252848, 0.02893044, 0.02579688,
       0.02306784, 0.02069109, 0.01862116, 0.01681844, 0.01524843,
       0.0138811 , 0.01269028, 0.01165319, 0.01074997, 0.00996335,
       0.00927828, 0.00868165, 0.00816203, 0.00770949, 0.00731538,
       0.00697214, 0.0066732 , 0.00641286, 0.00618613, 0.00598866,
       0.00581669, 0.00566692, 0.00553648, 0.00542288, 0.00532394,
       0.00523778, 0.00516274, 0.00509738, 0.00504046, 0.00499089,
       0.00494772, 0.00491013, 0.00487738, 0.00484887, 0.00482403,
       0.0048024 , 0.00478356, 0.00476716, 0.00475287, 0.00474043,
       0.00472959, 0.00472015, 0.00471193, 0.00470477, 0.00469854,
       0.00469311, 0.00468838, 0.00468426, 0.00468067, 0.00467755,
       0.00467483, 0.00467246, 0.0046704 , 0.0046686 , 0.00466703,
       0.00466567, 0.00466448, 0.00466345, 0.00466255, 0.00466177,
       0.00466108, 0.00466049, 0.00465997, 0.00465952, 0.00465913,
       0.00465878, 0.00465849, 0.00465823, 0.004658  , 0.0046578 ])
```

Fig. 1-12   Iteration loss results

After model training, a weight_history and bias_history set can be obtained; take the last value

as the optimal weight and bias values of the model, and draw the model.

```
1.  # Draw the final model
2.  line_x = np.linspace(X_train.min(),X_train.max(),500)
3.  line_y = [weight_history[-1]* xx +bias_history[-1]forxxinline_x]
4.  sns.scatterplot(X_train,y_train)
5.  sns.scatterplot(line_x,line_y)
6.  plt.show()
```

After running the code, see Fig. 1-13 for the results. The points in the figure represent the actual values of the sample, and the lines represent the predicted values of the model.

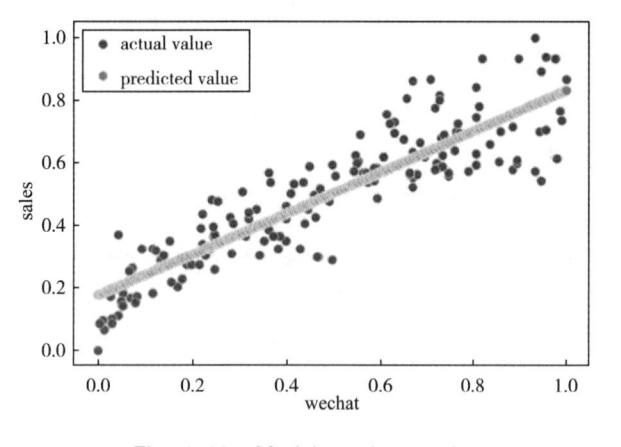

Fig. 1-13   Model running results

The weight_history [-1] value represents the weight value obtained after the last iteration, which is the weight of the linear model and also bias output. The code is as follows:

```
print ("Obtained parameters", "weight = ", weight_history [-1], "bias
= ", bias history [-1])
```

Code running result: The obtained parameter weight = 0.655 225 340 919 280 8, bias = 0.176 903 410 094 724 88, and these two values are the optimal weight and bias values of the model.

1.9   Simple Linear Model Loss

1. Draw a loss curve

During model training, iter = 100 is a preset parameter, whose range can be determined by plotting a loss curve to show the process of loss descent. The code is as follows:

```
1.  line_x = np.linspace(0,iter,iter)
2.  sns.scatterplot(line_x,loss_history)
3.  plt.show()
```

After running the code, see Fig. 1-14 for the output results. The horizontal axis in the figure represents the number of iterations, while the vertical axis represents the loss. From the figure, it can be seen that the loss changes rather slightly after 40 iterations, so the parameters of the iteration can be set to the values between 40 and 50.

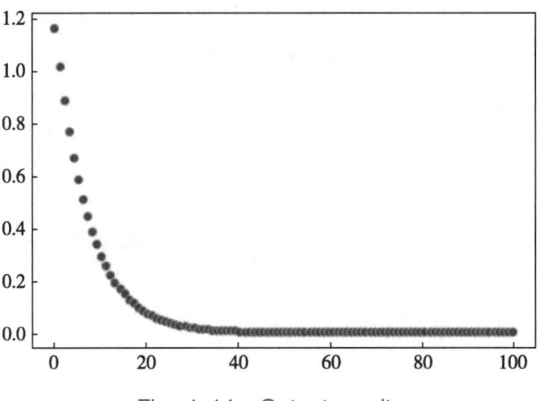

Fig. 1-14   Output results

2. Calculate current losses

The final loss of the training set can be calculated based on the obtained weight and bias. The code is as follows:

```
1.  print ('current loss', loss_function (X_train, y_rain, weight
    _history[-1], bias_history [-1]))
2.  print ('current weight', weight_history [-1])
3.  print ('current bias', bias_history [-1])
```

After running the code, the output is as follows: current loss 0. 004 657 804 055 314 04, current weight 0. 655 225 340 919 280 8, current bias 0. 176 903 410 094 724 88.

3. Calculate the test set loss

Based on the obtained weight and bias, use the loss_function ( ) function to calculate the loss of the test set. The code is as follows:

```
1.  print ('current loss', loss_function (X_test, y_test, weight
    _history[-1], bi-as_history [-1]))
2.  print ('current weight', weight_history [-1])
3.  print ('current bias', bias_history [-1])
```

After running the code, the output is as follows: current loss 0. 004 581 809 380 247 21, current weight 0. 655 225 340 919 280 8, current bias 0. 176 903 410 094 724 88.

1.10   Draw the Contour Plot of *w* and *b*

In machine learning, contour plot is often used. Here, a contour plot is used to represent the process of solving the bias value, that is, the dynamic process of solution based on the gradient_ descent ( ) function. Use Matplotlib-Animation to generate an animation and save it as a gif image. The code is as follows:

```
1.   # Design Contour Plot animation
2.   import matplotlib.animation as animation
3.
4.   theta0_vals = np.linspace( -2,3,100)
5.   theta1_vals = np.linspace( -3,3,100)
6.   J_vals = np.zeros((theta0_vals.size,theta1_vals.size))
7.
8.   for t1,element in enumerate(theta0_vals):
9.        for t2,element2 in enumerate(theta1_vals):
10.
11.  weight = element
12.  bias = element2
13.  J_vals[t1,t2] = loss_function(X_train,y_train,weight,bias)
14.
15.  J_vals = J_vals.T
16.  A,B = np.meshgrid(theta0_vals,theta1_vals)
17.  C = J_vals
18.
19.  fig = plt.figure(figsize = (12,5))
20.  plt.subplot(121)
21.  plt.plot(X_train,y_train,'ro',label = 'Trainingdata')
22.  plt.title('SalesPrediction')
23.  plt.axis([X_train.min() - X_train.std(),X_train.max() + X_train.
     std(),y_train.min() - y_train.std(),y_train.max() + y_train.std()])
24.  plt.grid(axis = 'both')
25.  plt.xlabel("WeChatAdsVolumn(X1)")
```

```
26. plt.ylabel("SalesVolumn(Y)")
27. plt.legend(loc = 'lowerright')
28.
29. line, =plt.plot([],[],'b - ',label = 'CurrentHypothesis')
30. annotation =plt.text( -2,3,'',fontsize =20,color = 'green')
31. annotation.set_animated(True)
32.
33. plt.subplot(122)
34. cp =plt.contourf(A,B,C)
35. plt.colorbar(cp)
36. plt.title('FilledContoursPlot')
37. plt.xlabel('Bias')
38. plt.ylabel('Weight')
39. track, =plt.plot([],[],'r - ')
40. point, =plt.plot([],[],'ro')
41.
42. plt.tight_layout()
43. plt.close()
44.
45. def init():
46.     line.set_data([],[])
47.     track.set_data([],[])
48.     point.set_data([],[])
49.     annotation.set_text('')
50.     returnline,track,point,annotation
51.
52. defanimate(i):
53.     fit1_X =np.linspace(X_train.min() -X_train.std(),X_train.max() +
    X_train.std(),1000)
54.     fit1_y =b_history[i] +w_history[i]* fit1_X
55.
56.     fit2_X =b_history.T[:i]
57.     fit2_y =w_history.T[:i]
58.
59.     track.set_data(fit2_X,fit2_y)
60.     line.set_data(fit1_X,fit1_y)
61.     point.set_data(b_history.T[i],w_history.T[i])
62.
63.     annotation.set_text('Cost =% .4f'% (l_history[i]))
```

```
64.        returnline,track,point,annotation
65.
66. anim = animation.FuncAnimation(fig,animate,init_func = init,
67.                            frames = 50,interval = 0,blit = True)
68.
69. anim.save('animation.gif',writer = 'imagemagick',fps = 500)
```

Load and display the gif image code:

```
1.   # Display ContourPlot animation
2.   import io
3.   import base64
4.   from IPython.display import HTML
5.
6.   filename = 'animation.gif'
7.
8.   video = io.open(filename, 'r + b').read()
9.   encoded = base64.b64encode(video)
10.  HTML(data = ''' < img src = "data:image/gif;base64,{0}" type = "gif"/ >
     '''.format(encoded.decode('ascii')))
```

Set weight = -2 and bias = -2 as the initialization values, with the running results as shown in Fig. 1-15. From the figure, it can be seen that as the parameters are fitted, the loss decreases continuously and ultimately reaches the center of the contour plot, which is namely the darkest part of the contour plot, which is the optimal solution. See Fig. 1-16 for status of the optimal point.

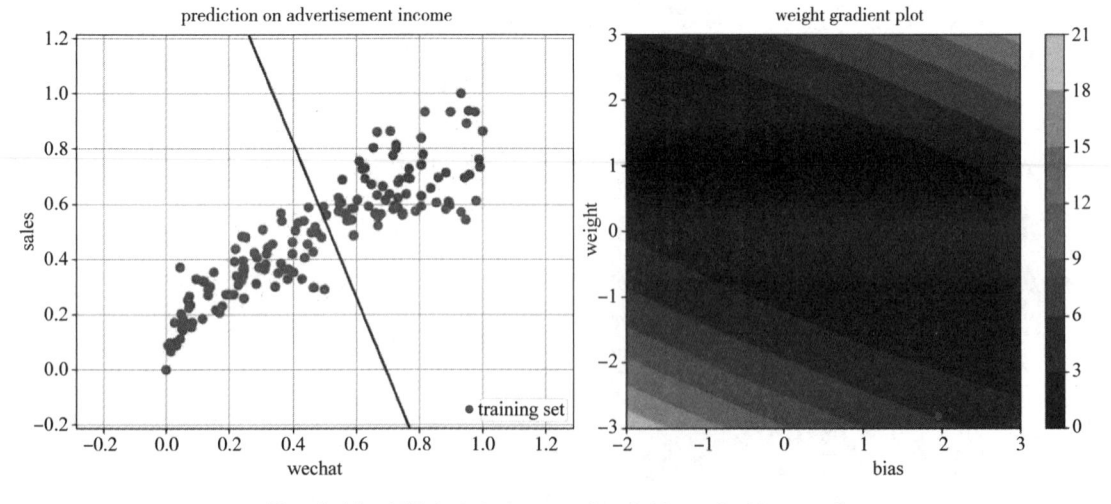

Fig. 1-15   Initial state image of weight = -2, bias = -2

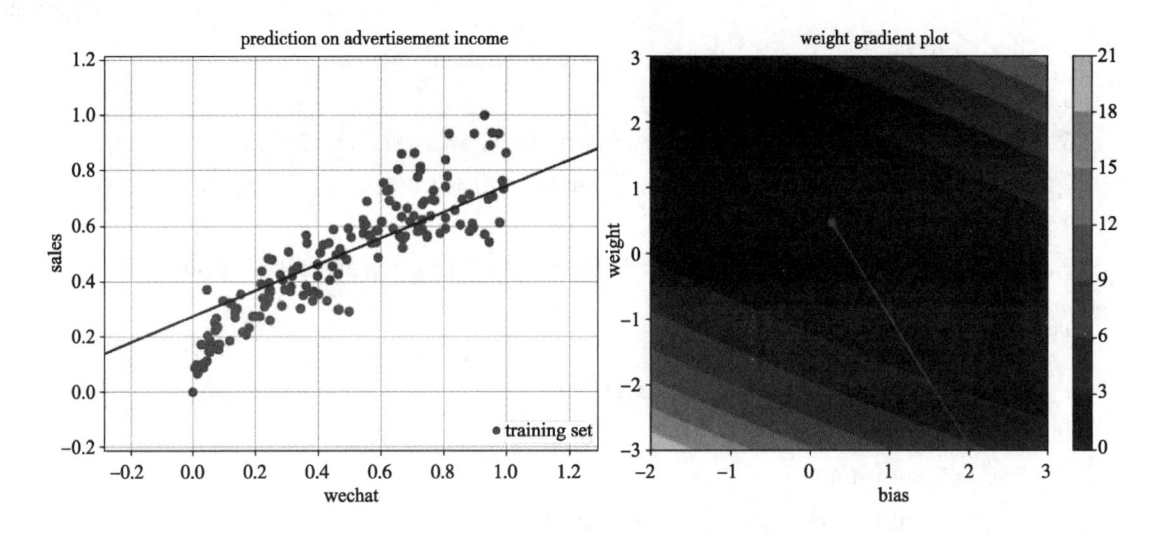

Fig. 1-16 State of approximate optimal point

Unit 2

Prediction on the Revenue of Advertising Investment via Multiple Linear Regression Model

In Unit 1, the wechat investment was used to predict the expected revenue of the company, but it has also invested funds in weibo and other media. To predict the returns of investing funds in all media, a multiple linear regression model is needed.

2.1　Introduction to Multiple Linear Regression

Multiple linear regression means that the regression model contains multiple independent variables. The multiple linear regression has its model as follows:

$$h(x) = w_1 x_1 + w_2 x_2 + w_3 x_3 + w_4 x_4 + w_5 x_6 + \cdots + w_n x_n + b$$

Where, x_1, x_2, x_3, $\cdots x_n$ are independent variables; b is a constant term.

In this unit, there are three features, "wechat", "weibo", and "others". Assuming $x_0 = 1$, we can obtain $w_0 x_0 = w_0$, and w_0 can be regarded as b, so the following formula can be used:

$$h(x) = w_0 x_0 + w_1 x_1 + w_2 x_2 + w_3 x_3$$

If a vector is used to represent $h(x)$, the weight can be expressed as $\boldsymbol{\theta}^{\mathrm{T}}$. If the matrix $\boldsymbol{x}$ is used to represent features, this formula can be written as follows:

$$h_\theta(x) = \sum_{i=1}^{n} \theta_i x_i = \boldsymbol{\theta}^{\mathrm{T}} \boldsymbol{x}$$

2.2 Dataset Reading

1. Import data load packages

Import such three packages as numpy, pandas and matplotlib in Jupyter Notebook to load data and analyze the data at the same time. The code is as follows:

```
1.  import pandas as pd
2.  import matplotlib.pyplot as plt
3.  import numpy as np
```

2. Load data

Use the pandas's read_csv method to open the dataset and use the head ( ) function to check the first five pieces of data. At the same time, the info ( ) function can be used to check the information of the data. The code is as follows:

```
1.  df_data = pd.read_csv("../data/advertising.csv")
2.  df_data.head()
```

2.3 Generate Training and Test Sets

1. Select data

Firstly, keep all features, then use numpy's delete to delete the label field "sales" and obtain the feature set, and then take out "sales" separately to obtain the label. Then, use the np. array ( ) and reshape( ) functions to convert the data into a tensor.

```
1.  X_data = np.array(df_data)
2.  X_data = np.delete(X_data,[3],axis =1)
3.  y_data = np.array(df_data.sales)
4.  #"the length of(% s)is% d"% ('runoob',len('runoob'))
5.  print("tensor X_data shape is:% s,dimension is% d"% (X_data.
    shape,X_data.ndim))
6.  print("tensor y_data shape is:% s,dimension is% d"% (y_data.
    shape,y_data.ndim))
7.  # print ("The shape of the tensor X_data is: % s, and the dimension
    is: % d"% (X_data. shape, X_data. ndim))
```

```
8.   #print ("The shape of the tensor y_data is: % s, and the dimension
     is: % d"% (y_data. shape, y_data. ndim))
```

Code run output result, tensor X_data shape is: (200, 3), dimension is 2; tensor y_data shape is: (200,), dimension is 1, and the shape is available after X_data and y_data have been converted into tensors.

At this point, the shape of y_data is a vector, and the reshape( ) function can be used to transform the vector into a matrix, namely a 2D tensor.

```
1.   y_data = y_data.reshape (-1,1)
2.   print ("tensor y_data shape is:% s, dimension is% d"% (y_data.
     shape, y_data.ndim))
```

Code run output result, tensor y_data shape is: (200, 1), dimension is 2.

2. Split training set and test set

Use the sklearn's model_selection function to split the x_data and y_data according to 80% of the training set and 20% of the test set to a training set and a test set.

```
1.   from sklearn.model_selection import train_test_split
2.   X_train,X_test,y_train,y_test = train_test_split(X_data,y_data,
     test_size =0.2,random_ state =0)
```

2.4 Denormalization

In machine learning, after model training, it is necessary to predict the test set data, namely, it requires denormalization operations. The denormalization function is used for data restoration. Moreover, it still need to save y_min, y_max and y_gap, calculate the maximum and minimum values of the training set, as well as their differences, for the subsequent denormalization process while keeping a copy of the original data. The code is as follows:

```
1.   #Denormalization
2.   def min_max_gap (train): #Calculate the maximum and minimum values
     of the training set, as well as their differences, for the
     subsequent denormalization process
3.       min = train.min (axis =0) #The minimum value of the training set
4.       max = train.max (axis =0) #The maximum value of the training set
5.       gap = max -min #Difference between maximum and minimum values
```

```
6.       return min,max,gap
7.
8.       y_min, y_max, y_gap =min_max_gap(y_train)
```

Keep the original data. The code is as follows:

```
1.   # Keep a copy of the original data
2.   X_data_original =X_train.copy()
```

Use normalization functions to process data. The code is as follows:

```
1.   X_train,X_test =scaler(X_train,X_test) # Normalize features
2.   y_train,y_test =scaler(y_train,y_test) # Normalize labels as well
```

Add a value of x_0, where all x_0 values are 1. The code is as follows:

```
1.   train0 =np.ones((len(X_train),1))
2.   print ("The shape of the tensor X_train is: %  s, and the dimension
     is: % d"% (X_train. shape, X_train. ndim))
3.   X_train =np.append(X_train0,X_train,axis =1)
4.   X_test0 =np.ones((len(X_test),1))
5.   X_test =np.append(X_test0,X_test,axis =1)
6.   print ("The shape of the tensor X_data is: %  s, and the dimension
     is: % d"% (X_train. shape, X_train. ndim))
```

After running the code, the shape of the output tensor X_train is: $(160, 3)$, and the dimension is: 2; the shape of the tensor X_data is: $(160, 4)$, and the dimension is: 2.

2.5　Define Loss Function and Mean Square Error (MSE) Loss Function

Like simple linear regression, multiple linear regression uses MSE as the loss function. The formula is as follows:

$$MSE = J(\theta) = \frac{1}{2m} \sum_{i=1}^{m} (h_\theta(x^{(i)}) - y^{(i)})^2$$

Usually represented by a matrix:

$$MSE = J(\theta) = \frac{1}{2m}(X\theta^T - y)^2$$

Where, $J(\theta)$ represents the loss function; $h_\theta(x^{(i)}) = y' = X\theta^T$ represents the predicted value of the model; y represents the label value.

It can be calculated with vector inner product. In Python, np. dot is used to accomplish such an

operation, X. dot (θ. T), in which T undertakes to change the shape of θ from $(1,4)$ to $(4,1)$, and then subtract the corresponding label value y to obtain $h_\theta(x^{(i)}) - y^{(i)}$.

For example, when the feature is a 3×3 matrix, labeled as a 3×1 matrix, the value of $J(\theta)$ can be calculated based on the following equation:

$$
\boldsymbol{y'} - \boldsymbol{y} = \begin{pmatrix} x_0^{(1)}\theta_0 + x_1^{(1)}\theta_1 + x_2^{(1)}\theta_2 \\ x_0^{(2)}\theta_0 + x_1^{(2)}\theta_1 + x_2^{(2)}\theta_2 \\ x_0^{(3)}\theta_0 + x_1^{(3)}\theta_1 + x_2^{(3)}\theta_2 \end{pmatrix} - \begin{pmatrix} y^{(1)} \\ y^{(2)} \\ y^{(3)} \end{pmatrix}
$$

At this point, you can get the shape of $[\boldsymbol{y'} - \boldsymbol{y}]$ as $(\text{len}(x), 1)$, and then calculate $(\boldsymbol{y'} - \boldsymbol{y})^2$ and finally use np. sum to sum the values in the matrix to obtain the loss value. The code is as follows:

```
1.   def loss_function(X,y,w):
2.       y_hat =X.dot(w.T)
3.       loss =y_hat.reshape((len(y_hat)),1) -y
4.       cost =np.sum(loss* * 2)/(2* len(X))
5.       return cost
```

2.6 Define Gradient Descent Function

The loss function was previously defined:

$$
J(\theta) = \frac{1}{2m} \sum_{i=1}^{m} (h_\theta(x^{(i)}) - y^{(i)})^2
$$

Where, this function $h_\theta(x) = \theta_0 x_0 + \theta_1 x_1 + \cdots + \theta_{n-1} x_{n-1} + \theta_n x_n$ is a multivariate function. According to the rule for gradient solution, the gradient $\nabla j(\theta)$ of the multivariate function is a column vector consisting of partial derivative.

$$
\nabla j(\theta) = \begin{pmatrix} \dfrac{\partial j(\theta)}{\partial \theta_0} \\[2ex] \dfrac{\partial j(\theta)}{\partial \theta_1} \\[2ex] \dfrac{\partial j(\theta)}{\partial \theta_2} \\[1ex] \vdots \\[1ex] \dfrac{\partial j(\theta)}{\partial \theta_n} \end{pmatrix}
$$

The partial derivative can be calculated as follows according to the chain derivative rule:

$$
\frac{\partial j(\theta)}{\partial \theta_0} = \frac{1}{m} \sum_{i=1}^{m} (h_\theta(x^{(i)}) - y^{(i)}) x_0^{(i)}
$$

$$\frac{\partial j(\theta)}{\partial \theta_1} = \frac{1}{m} \sum_{i=1}^{m} (h_\theta(x^{(i)}) - y^{(i)}) x_1^{(i)}$$

$$\frac{\partial j(\theta)}{\partial j} = \frac{1}{m} \sum_{i=1}^{m} (h_\theta(x^{(i)}) - y^{(i)}) x_j^{(i)}$$

$(h_\theta(x^{(i)}) - y^{(i)}) = X\theta - y$ and $x^{(i)}$ can be partially converted into a $1 \times m$ row vector (m is the number of data), so the above equation can be written as

$$\frac{\partial j(\theta)}{\partial j} = \frac{1}{m} \sum_{i=1}^{m} (h_\theta(x^{(i)}) - y^{(i)}) x_j^{(i)} = x_j^{\mathrm{T}} (X\theta - y)$$

Then

$$\nabla j(\theta) = \begin{pmatrix} \dfrac{\partial j(\theta)}{\partial \theta_0} \\ \dfrac{\partial j(\theta)}{\partial \theta_1} \\ \dfrac{\partial j(\theta)}{\partial \theta_2} \\ \vdots \\ \dfrac{\partial j(\theta)}{\partial \theta_n} \end{pmatrix} = \begin{pmatrix} x_1^{\mathrm{T}}(X\theta - y) \\ x_2^{\mathrm{T}}(X\theta - y) \\ x_3^{\mathrm{T}}(X\theta - y) \\ \vdots \\ x_n^{\mathrm{T}}(X\theta - y) \end{pmatrix} = \begin{pmatrix} x_1^{\mathrm{T}} \\ x_2^{\mathrm{T}} \\ x_3^{\mathrm{T}} \\ \vdots \\ x_n^{\mathrm{T}} \end{pmatrix} \times (X\theta - y)$$

$$= (x_1, x_2, \cdots, x_n)^{\mathrm{T}} (X\theta - y)$$

Where, $(x_1, x_2, \cdots, x_n)$ is an $m \times n$ matrix while $(X\theta - y)$ is an $m \times 1$ column vector, so the gradient function of the multivariate linear model is

$$\nabla j(\theta) = \frac{1}{2m} X^{\mathrm{T}} (X\theta - y)$$

In this formula, it can be implemented step by step:

Step 1: Use the numpy dot method to implement $(wx_i' + b)$. The code is as follows:

```
y_hat = X.dot(w)
```

Here, the shape of X is $(160, 4)$, and the shape of w is $(4,)$. After dot operation, the shape of y_hat is $(160,)$.

Step 2: Implement $(wx_i' + b) - y_i'$. The code is as follows:

```
loss = y_hat.reshape((len(y_hat),1)) - y
```

After y_hat. reshape, the shape of y_hat is $(160, 1)$ and the shape of y is $(160,1)$, which then is subtracted from y to obtain the shape of loss $(160,1)$.

Step 3: Implement $\frac{1}{2m} \sum_{i=1}^{m} [(wx_i' + b) - y_i'] x_i'$. The code is as follows:

```
derivative = X.T.dot(loss)/2(len(X))
```

The shape of X is $(160, 4)$. By performing a matrix transpose operation on X, the shape of X is changed to $(4, 160)$; then performing a dot operation with loss $(160, 1)$, a matrix with the shape of $(4, 1)$ can be obtained, which is the gradient matrix (partial derivative matrix) to be obtained.

Step 4: derivative_w = derivative_w. reshape(len(w)).

After performing derivative_w. reshape (4) operations, the shape of derivative_w will be changed to $(4,)$, and then perform w = w − lr * derivative_w to obtain the updated w value. The code is as follows for defining the gradient descent function:

```
1.   def gradient_descent(X,y,w,lr,iter):
2.       l_history = np.zeros(iter)
3.       w_history = np.zeros((iter,len(w)))
4.       for iter in range(iter):
5.           y_hat = X.dot(w)
6.           loss = y_hat.reshape((len(y_hat),1)) - y
7.           derivative_w = X.T.dot(loss)/(2* len(X))
8.           derivative_a = derivative_w
9.           derivative_w = derivative_w.reshape(len(w))
10.          w = w - lr* derivative_w
11.          l_history[iter] = loss_function(X,y,w)
12.          w_history[iter] = w
13.          #print(X.shape,y_hat.shape,loss.shape,derivative_a.shape)
14.      return l_history,w_history
```

2.7   Model Definition and Training

1. Parameter initialization

Give an initial value for iter, lr, and weight, and calculate the loss value for the current value. The code is as follows:

```
1.   iter = 300
2.   lr = 0.15
3.   weight = np.array([0.5,1,1,1])#weight[0] = bias
4.   print ('Current loss:', loss_function (X_train, y_train, weight))
5.   print(weight.shape)
```

After running the code, the current loss output is 0. 803 918 373 360 485 7.

2. Define a linear model

It mainly includes two parts: calling the gradient descent function to start iterative operation and calculating accuracy. The code is as follows:

```
1.  def linear_regression(X,y,weight,lr,iter):
2.      loss_history,weight_history = gradient_descent(X,y,weight,lr,
        iter)
3.      print ("Training final loss:", loss_history[-1])
4.      y_pred = X.dot(weight_history[-1])
5.      training_acc = 100 - np.mean(np.abs(y_pred - y)) * 100
6.       print ("Accuracy of linear regression training: {:.2f}% ".
        format(training_acc))
7.      return loss_history,weight_history
8.
9.  loss _history,weight _history = linear _regression (X _train,y _
    train, weight,lr,iter)
```

Calculate the difference between the predicted value y_pred and the actual value y first, use np. abs () to find the absolute value, and then use np. mean () to find the average of all distances. This value is the "distance" between y_pred and the actual value y, and the distance can be used as the accuracy of the model.

3. Model training

Call and run the model function. The code is as follows:

```
loss_history, weight _history = linear_regression (X_train, y_train,
weight,lr,iter)
```

After running the code, the final loss of output training is 0. 004 334 018 335 12, and the accuracy of linear regression training is 74. 52%.

4. Use model prediction

It is necessary to denormalize the predicted results. The code is as follows:

```
1.  print(y_min,y_max,y_gap)
2.  X_plan = [250,50,50]
3.  X_train,X_plan = scaler(X_data_original,X_plan)
4.  X_plan = np.append([1],X_plan)
5.  y_plan = np.dot(weight_history[-1],X_plan)
6.  y_value = y_plan* y_gap + y_min
7.  print ("Estimated sales revenue of goods:", y_value, "thousand yuan")
```

After running the code, the expected sales revenue of the product is output as $[6.08890958]$ thousand yuan.

2.8   Loss Calculation

1. Draw a loss curve and output a linear equation

Use the np. linspace ( ) function to generate x coordinate points with y coordinate as loss_ history. The code is as follows:

```
1.  line_x =np.linspace(0,iter,iter)
2.  sns.scatterplot(line_x,loss_history)
3.  plt. xlabel ("iterations")
4.  plt. ylabel ("loss")
5.  plt.show()
```

After running the code, as shown in Fig. 2-1, when the value of iter is greater than 25, the loss value basically reaches a minimum value.

After the model training is completed, the trained θ value, also known as the weight, can be output via the following code:

```
1.  print(weight_history[-1])
2.  print ("The predicted linear model is: y =% .2f* x1 +% .2f* x2 +
    % .2f* x3% .2f"% (weight_history[-1][1],weight_history[-1][2],
    weight_history[-1][3],weight_history[-1][0]))
```

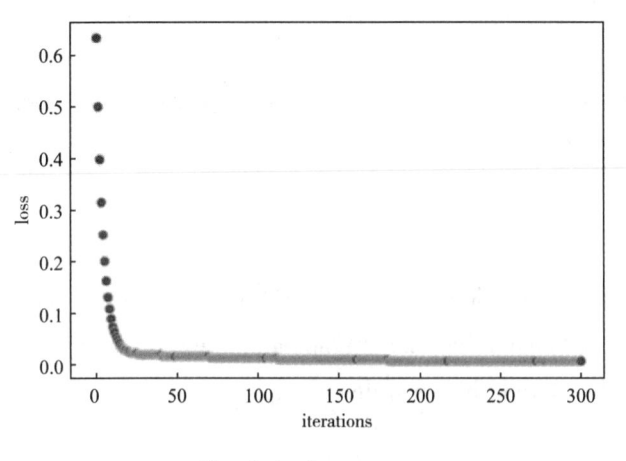

Fig. 2-1   Loss curve

After running the code, the output weight value is $[-0.04161205 \quad 0.6523009$

0. 246 867 67　0. 377 415 12], and the predicted linear model is: y = 0. 65 ∗ x1 + 0. 25 ∗ x2 + 0. 38 ∗ x3 − 0. 04.

2. Draw a fitting curve

After obtaining the model, predictions can be made. In order to observe the effect of the model more intuitively, the predicted and actual values can be plotted on the same chart for visual comparison. First obtain the length of X_train and output it, and then use np. ones to generate the weight θ of $\theta_0 x_0$ in the model, and then concatenate θ_0 and X_train to obtain the training set data. The code is as follows:

```
1.  x = np.arange(len(X_train))
2.  print(x.shape)
3.  y_train = y_train.reshape(len(y_train))
4.  X_train0 = np.ones((len(X_train),1))
5.  X_train = np.append(X_train0,X_train,axis =1)
```

Use the weight_history [− 1] parameter, namely the final model weight coefficient, and then use dot to calculate y_pred to obtain the predicted value of the model. The code is as follows:

```
1.  y_pred = X_train.dot(weight_history[ -1])
2.  y_pred
```

Finally, use lineplot to draw a fitting curve and output the label value (y) and predicted value (y_pred) in the same chart. The code is as follows:

```
1.  print(y_train.shape,y_pred.shape)
2.  data = np.append(y_train,y_pred,axis =0)
3.  data = data.reshape(160,2)
4.  print(data.shape)
5.  w_d = pd.DataFrame(data, x, ["actual value", "predicted value"])
6.  plt.figure(figsize = (15,5))
7.  sns.lineplot(data = w_d)
8.  plt.show()
```

After running the code, see Fig. 2-2 for the results. As can be seen, there is still a certain difference between the predicted value and the actual value. To reduce the error, several methods can be used, for example, adding higher order terms.

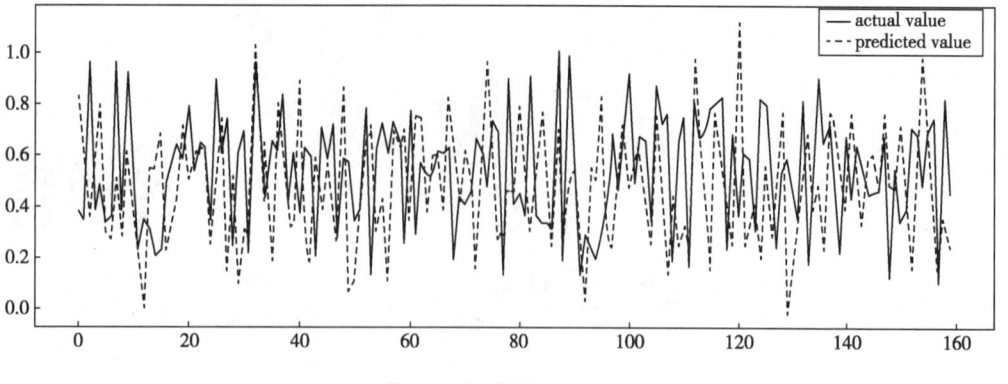

Fig. 2-2   Fitting curve

Unit 3

Using the Scikit-learn Library to Achieve Regression

In the previous instances, when using simple linear regression and multiple linear regression to predict advertising investment, the model code, loss function and gradient descent function, etc. were written in Jupyter Notebook. In actual projects, we usually use the open source Scikit-learn library to complete the above work. This unit mainly discusses how to use the Scikit-learn library to build a model for predicting advertising investment.

3.1 Scikit-learn ABC

Scikit-learn (formerly known as scikits learn, abbreviated as sklearn) is a free software machine learning library for the Python programming language. It is a Python machine learning project and a simple and efficient tool for data mining and analysis. It is built based on NumPy, SciPy and matplotlib. It has various functions such as classification, regression and clustering, dimension reduction, model selection and preprocessing. It is open source and can be used commercially.

3.2 Dataset Loading

1. Import data load packages

Import such three packages as numpy, pandas and matplotlib in Jupyter Notebook to load data and analyze the data at the same time. The code is as follows:

```
1.  # import data processing library
2.  import numpy as np
3.  import pandas as pd
4.  from matplotlib import font_manager as fm,rcParams
5.  import matplotlib.pyplot as plt
6.  from sklearn.model_selection import train_test_split
7.  from sklearn import datasets
8.  plt.rcParams['font.sans-serif'] = ['SimHei']
9.  plt.rcParams['axes.unicode_minus'] = False
```

2. Dataset visualization display

The main function of distplot in seaborn is to plot a single variable histogram. It can also add some contents of kdeplot and rugplot (histogram + kernel density estimation) to the histogram. It is a function fairly powerful and practical. The following code demonstrates how to use seaborn to draw a distplot image.

```
1.  import seaborn as sns
2.  f,ax =plt.subplots(1,1)
3.  sns.distplot(data['weibo'],color = 'r',label = "weibo",ax =ax)
4.  sns.distplot(data['wechat'],color = 'b',label = "wechat",ax =ax)
5.  sns.distplot(data['others'],color = 'g',label = "others",ax =ax)
6.  plt.ylabel ('kernel density')
7.  plt.xlabel ('advertising investment')
```

After running the code, a kernel density function of weibo (dashed line), wechat (solid line) and others (dotted line) was drawn. As can be seen, both wechat and weibo are distributed leftward, as shown in Fig. 3-1.

At the same time, pairplot was also used to visualize the data. In pairplot, pair means paired, and pairplot mainly displays the relationship between two variables (linear or nonlinear, with or without obvious correlation). The code is as follows:

```
1.  # Scatter chart showing sales and various advertising campaigns.
2.  sns.pairplot(data,x_vars = ['wechat','weibo','others'],
3.                          y_vars = 'sales',
4.                          height =4,aspect =1,kind = 'scatter')
5.  plt.show()
```

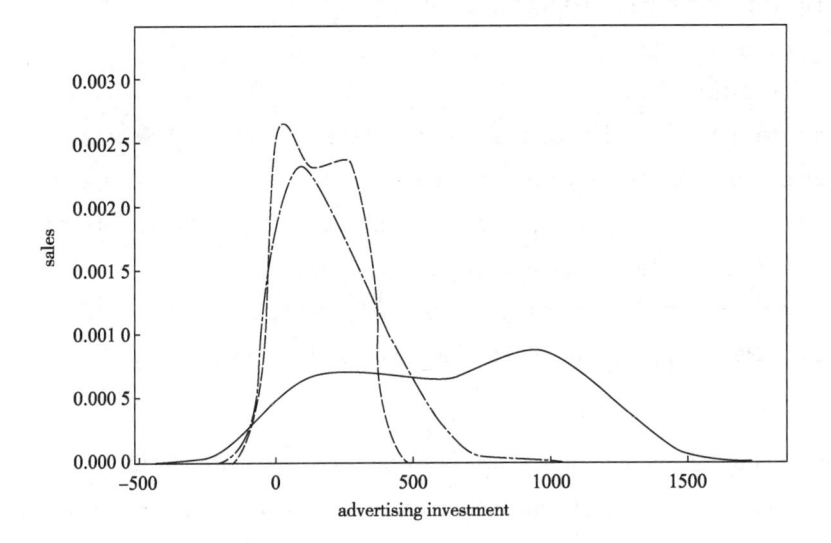

Fig. 3-1　Kernel density chart for advertising investment

3.3　Data Normalization

1. Z-score standardization

Z-score standardization is a common method for data processing, which uses the mean x_{mean} and standard deviation σ of the original data to standardize. The processed data conform to the standard normal distribution, that is, the mean value is 0 and the standard deviation is 1. The conversion function is

$$x_{scale} = \frac{x - x_{mean}}{\sigma}$$

Where, x_{mean} is the mean of all sample data; σ is the standard deviation of all sample data; x_{scale} is the standardized value.

The standardized code is as follows:

```
1.  # Feature scaling (x - data.mean ())/data.std ()
2.  data = (data - data.mean ())/data.std ()
3.  # Check scaled data
4.  data.head (10)
```

After the code is run, it will output the value after data normalization as shown in Fig. 3-2.

2. Normalization, standardization, centralization

In machine learning, normalization, standardization, and centralization are commonly used to preprocess data.

Normalization: Convert data into decimals between (0, 1) or (− 1, 1) mainly for the convenience of data processing. Mapping data to the range of 0-1 for processing is more convenient and fast. Normalization is a simplified method of calculation, which can transform dimensional expressions into dimensionless expressions to achieve pure quantities, making it easier for indicators of different units or magnitudes to be compared and weighted.

	wechat	weibo	others	sales
0	-0.933464	-0.778888	0.286795	-1.027746
1	0.877676	-1.277311	0.883711	0.297035
2	1.080421	-1.290782	0.291387	0.410587
3	-1.493829	-0.105343	0.911261	-1.538733
4	0.967529	0.979066	1.774493	1.319009
5	0.695922	-1.216692	-0.512154	0.391662
6	1.556821	-0.630708	0.295978	0.183482
7	-1.614913	0.265107	-1.306511	-1.784764
8	-0.835675	-0.199639	0.089354	-0.724939
9	-0.822876	-1.513052	-0.723370	-1.084522

Fig. 3-2   Results from data normalization

Standardization: In machine learning, different types of data need to be processed and the data with different feature dimensions needs scaling transformations (with the data scaled proportionally to fit into a small specific interval), with the aim of making features between different measures comparable. At the same time, the distribution state of the original data will not be changed. After being converted into dimensionless pure numerical values, it is convenient for indicators of different units or magnitudes to be compared and weighted.

Standardization is widely used in many machine learning algorithms (such as support vector machines, logistic regression, and nerve-like networks).

Basic principle: Subtract the sample average from the sample value, and then divide it by its sample standard deviation to get the data subject to the standard normal distribution with the mean value of 0 and the standard deviation of 1.

For example, predicting a person's health index based on his/her height and weight. Assuming the following data is available, see Table 3-1.

Table 3-1   Height-age-weight data

Sample	Height/m	Age	Weight/g
Sample 1	1.7	20	70 000
Sample 2	1.8	18	80 000
Sample 3	1.6	13	60 000

In the example, the weights of height, age, and weight are the same. When conducting feature analysis, it is evident that weight has the greatest impact on the calculation results. Therefore, standardized methods are used to process numerical values within the range of 0-1. The standard-scaler class is provided in Python to standardize data samples.

Centralization: The average value is 0, with no requirement for standard deviation.

Difference between normalization and standardization: Normalization is to convert the eigenvalues of a sample to the same dimension and map the data to the interval $(0, 1)$ or $(-1, 1)$. Standardization is to process data according to the columns of the characteristic matrix and convert the data into standard normal distribution by solving the standardization calculation method. Standardization is related to the overall sample distribution, and each sample point can have an impact on standardization. Similarly, they both can eliminate the errors caused by different dimensions; both are linear transformations that compress the vector X proportionally and then translate it.

Difference between standardization and centralization: Standardization is the original data minus the average, and then divided by the standard deviation. Centralization is the original data minus the average. So in general, standardization follows centralization.

3. When will normalization be used? When will standardization be used?

If there are requirements for the range of output results, use normalization; if the data is relatively stable and there is no extreme maximum or minimum value, use normalization; if there are outliers and more noise in the data, standardization can be used to avoid the impact of outliers and extreme values indirectly through centralization.

3.4   Data Preprocessing

1. Data selection

Here, iloc is used to retrieve data. iloc takes values through indexes (loc takes values through column names, both before and after). Here the slicing operation is used, which is a front-closed and back-open operation.

First, use shape [1] to return the number of columns, and then use data. iloc [:, 0: cols − 1] to take the first cols − 1 column, namely the input vector. data. iloc [:,cols − 1:cols], with the last column taken, which is the target variable, also known as the label y value. The following code is provided for using the iloc function to intercept the data:

```
1.   # variable initialization
2.   # the last column is y, and the others are x.
3.   cols = data. shape [1]
4.   X = data.iloc[:,0:cols - 1] # take the first cols - 1 column, which is
     the input vector
```

```
5.  y = data.iloc[:,cols -1:cols] #take the last column, which is the
    target variable
6.  X.head(10)
```

After the code runs, output the data captured by iloc. There is also a function loc in Python that can be used to select data for rows and columns, which differs from iloc as follows:

The loc function retrieves row data through specific values in the row index Index (for example, take rows with Index as A); the iloc function retrieves row data through row numbers (for example, take the data from the second row). iloc starts counting from 0 based on the position of the label, selecting rows first and then columns; loc selects rows and columns based on the specific DataFrame labels, which also takes the row label first and then the column label.

2. Divide training and test sets

Use the sklearn's model_selection function to split x_data and y_data into training and test sets according to 80% of the training set and 20% of the test set to obtain a training set and a test set. The code for splitting the training and test sets is as follows:

```
1.  # Split training and test sets
2.  X_train,X_test,y_train,y_test = train_test_split(X,y,test_size =
    0.2)
```

Use the np. matrix ( ) function to convert data into a numpy matrix, with the following code:

```
1.  # convert data into a numpy matrix
2.  X_train = np.matrix(X_train.values)
3.  y_train = np.matrix(y_train.values)
4.  X_test = np.matrix(X_test.values)
5.  y_test = np.matrix(y_test.values)
6.  theta = np.matrix([0,0,0,0])
7.  X_train.shape,X_test.shape,y_train.shape,y_test.shape
```

3.5   Use sklearn to Establish a LinearRegression Model

1. Import the sklearn library and use LinearRegression for linear regression

Import the linear_model library of the linear models in sklearn. The linear_model contains a variety of classes and functions, and LinearRegression is such a model in the linear_model library used to complete ordinary linear regression.

The performance of linear regression often depends on the data itself, rather than the ability to

adjust parameters, and therefore, linear regression has high requirements for data. In reality, there are more or less linear connections between most continuous variables. So although linear regression is simple, its functionality is very powerful. Linear regression in sklearn can handle multi-label problems by simply inputting multi-dimensional labels during model training. The code for importing the sklearn library to establish a linear regression model is as follows:

```
1.  from sklearn.linear_model import LinearRegression
2.  reg = LinearRegression()
3.  model = reg.fit(X_train,y_train)
4.  print(model)
5.  print(reg.coef_)
6.  print(reg.intercept_)
```

After the code runs, it will output the LinearRegression information established using sklearn, and also output coef_, namely, the weight value and intercept_ (bias) value of the model as shown in Fig. 3-3.

```
LinearRegression(copy_X=True, fit_intercept=True, n_jobs=None, normalize=False)
[[0.88775328  0.30866359  0.00554074]]
[0.01322183]
```

Fig. 3-3   Code running results

2. Usage of linear_model. LinearRegression ( )

LinearRegression( ) is used to establish a linear model, with the function and parameter code as follows:

```
1.  LinearRegression(
2.      fit_intercept = True,
3.      normalize = False,
4.      copy_X = True,
5.      n_jobs = None,
6.  )
```

The LinearRegression( ) function has such main parameters and usage as follows:

fit_intercept: boolean, optional, default True. Calculate the intercept or not? Calculate it by default. If using centralized data, you can consider setting it to False without considering intercept. Note that it is just considered and that usually the intercept should be considered.

normalize: boolean, optional, default False. Standardization switch, OFF by default; this parameter will be ignored automatically when fit_intercept is set to False. If True, the regression will standardize the input parameters. It is recommended to place standardization work before model

training. If False, sklearn. reprocessing. StandardScaler can be used for standardization before model training.

copy_X: boolean, optional, default True. True by default; otherwise X will be rewritten.

n_jobs: int, 1 by default. When it is -1, all CPUs are used by default.

Return values and properties of the LinearRegression( ) function:

coef_array type variable, with the shape of (n_features,) or (n_targets, n_features). In this example, the weight coefficient ($w_1 - w_n$) is returned. For linear regression problems, the feature coefficient is calculated. If the input is a multi-objective problem, it will return a two-dimensional array (n_targets, n_features); if it is a uni-objective problem, it will return a one-dimensional array (n_features,).

intercept_array type variable, in this example, the value of bias/w_0 returns, but if fit_intercept = False is set in LinearRegression, then the intercept_ is 0.0.

How to use the LinearRegression( ) function:

fit(self,X,y[,sample_weight]) method, training model, sample_weight is the weight value of each sample, None by default.

predict(self,X) method, model prediction, predicted values return.

score(self,X,y[,sample_weight]) method, it will evaluate the model and return its evaluation score. The optimal value is 1, indicating that all data are predicted correctly.

3.6　LinearRegression Model Prediction and Error

1. Use the predict( ) function of LinearRegression to predict the y_test value

After defining the linear regression model, call the predict( ) function to predict the y_test value, then use mean_squared_error ( ) function to calculate the mean squared error (MSE). The code is as follows for MSE calculation:

```
1.  from sklearn.metrics import mean_squared_error
2.  y_hat = reg.predict(X_test)
3.  mean_squared_error(y_test,y_hat)
```

2. Evaluation indexes of linear regression algorithm

The linear regression algorithm includes such evaluation indicators as MSE, RMSE, MAE, R2_score. The indexes meaning as follows:

(1) MSE (mean squared error)

Example: Assuming two models are established, one model uses 10 samples and the loss value calculated is 100; the other model uses 50 samples and the loss value calculated is 200,

making it difficult to determine which model is better. So what should we do? In fact, it's easy! Only if the loss value is divided by the number of samples, the impact of the sample size can be eliminated.

The MSE of the two models shown above is 10 (100/10) and 4 (200/50), so the latter performs better.

(2) RMSE (root mean squared error)

There is a problem with the MSE formula (see Section 1.6), as it changes dimensions by taking the square. For example, the unit of the y value is ten thousand yuan, and MSE calculates the square of ten thousand yuan, which is difficult to explain its meaning. So in order to eliminate the influence of dimensionality, we can extract the root of this MSE. As can be seen, MSE and RMSE are positively correlated, namely, a higher MSE value will lead to a higher RMSE value.

The following formula is provided for RMSE calculation:

$$RMSE = \sqrt{\frac{1}{2n} \sum_{i=1}^{n} (y_i - y_i')^2}$$

Where, y_i represents the actual value; y_i' represents the predicted value; n represents the number of samples.

For example, to predict a house price, the selling price per square meter is ten thousand yuan, and the prediction result is also ten thousand yuan. So the square unit of the difference should be in the tens of millions level. It is not easy to describe the effect of the model made at this time. So we extracted the root of the result. At this point, the result of the error is at the same level as the data. When describing the model, it is said that the error of the model is tens of thousands of yuan.

(3) MAE (mean absolute error)

In MSE and RMSE, in order to avoid positive and negative offset of the errors, the square of the difference is calculated. There is another formula that can also have the same effect, which is to calculate the absolute value of the difference. At this time, the mean absolute error (MAE) can be obtained. The MAE calculation formula is as follows:

$$MAE = \frac{1}{n} \sum_{i=1}^{n} |(y_i - y_i')|$$

Where, y_i represents the actual value; y_i' represents the predicted value; n represents the number of samples.

The above three methods eliminate the influence of the sample size m and dimensionality. But they all have the same problem: When the dimensionality is different, it's difficult to measure the effectiveness of the model.

For example, the error RMSE predicted by the model on one housing price dataset is 50 000

yuan, while the error obtained on another student performance dataset is 10 points. Based on these two values, it is difficult to know which dataset the model performs better on.

(4) Linear regression coefficient of determination R square value (R2_score)

The coefficient of determination is also called goodness of fit. It represents to what extent the regression equation explains the changes in the dependent variable, or how well the equation fits the observed values.

If the residual sum of squares is used only, it will be affected by the absolute values of dependent and independent variables, which is not conducive to the relative comparison between different models. However, if goodness of fit is provided, this problem can be solved. For example, if the dependent variables in one model are 10 000, 20 000⋯, while 1, 2⋯in the other model, then the residual sum of squares of the first model in these two models may be very great, while rather small in the second model, but this does not mean that the first model is inferior to the second one. At this time, R2_score can be used to determine the quality of the model. The R2_score calculation formula is as follows:

$$R2\_score = 1 - \frac{MSE(\hat{y}, y)}{Var(y)}$$

Where, the numerator is the mean square error, and the denominator is the variance, both of which can be worked out directly, thus quickly calculating R2_score value. The R2_score value can have such three situations as follows:

R2_score = 1, up to the maximum value. The numerator is 0, which means that the predicted and true values in the sample are completely equal without any error. That is to say, the established model perfectly fits all real data and is most effective. But usually a model is not so perfect and there will always be some error. When the error is very small, the numerator is smaller than the denominator, and the model will approach 1, still a good model. As the error increases, R2_score will also move further and further away from the maximum value of 1 until the second situation occurs.

R2_score = 0. At this point, the numerator is equal to the denominator, and each predicted value of the sample is equal to the mean value. That is to say, the trained model is exactly the same as the mean model mentioned earlier. The third situation arises as the error increases.

R2_score < 0. If the numerator is greater than the denominator, the error generated by model training is greater than that generated by using the mean, which means that the model training is not as effective as direct using the mean value. This situation usually occurs when the model itself is not linear and a linear model is mistakenly used, resulting in significant errors.

The functions in sklearn can be directly used to calculate the MSE, MAE, and R2 values of the model. The code is as follows:

```
1.  from sklearn.metrics import mean_squared_error
2.  from sklearn.metrics import mean_absolute_error
3.  from sklearn.metrics import r2_score
4.  y_hat = reg.predict(X_test)
5.  print(mean_squared_error(y_test,y_hat)) # mean squared error
6.  print(mean_absolute_error(y_test,y_hat)) # squared error
7.  print(r2_score(y_test,y_hat)) # R2 value
```

Run the program and it will output the MSE, MAE and R2 values of 0. 085 430 660 95, 0. 226 210 669 0 and 0. 903 921 966 4, respectively.

3.7 Plot a Fit Curve

Take the X_test value, use the model prediction to obtain y_hat and plot y_hat and y_test in the same plt to intuitively display the relationship between the predicted value and the actual value.

Here, the lineplot ( ) function is used to draw a line chart. First, generate the data of the x axis, and then put the y_hat and y_test data together, and finally convert the appended data into a dataframe subject to the code as follows:

```
1.  #y = y_test.reshape((40,1)).ravel()
2.  x = np.arange(len(X_test)) # generate x - axis data
3.  print(y_test.shape,y_hat.shape)
4.  df = np.append(y_test, y_hat, axis = 1) # append to generate y -
    axis data
5.  df = df.reshape(40,2)
6.  print(df.shape)
7.  w_d = pd.DataFrame(df,x,["actual value", "predicted value"])
    # generate dataframe data
8.  plt.figure(figsize = (15,5)) # set the figure size
9.  sns.lineplot(data = w_d) # Plot a line chart
10. plt.show()
```

After running the code, the predicted and actual values were plotted, as shown in Fig. 3-4. As can be seen from the figure, the predicted and actual values have a high degree of overlap, and the model is high in accuracy.

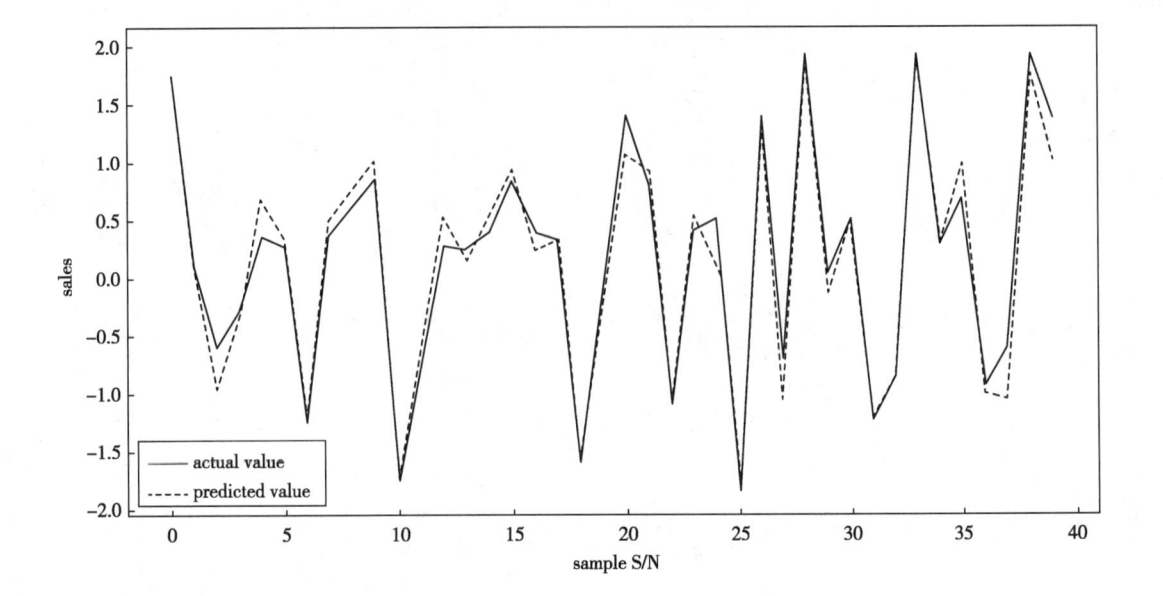

Fig. 3-4   Fit curve of predicted and actual values

Unit 4

Use the Scikit-learn Library for Polynomial Regression

Previously, simple and multiple linear regressions were used to fit advertising investment. However, in real life, many data are mutually nonlinear. Although linear regression can also be used to fit nonlinear regression, the effect is rather poor. At this time, it is necessary to improve the linear regression model and add higher-order terms in the model, that is, the amount of nonlinearity should be increased, so that it can fit nonlinear data.

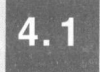 4.1 Introduction to Polynomial Regression

In practical projects, nonlinear problems are often encountered. The data shown in Fig. 4-1 shows a linear relationship, and good fitting results can be obtained using linear regression. The data shown in Fig. 4-2 shows a nonlinear relationship, and a polynomial regression model is required. Polynomial regression is an improvement on the basis of linear regression, which is equivalent to adding feature items to the sample. As shown in Fig. 4-2, x^2 feature term was added to the sample, which can better fit nonlinear data.

The above data points in Fig. 4-2 can be fitted using a quadratic curve, with the corresponding equation $y = ax^2 + bx + c$. At this point, the data model is a polynomial, where x^2 can be seen as an added feature.

Polynomial regression solves the problem of nonlinearity, fitting data that are not linear relationships but other curve relationships. Similar to linear regression, assuming that the

relationship between feature (x) and value (y) is a polynomial mathematical model, in specific operation, it can be seen as adding polynomial ax^2 to the linear relationship $(y = bx + c)$. If seen from a regression perspective, this equation is a linear regression equation; from the perspective of the x expression, it is a quadratic equation.

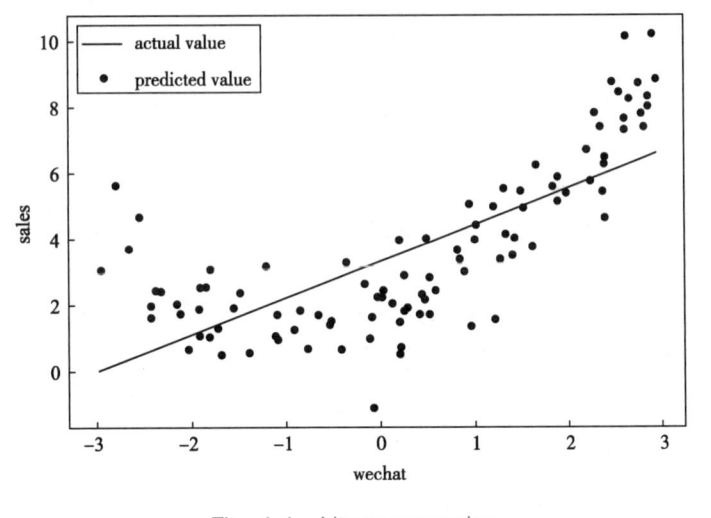

Fig. 4-1　Linear regression

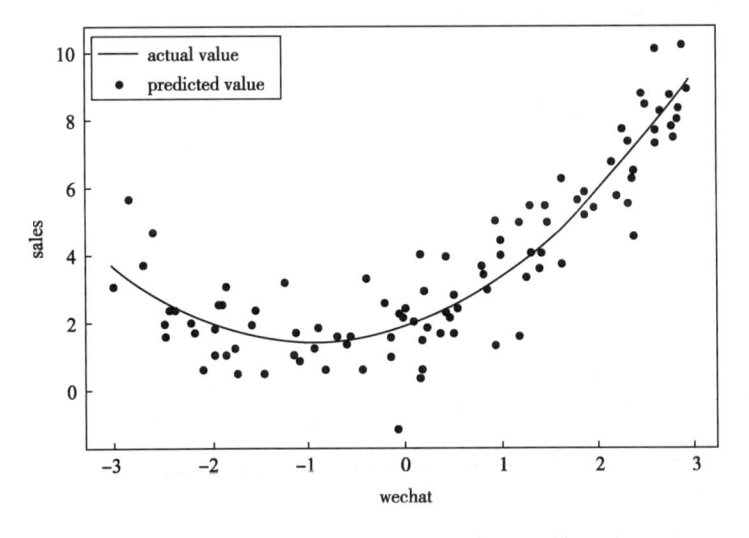

Fig. 4-2　Quadratic polynomial regression

Polynomial regression is the method for regression analysis of polynomials between a dependent variable and one or more independent variables. If there is only one independent variable, it is called unary polynomial regression; if there are multiple independent variables, it is called multivariate polynomial regression. Since any function can use polynomial approximation, polynomial regression has been widely applied.

Unary n-degree polynomial regression: $y = b_0 + b_1 x + b_2 x^2 + \cdots + b_m x^m$.

Binary quadratic polynomial regression: $y = b_0 + b_1 x_1 + b_2 x_2 + b_3 x_1^2 + b_4 x_2^2 + b_5 x_1 x_2$.

4.2   Data Preprocessing

① Firstly, import the package into Jupyter Notebook subject to the following code:

```
1.  # import data processing library
2.  import numpy as np
3.  import pandas as pd
4.  from matplotlib import font_manager as fm, rcParams
5.  import matplotlib.pyplot as plt
6.  from sklearn.model_selection import train_test_split
7.  from sklearn import datasets
8.  plt.rcParams['font.sans-serif'] = ['SimHei']
9.  plt.rcParams['axes.unicode_minus'] = False
```

② Read the data subject to the following code:

```
1.  data = pd.read_csv('../data/advertising.csv')
2.  data.head()
```

③ Generate features and labels subject to the following code:

```
1.  # variable initialization
2.  # the last column is y, and the others are x
3.  cols = data.shape[1]
4.  X = data.iloc[:,0:cols-1] # take the first cols-1 column, namely
    the input vector
5.  y = data.iloc[:,cols-1:cols] # take the last column, namely the
    target variable
6.  X.head(10)
```

④ Split to obtain the training set and test set, with the following codes:

```
1.  from sklearn.preprocessing import PolynomialFeatures
2.  from sklearn.linear_model import LinearRegression
3.  from sklearn.metrics import r2_score
4.  from sklearn.pipeline import Pipeline
5.  from sklearn.model_selection import train_test_split
```

```
6.
7.  X_train,X_test,y_train,y_test =train_test_split(X,y,test_size =0.2)
8.
9.  X_train =np.array(X_train)
10. X_test =np.array(X_test)
11. y_train =np.array(y_train)
12. y_test =np.array(y_test)
```

Use Polynomial Linear Regression to Predict the Revenue of Advertising Investment

1. Use PolynomialFeatures ( ) to expand features

The PolynomialFeatures ( ) function in the sklearn library can be used to expand features, generate x^m features, and then use pipeline to achieve polynomial regression.

(1) PolynomialFeatures() function

The function format is as follows:

```
10.  sklearn.preprocessing.PolynomialFeatures(
11.          degree =2,* ,interaction_only =False,
12.          include_bias =True,order = 'C')
```

(2) Common parameters and usage of PolynomialFeatures()

degree: It controls the degree of the polynomial.

interaction_only: False by default. If specified as True, there will be no x^2 item. If degree $=2$ is set, there will be no x^2 in the combined features.

include_bias: True by default, indicating that there will be a zero-power term in the result, namely all are 1 in this column.

Here, PolynomialFeatures() is used to append data. Set degree $=2$ to append features $x_1x_2x_3$, and generate such 10 features as follows:

$$x_0,x_1,x_2,x_3,x_1x_2,x_1x_3,x_2x_3,x_1^2,x_2^2,x_3^2$$

The expand code is as follows:

```
1.  polyCoder = PolynomialFeatures (degree = 2, include_bias = True,
    interac - tion_only =False)
2.  df =polyCoder.fit_transform(X)
3.  print(df)
4.  print(df.shape)
```

The fit _ transform ( ) function in the above code undertakes to append data features. After running the code, it will output 10 features newly generated, as shown in Fig. 4-3.

```
[[1.0000000e+00 3.0440000e+02 9.3600000e+01 ... 8.7609600e+03
  2.7555840e+04 8.6671360e+04]
 [1.0000000e+00 1.0119000e+03 3.4400000e+01 ... 1.1833600e+03
  1.3704960e+04 1.5872256e+05]
 [1.0000000e+00 1.0911000e+03 3.2800000e+01 ... 1.0758400e+03
  9.6825600e+03 8.7143040e+04]
 ...
 [1.0000000e+00 1.2190000e+02 2.6400000e+02 ... 6.9696000e+04
  4.0761600e+04 2.3839360e+04]
 [1.0000000e+00 3.4350000e+02 8.6400000e+01 ... 7.4649600e+03
  4.1472000e+03 2.3040000e+03]
 [1.0000000e+00 7.9670000e+02 1.8000000e+02 ... 3.2400000e+04
  4.5360000e+04 6.3504000e+04]]
(200, 10)
```

Fig. 4-3   Feature results after appended

2. Use pipeline to achieve polynomial linear regression

Define the polynomial _ regression ( ) function and set the pipeline for quick polynomial regression. The code for appending features using pipeline is as follows:

```
1.  def polynomial_regression(degree =1):
2.      polyCoder = PolynomialFeatures(degree = degree,include_bias =
        True,interaction_only = False)
3.      linear_regression_model = LinearRegression(normalize = True)
        # data normalization
4.      new_model = Pipeline([("polynomial_features",polyCoder),
5.      ("linear_regression",linear_regression_model)])
6.      return new_model
```

In the polynomial_regression( ) function, introduce the parameter degree with the default value of 1. When calling the function, different degree values can be introduced to obtain n-degree polynomials, and finally return the generated polynomial model.

3. Usage of pipeline function

The pipeline in Python is similar to the pipeline in Linux. Connect several commands together, with the output of the previous command being the input of the latter command, ultimately completing a pipeline-like function.

You can use sklearn. pipeline. Pipeline to implement pipelines.

Parameter: steps is a list with the elements of (name, transform) tuples, where name is the name of the learner used for output and logs; transform is a learner and it is called a transform because this learner (except for the last one) must provide a transform method.

Property of function: named_steps is a dictionary, where the key is the name element of each

tuple in steps, and the value of the dictionary is the transform element of each tuple in steps.

Common methods:

fit (X [, y]): Start the pipeline, execute the fit and transform methods sequentially to convert data for each learner (except for the last learner), and execute the fit method to train the learner for the last learner.

transform (X): Start the pipeline and execute the fit and transform methods sequentially to convert data for each learner. Each learner is required to implement the transform method.

fit_transform(X[,y]): Start the pipeline, execute the fit and transform methods to convert data for each learner (except for the last learner) in sequence, and execute the fit_transform method for the last learner to convert data.

inverse_transform (X): Invert the transformed data into the original data. Each learner is required to implement the inverse_transform method.

predict(X)/predict_log_proba(X)/predict_proba(X): After converting X to data, use the last learner for prediction.

score(X,y): After converting X to data, use the last learner for a predictive score.

When executing pipe.fit(X_train,y_train), it is up to StandardScaler to execute the fit and transform methods on the training set at first, and the transformed data is then passed to the next step of the pipeline object, PCA(). Like StandardScaler, PCA also executes fit and transform methods, ultimately passing the converted data to LogisticRegression. The code is as follows:

```
1.  # use the pipeline mechanism
2.  from sklearn.preprocessing import StandardScaler # normalize to
    make the mean of each feature 1 and the variance 0
3.  from sklearn.decomposition import PCA
4.  from sklearn.linear_model import LogisticRegression
5.
6.  from sklearn.pipeline import Pipeline
7.  pipe = Pipeline ([('sc',StandardScaler()), # data standardization
8.                    ('pca',PCA(n_components =2)),
                 # PAC data dimensionality reduction
9.                    ('clf',LogisticRegression(random_state =666))
10.
11. pipe.fit(x_train, y_train)
12. print('Test accuracy is %.3f' % pipe.score(x_test, y_test))
```

4.4 Model Training

Use the fit ( ) function to train the model, and call the score( ) function to output the coefficient of determination R2 of the model. The code is as follows:

```
1.  model = polynomial_regression(2)
2.  model.fit(X_train,y_train)
3.  train_score = model.score(X_train,y_train)
4.  test_score = model.score(X_test,y_test)
5.  print(model.named_steps['linear_regression'].coef_)
6.  print(train_score)
7.  print(train_score)
8.  print(model)
```

After running the code, it will output the weight of the model, the score value of the training set, and the score of the test set, as shown in Fig. 4-4.

```
[[ 0.00000000e+00  1.66223817e-02 -3.03453877e-03 -2.56333282e-03
  -5.34773625e-06  1.15973927e-05  3.10131074e-08  2.24489815e-05
   4.97254102e-06  2.16733893e-06]]
0.9304880804198212
0.9304880804198212
Pipeline(memory=None,
      steps=[('polynomial_features', PolynomialFeatures(degree=2, include_bias=True, interaction_only=False)),
 ('linear_regression', LinearRegression(copy_X=True, fit_intercept=True, n_jobs=1, normalize=True))])
```

Fig. 4-4   Code running results

4.5 Plot a Fit Curve

Use lineplot to plot a fit curve between predicted and actual values, with the following code:

```
1.  x = np.arange(len(X_test)) # generate X - axis data
2.  print(y_test.shape,y_hat.shape)
3.  df = np.append(y_test,y_hat,axis =1)# append to generate y - axis data
4.  df = df.reshape(40,2)
5.  print(df.shape)
6.  w_d = pd.DataFrame (df, x, ["actual value", "predicted value"])
    # generate dataframe data
7.  plt.figure(figsize = (15,5)) # set the figure size
```

```
8.   sns.lineplot(data = w_d) # plot a line chart
9.   plt.title ("multiple Linear regression prediction on advertising
     revenue", fontsize =18)
10.plt.show()
```

After running the code, it will output the result as shown in Fig. 4-5. As can be seen, the degree of fitting between the predicted value and the actual value in the polynomial model is higher than that in multiple linear regression.

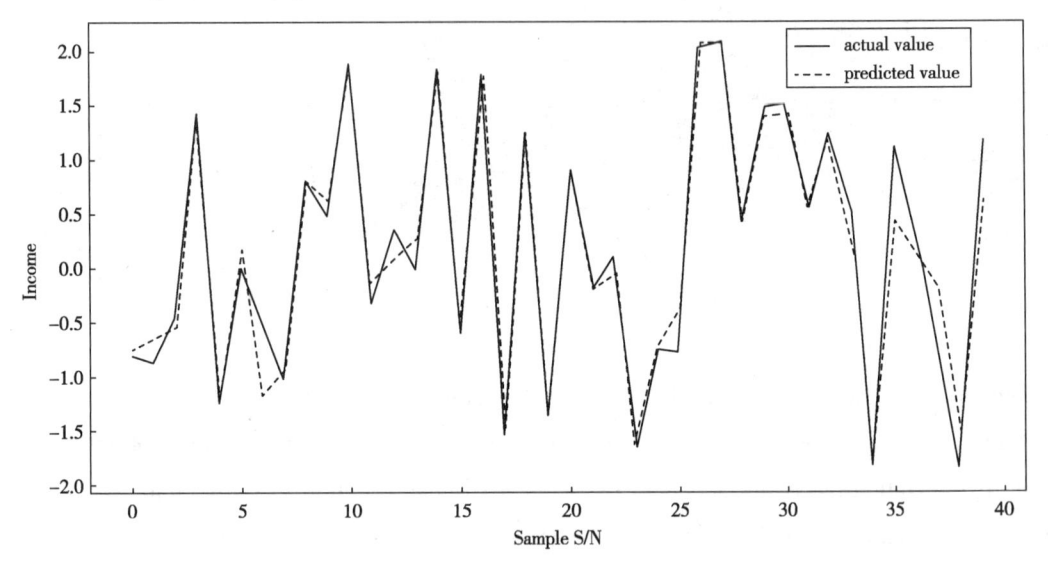

Fig. 4-5   Fit run results

Unit 5

Use Ridge Regression to Predict House Prices

The housing price prediction dataset is a high-dimensional data (with 80 columns), and overfitting occurs with the increase of features in linear regression. In order to prevent overfitting, L1 and L2 regularization terms are added to the standard linear equation loss function. What is added with L1 regularization terms is called Lasso regression, while what is added with L2 regularization terms is called ridge regression.

5.1　Overfitting and Underfitting

1. Introduction to overfitting and underfitting

Overfitting and underfitting are common problems encountered in artificial intelligence algorithm training. Underfitting means that the model is weak in learning and cannot learn the "general rules" in the sample data, resulting in weak generalization ability of the model. On the contrary, overfitting means that the model is strong in learning, to the extent that the "individual characteristics" in the sample data are also regarded as "general rules".

Underfitting: If a hypothesis cannot achieve better fitting on the training data and cannot fit the data well on the test dataset, then it is considered to have underfitting, as shown in Fig. 5-1. (The model is too simple)

Due to the limited number of swan features learned by the machine, the differentiation criteria are too rough to accurately identify swans.

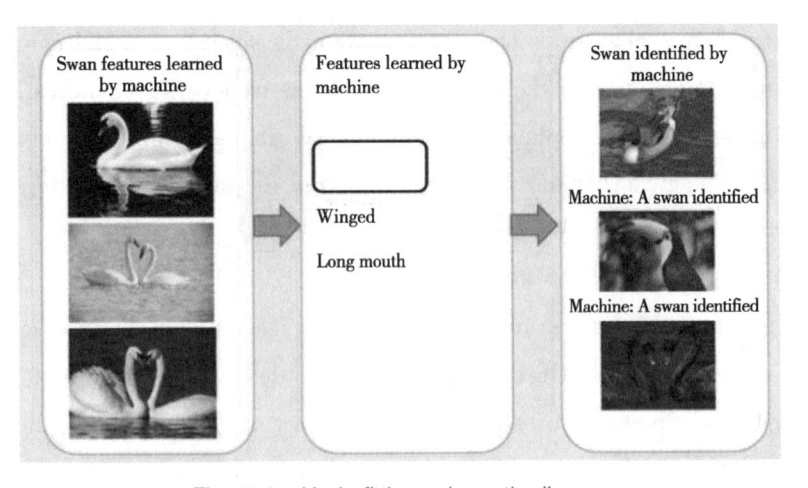

Fig. 5-1　Underfitting schematic diagram

Overfitting: If a hypothesis can achieve better fitting than other hypotheses on training data, but cannot fit the data well on the test dataset, then it is considered to have overfitting, as shown in Fig. 5-2. (The model is too complex)

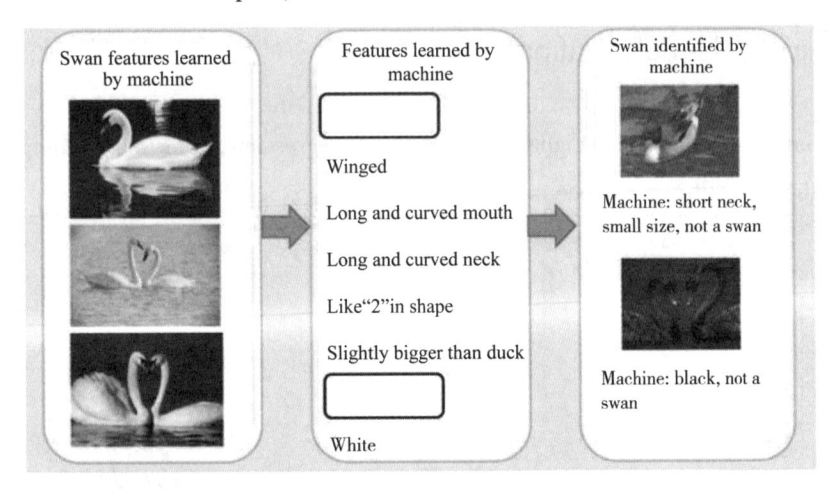

Fig. 5-2　Overfitting schematic diagram

2. Dataset accuracy in case of overfitting and underfitting

For a training dataset, the more complex the model, the higher the accuracy of the model, and the better the fitting of the training dataset. For a test dataset, when the model is very simple, the accuracy of the model is relatively low. As the model gradually becomes complex, the accuracy of the test dataset will improve gradually. If the model continues to become complex after improved to a certain extent and as the model increases in complexity, the accuracy of the model on the test set will decrease accordingly, it indicates that the model is in a state of overfitting.

3. Model generalization

Essentially, the artificial intelligence algorithm is to use algorithm models to fit sample data and

effectively predict unknown new data. The more accurate the algorithm model can predict the data beyond the training set, the stronger the generalization of the model. This requires learning as much as possible the "universal rules" applicable to all potential samples from the training samples. Both overfitting and underfitting can weaken the generalization of the model. Overfitting and underfitting should be two common reasons for low generalization of models.

4. Solutions to overfitting and underfitting

The model learns too few features from the data, which leads to underfitting, so we can increase the number of features in the data. Overfitting occurs due to the excessive number of original features and the presence of some noisy features. The model is too complex because it attempts to balance various test data points, thereby learning some features beyond the "general rules".

Here, regularization is chosen for regression. However, for other machine learning algorithms such as classification algorithms, such problems also occur. In addition to the role of some algorithms (decision trees, neural networks), more attention should be paid to feature selection.

5.2 L1, L2 Regularization

For polynomial models, the higher the order of regression, the more complex the model becomes, and then overfitting will occur, as shown in Fig. 5-3.

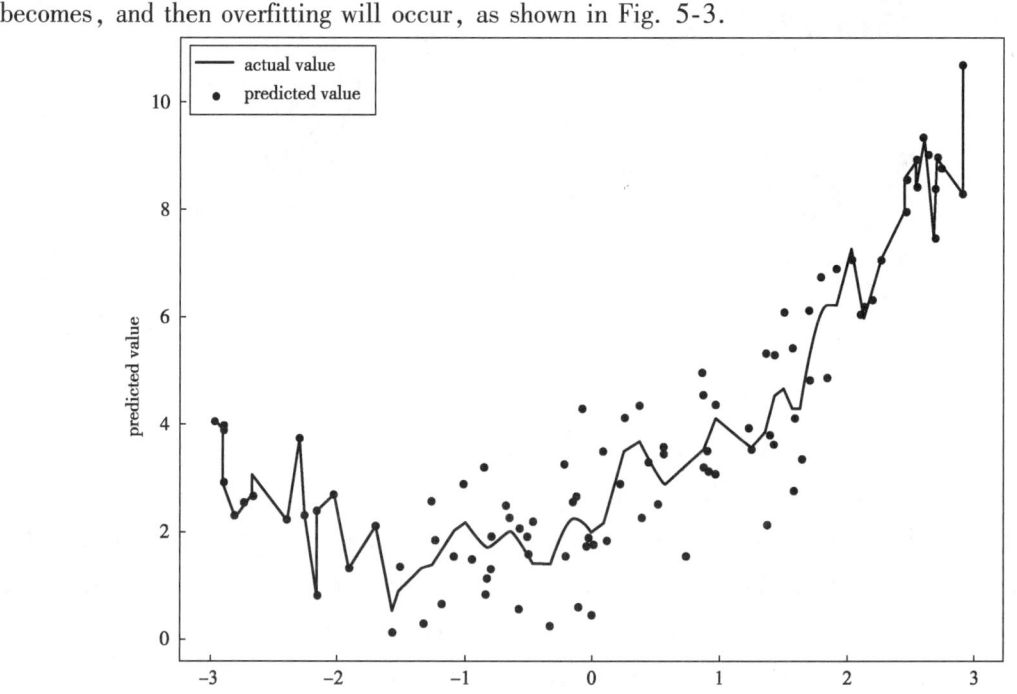

Fig. 5-3 Overfitting

The reason for overfitting is that the values of higher-order terms have a particularly significant impact on the predicted values, which requires minimizing the impact of higher-order terms on features as much as possible. Therefore, a penalty term is added to the loss function. The penalty term refers to the restriction on some parameters in the loss function. The common practice is to add L1 and L2 penalty terms to the model.

L1 regularization (Lasso regression) refers to adding the absolute value of the weight parameter on the basis of the original loss function. The formula is as follows:

$$J(\theta) = \frac{1}{2m}\sum_{i=1}^{m} (y_i - y_i')^2 + \frac{\lambda}{2m}\sum_{j=1}^{n} |\theta_j|$$

Where, y_i represents the actual label value; y_i' represents the predicted value; θ_j represents the weight coefficient; m represents the number of samples; $\frac{\lambda}{2m}\sum_{j=1}^{n} |\theta_j|$ represents the L1 penalty term.

L2 regularization (ridge regression) means to add the sum of squares of weight parameters on the basis of the original loss function. The formula is as follows:

$$J(\theta) = \frac{1}{2m}\sum_{i=1}^{m} (y_i - y_i')^2 + \frac{\lambda}{2m}\sum_{j=1}^{n} \theta_j^2$$

The addition of the penalty term $\frac{\lambda}{2m}\sum_{j=1}^{n} \theta_j^2$ guarantees the invertibility of the matrix, and the diagonal of the identity matrix I is dotted with 1, like a mountain ridge, which is the origin of the name of ridge regression.

5.3   Housing Price Dataset Reading

The housing price prediction dataset is a high-dimensional data (with 80 columns) and can be used to establish a ridge regression model to predict housing prices.

1. Import data load packages and open the csv data file

Import numpy, pandas and matplotlib packages to load and analyze the data at the same time. The index of the source data can be used as the index of the pandas dataframe. When the read_csv() function is used to read data, the index_col = 0 parameter can be added, with the first column of data used as the serial number, subject to the code as follows:

```
1.  train_df = pd.read_csv('../input/train.csv',index_col =0)
2.  test_df = pd.read_csv('../input/test.csv',index_col =0)
```

After running the code, open the housing price test and training sets respectively, and display the training set. The data contains a total of 80 columns, and the output results are shown in Fig. 5-4.

	MSSubClass	MSZoning	LotFrontage	LotArea	Street	Alley	LotShape	LandContour	Utilities	LotConfig	...	PoolArea	PoolQC	Fence	MiscFeature	Misc
Id																
1	60	RL	65.0	8450	Pave	NaN	Reg	Lvl	AllPub	Inside	...	0	NaN	NaN	NaN	
2	20	RL	80.0	9600	Pave	NaN	Reg	Lvl	AllPub	FR2	...	0	NaN	NaN	NaN	
3	60	RL	68.0	11250	Pave	NaN	IR1	Lvl	AllPub	Inside	...	0	NaN	NaN	NaN	
4	70	RL	60.0	9550	Pave	NaN	IR1	Lvl	AllPub	Corner	...	0	NaN	NaN	NaN	
5	60	RL	84.0	14260	Pave	NaN	IR1	Lvl	AllPub	FR2	...	0	NaN	NaN	NaN	

5 rows × 80 columns

Fig. 5-4　Data display results

Next, it is necessary to observe the data and determine whether there is "NaN" in the data, which fields are numerical and which are categorical. These data need to be processed separately.

2. Use the info() function to view data information

The running results are shown in Fig. 5-5.

From the running results, the types of all features and the types of each feature can be seen, for example, MSSubClass- represents the type of residence, with 1460 values and type int64.

```
train_df.info()

<class 'pandas.core.frame.DataFrame'>
Int64Index: 1460 entries, 1 to 1460
Data columns (total 80 columns):
 #   Column         Non-Null Count   Dtype
---  ------         --------------   -----
 0   MSSubClass     1460 non-null    int64
 1   MSZoning       1460 non-null    object
 2   LotFrontage    1201 non-null    float64
 3   LotArea        1460 non-null    int64
 4   Street         1460 non-null    object
 5   Alley          91 non-null      object
 6   LotShape       1460 non-null    object
 7   LandContour    1460 non-null    object
 8   Utilities      1460 non-null    object
 9   LotConfig      1460 non-null    object
 10  LandSlope      1460 non-null    object
 11  Neighborhood   1460 non-null    object
 12  Condition1     1460 non-null    object
```

Fig. 5-5　Running results

5.4　Handling Label Values

1. Visualization of label value distribution

Firstly, check the label house price SalePrice, extract the SalePrice from the train, and plot a

chart subject to the following code:

```
1.  import seaborn as sns
2.  sns.histplot(train_df['SalePrice'])
```

After the program runs, the horizontal axis SalePrice represents the housing price, and the vertical axis Count represents the corresponding statistical quantity of the housing price. It can be seen that the data are distributed rightward, as shown in Fig. 5-6.

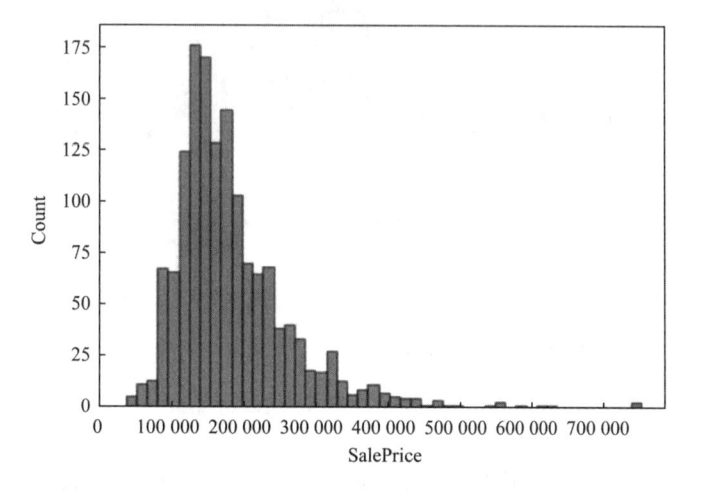

Fig. 5-6   Housing price histogram

2. Use log1p handling label

In machine learning, probability normal distribution is frequently used, so first you need to adjust the data to normal distribution, and then it will output a normal distribution after such processing. The SalePrice itself in the figure is not smooth. In order to make the learning of the classifier more accurate, the label will first be "smoothed" (normalized). The log1p( ) used here, also known as log $(x + 1)$, can avoid complex values.

If the housing price SalePrice data is smoothed with log1p here, then when calculating the final results, remember to change back the smoothed data already predicted. According to the principle of "how to come and how to go", log1p( ) needs expm1( ); similarly, log( ) needs exp( ). The code processed with log1p( ) is as follows:

```
sns.histplot(np.log1p(train_df['SalePrice']))
```

After running the code, it can be seen that the data after processed with log1p( ) basically conforms to normal distribution, as shown in Fig. 5-7.

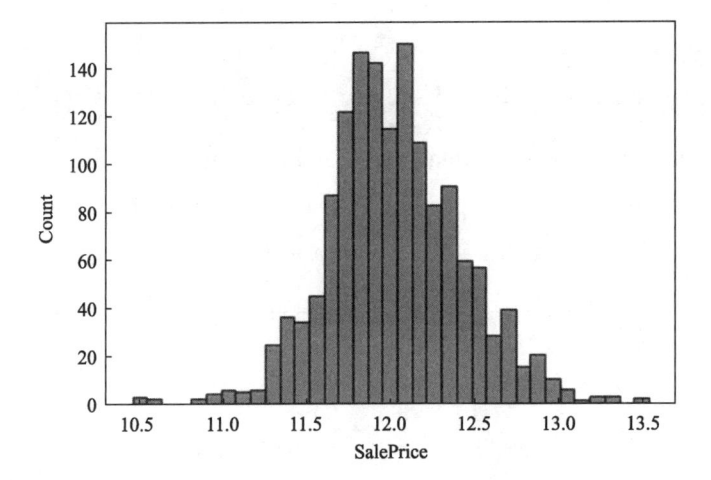

Fig. 5-7   Smoothed housing price histogram

5.5 Type Feature One-hot Code

1. Combine train and test sets

The project provides training and test datasets, and the test is the data used for testing. After the model is trained, it is necessary to use it to predict the housing prices in the test. Therefore, it is necessary to first combine train and test, mainly to make data preprocessing more convenient. After all the necessary preprocessing is completed, separate them subject to the code as follows:

```
1.  y_train = np.log1p(train_df.pop('SalePrice'))
2.  all_df = pd.concat((train_df,test_df),axis =0)
3.  all_df.shape
4.  y_train.head()
```

2. One-hot coding

(1) One-hot code

In machine learning algorithms, classification features are often encountered, such as the gender of a person being male or female, and nationality being Chinese, American, French, etc. These eigenvalues are not continuous, but discrete and unordered. This type of data is category/nominal (category) data. If there is no sense of size between category discrete type data, one-hot coding conversion can be used. If there is a sense of size between category discrete type data, LabelEncoder label coding can be used for conversion.

For example: ["China", "United States", "France"] (N = 3):

China = > 100   US = > 010   France = > 001

In the above example, the categorical value is mapped to an integer value; then, each integer value is represented as a binary vector. This processing method is one-hot coding.

One-hot coding is also known as one-bit valid coding, namely, use an N-bit status register to encode N states, each state has its own independent register bit, and at any time, only one bit is valid.

(2) MSSubClass one-hot coding

For example, the value of MSSubClass (Residence Type) in the housing price prediction dataset is a category type, and in DF it is an int type value. As can be seen with all_df ['MSSubClass']. value_counts(), MSSubClass has a total of 16 values, which have no specific meaning and need to be processed through one-hot coding.

Firstly, you need to convert it back to a string, and then use the get_dummies method that comes with pandas to convert to one-hot coding, with the following code:

```
1.  all_df['MSSubClass'].dtypes
2.  all_df['MSSubClass'] = all_df['MSSubClass'].astype(str)
```

After running the code, perform one-hot coding according to the type of MSSubClass, as shown in Fig. 5-8.

Id	MSSubClass_120	MSSubClass_150	MSSubClass_160	MSSubClass_180	MSSubClass_190	MSSubClass_20	MSSubClass_30	MSSubClass_40	MSSubClass_
1	0	0	0	0	0	0	0	0	
2	0	0	0	0	0	1	0	0	
3	0	0	0	0	0	0	0	0	
4	0	0	0	0	0	0	0	0	
5	0	0	0	0	0	0	0	0	
6	0	0	0	0	0	0	0	0	
7	0	0	0	0	0	1	0	0	
8	0	0	0	0	0	0	0	0	
9	0	0	0	0	0	0	0	0	
10	0	0	0	0	1	0	0	0	

Fig. 5-8   Results after converted to characters

As shown in the row results, MSSubClass is now divided into 16 columns, each representing a category. For the corresponding category, it is represented by 1, but against the category, it is represented by 0. A vector can be used to represent a category. If the first category, it can be represented as [1,0,0,0,0,0,0,0,0,0,0,0,0,0,0,0,0,0,0,0,0,0,0,0,0,0,0,0,0,0,0], with 1 for the corresponding value and 0 for others.

3. Convert the data in the object into one-hot coding

Apply one-hot coding to all category data. In this way, the category data can be converted into numerical data. Here get_dummies() is used to convert data into a one-hot coding format. Namely, apply one-hot encoding to object and str data. The code is as follows:

```
1.  all_dummy_df =pd.get_dummies(all_df,drop_first =True)
2.  all_dummy_df.head()
```

After running the code, it can be seen that after one-hot coding, the dataset has become 259 columns with the output result as shown in Fig. 5-9.

Id	LotFrontage	LotArea	OverallQual	OverallCond	YearBuilt	YearRemodAdd	MasVnrArea	BsmtFinSF1	BsmtFinSF2	BsmtUn
1	65.0	8450	7	5	2003	2003	196.0	706.0	0.0	1٤
2	80.0	9600	6	8	1976	1976	0.0	978.0	0.0	2٤
3	68.0	11250	7	5	2001	2002	162.0	486.0	0.0	4٤
4	60.0	9550	7	5	1915	1970	0.0	216.0	0.0	5٤
5	84.0	14260	8	5	2000	2000	350.0	655.0	0.0	4٩

5 rows × 259 columns

Fig. 5-9　One-hot coding running results

5.6　Missing Data Processing in the Housing Price Dataset

1. Missing values in statistical data

If there are missing values in the data, anomalies will occur during model training, so it is necessary to fill in the missing data. For numerical type data, it is necessary to first check the missing values, then fill in the missing values, and finally perform normalization operations.

Here, the isnull() method is first used to detect which features in the dataset have NULL values, and then use sort_values for sorting and calculate the proportion of missing values. Here 35 features have missing values with the code as follows:

```
1.  conut =all_dummy_df.isnull().sum().sort_values(ascending =False)
2.  percent = (all_dummy_df.isnull().sum()/all_df.isnull().count
    ()).sort_values(ascending =False)
3.  miss_df = pd.concat([conut,percent],axis =1,keys = ['Total','
    precent'],sort =False)
4.  miss_df.head(40)
```

Processing these missing data requires analysis based on the feature description file provided by the topic and the description can clarify the meaning represented by such missing data.

2. Fill in missing data

After finding any missing data, it is necessary to fill in the data. There are many methods for data filling, such as median filling, mean filling and random forest filling. If a feature has too many values missing, it can also be deleted. Here, the mean filling is used with the code as follows:

```
1.  mean_clos = all_dummy_df.mean()
2.  mean_clos.head()
3.  mean_clos.shape
```

Use fillna to fill in the mean and then carry out statistics with the code as follows:

```
1.  all_dummy_df = all_dummy_df.fillna(mean_clos)all_dummy_df.
    isnull().sum().sum()
2.  all_dummy_df.isnull().sum().sum()
```

5.7   Normalization of the Housing Price Dataset

1. Data normalization processing

Here standard distribution is used to normalize the data, which can make the data smoother and easier to calculate. There are two steps here: firstly, identify numerical features; secondly, use the standard deviation normalization formula to normalize the data. Let's first see which are numerical, and the code is as follows:

```
1.  numeric_cols = all_df.columns[all_df.dtypes! = 'object']
2.  numeric_cols
```

Use standard deviation normalization to normalize numerical data, with the following code:

```
1.  numeric_cols_means = all_dummy_df.loc[:,numeric_cols].mean()
2.  numeric_cols_std = all_dummy_df.loc[:,numeric_cols].std()
3.  all_dummy_df.loc[:,numeric_cols] = (all_dummy_df.loc[:,numeric_
    cols] - numeric_cols_means)/numeric_cols_std
```

2. Split to get train and test sets

Use the loc() function to obtain the train and test sets from the data, and output the train and

test shapes at the same time. The code is as follows:

```
1.  dummy_train_df = all_dummy_df.loc[train_df.index]
2.  dummy_test_df = all_dummy_df.loc[test_df.index]
3.  dummy_train_df.shape,dummy_test_df.shape
```

5.8 Establishment of a Ridge Regression Model for Housing Price Prediction

1. Import the sklearn package to establish a ridge regression model

Import the sklearn package to implement model construction and hyperparameter selection. Moreover, it is high-dimensional data, so ridge regression is chosen as the basic model.

2. Use cross validation sets to obtain optimal parameters for ridge regression

np. logspace (-2,2,50) constructs a geometric progression from the -2 power of 10 to the 2 power of 10. This geometric progression is 50 elements in length, and this value corresponds to the L2 regularization coefficient λ in the ridge regression.

Then, through a loop, save the test_score value of each corresponding numerical value and then save the test_score value corresponding to each λ with the code as follows:

```
1.  test_scores = []
2.  for alpha in alphas:
3.      clf = Ridge(alpha)
4.      test_score = np.sqrt( - ross_val_score(clf, X_train, y_train,
        cv =10, scoring = 'neg_mean_squared_error'))
5.      test_scores.append(np.mean(test_score))
6.  print(test_score)
```

Then use sns. lineplot to plot a test_score chart with the code as follows:

```
1.  import seaborn as sns
2.  sns.lineplot(alphas,test_scores)
3.  plt.xlabel("Iterations")
4.  plt.ylabel("loss")
5.  ax.legend()
```

See Fig. 5-10 for the running results. From Fig. 5-10, it can be seen that when the iteration (alpha) value is 10-20, the loss (R-score) will reach around 0. 135 5. The value of λ can be selected as 15.

3. Cross validation

Divide all samples into k equally sized subsets of samples; traverse through these k subsets in sequence, using the current subset as the validation set and all other samples as the train set for model training and evaluation each time; finally, use the average of k evaluation indicators as the final evaluation indicator. In practical experiments, k is usually taken as 10.

For example, $k = 10$ is taken here, as shown in Fig. 5-11.

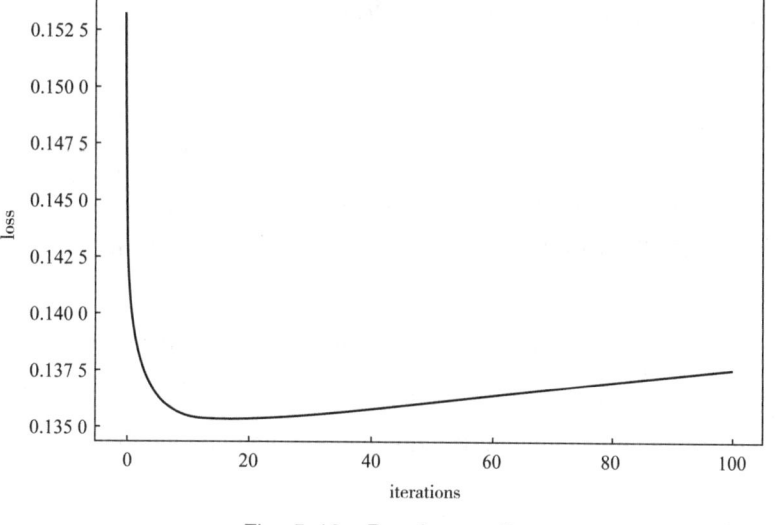

Fig. 5-10   Running results

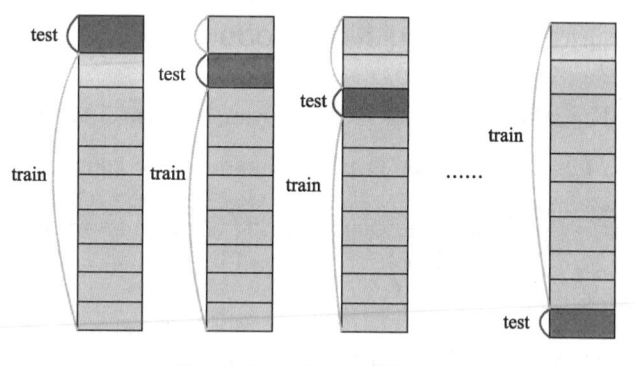

Fig. 5-11   Cross validation

① Firstly, divide the original dataset into 10 parts.

② Take one of them as the test set and the remaining 9 as the training set; now, the training set has become $k \times D$ (D represents the number of data samples included in each part).

As shown in Fig. 5-11, each time one data is taken as the test set, and other data are used as the train set. The train model can get 10 model evaluation indicators, and the average value is used as the final indicator for model evaluation.

The cross_val_score ( ) function can be used to implement the function of cross validation, and the function can be specifically used as follows:

```
sklearn.model_selection.cross_val_score(estimator,X,y = None,groups =
None,scoring =None,cv = 'warn',n_jobs = None,verbose =0,fit_params =None,pre_
dispatch = '2* n_jobs',error_score = 'raise -deprecating')
```

Parameters:

estimator: It requires using cross validation algorithms.

X: Enter the sample data.

y: Sample labels.

groups: The group labels of the samples used when the dataset is divided into train/test sets.

cv: Cross validation folds or number of iterations.

n_ jobs: Number of CPUs working simultaneously (-1 represents all).

verbose: Level of detail.

fit_params: The parameter of the fitting method passed to the estimator (validation algorithm).

pre_ dispatch: The number of tasks dispatched in parallel by the control program. If this quantity is reduced, is useful for avoiding the expansion of CPU memory consumption when the CPU sends more jobs.

5.9 Use a Cross-validation Set to Obtain the Optimal Parameters of the Housing Price Prediction Model

Import the RandomForestRegressor package in Jupyter Notebook, set a max_features list and use loop and cross validation to solve optimization parameters with the code as follows:

```
1.  from sklearn.ensemble import RandomForestRegressor
2.  max_features = [.1,.3,.5,.7,.9,.99]
3.  test_scores = []
4.  for max_feat in max_features:
5.      clf = RandomForestRegressor(n_estimators =200,max_features =max_
    feat)
6.      test_score =np.sqrt( -cross_val_score(clf,X_train,y_train,cv =5,
    scoring = 'neg_mean_squared_error'))
7.      test_scores.append(np.mean(test_score))
8.  sns.lineplot(max_features,test_scores)
```

After running the code, it can be seen that when max_features $=0.3$, test_scores has achieved

the optimal value of 0. 137, as shown in Fig. 5-12.

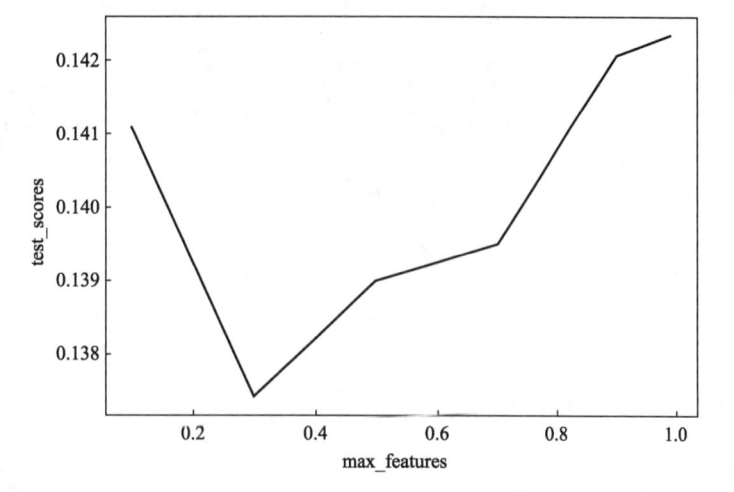

Fig. 5-12   Solving for optimal parameters

Unit 6

Use Logistic Regression to Forecast Shared Bike Usage

The "last mile" of public transportation is the main obstacle for urban residents to use public transportation for trip, and it is also the main challenge faced in the process of building green and low-carbon cities. Shared bike enterprises provide services on campuses, subway stations, bus stops, residential areas, commercial areas, public service areas, etc., completing the final "puzzle" of the transportation industry and driving residents' enthusiasm for using other public transportation tools. In this way, a synergistic effect can be stimulated with other public transportation modes. Shared bike is a time-sharing rental model, and also a new, green and environmentally friendly sharing economy.

At the same time as its rapid development provides convenient transportation for citizens, due to the tidal nature of transportation and untimely vehicle scheduling, it also faces such problems as imbalance in vehicle supply and demand between stations during peak hours. Therefore, it is necessary to accurately and efficiently predict the number of shared bikes between stations, and establish a prediction model to accurately predict vehicle scheduling and predict the demand and status of shared bikes between stations. The demand for shared bike stations is not only dynamically related to time and space, but also to complex conditions such as weather, holidays, temperature and wind power. Its model is actually a nonlinear time series problem. In this unit, a data model is established by analyzing the past shared bike data to predict the demand of shared bikes at stations, thereby achieving accurate scheduling of shared bikes.

6.1 Data Loading and Display

1. Import data load packages

Import numpy, pandas and matplotlib packages into Jupyter Notebook to load and analyze the data. The code is as follows:

```
1.  # import the required package
2.  import pandas as pd
3.  #import seaborn as sns
4.  import matplotlib.pyplot as plt
5.  import numpy as np
```

2. Open a csv data file to import data

Use pandas read_csv to open the csv file to import data and return a DataFrame object.

Comma-separated values (CSV) are also called character-separated values (because not only commas can separate values as character). Its file stores table data (numbers and texts) in plain text format. A plain text means that the file is a sequence of characters and does not contain data that must be interpreted like binary digits. A CSV file consists of any number of records separated by a certain line break; each record consists of fields, and the separators between fields are other characters or strings, the most common being commas or tabs. Usually, all records have an identical field sequence. Usually it is a plain text file. It is recommended to use WORDPAD or Notepad to open it, or save it as a new document and then use EXCEL to open it. The code is as follows:

```
1.  # import data
2.  data =pd.read_csv('kaggle_bike_competition_train.csv',header =
    0,skip_blank_lines =True)
3.  # display data
4.  data.head()
```

Note: When skip_blank_lines = True, this parameter ignores comment and blank lines. So header = 0 indicates that the first line is data rather than the first line of the file.

After the code runs, it will display the shared bike prediction data, as shown in Fig. 6-1.

3. Display data and check basic data information

Use the data. info ( ), data. isnull ( ) and data. describe ( ) functions to display data basic information, including row count, column count, column index, number of non-null values, column type, memory usage, and so on, as shown in Fig. 6-2.

	datetime	season	holiday	workingday	weather	temp	atemp	humidity	windspeed	casual	registered	count
0	2011/1/1 0:00	1	0	0	1	9.84	14.395	81	0.0	3	13	16
1	2011/1/1 1:00	1	0	0	1	9.02	13.635	80	0.0	8	32	40
2	2011/1/1 2:00	1	0	0	1	9.02	13.635	80	0.0	5	27	32
3	2011/1/1 3:00	1	0	0	1	9.84	14.395	75	0.0	3	10	13
4	2011/1/1 4:00	1	0	0	1	9.84	14.395	75	0.0	0	1	1

Fig. 6-1   Basic information of data[①]

From the execution results of the info( ) function, it can be seen that the data types for shared bikes are int64 and float64. However, there is a special type of data—datetime, which is of the object type and needs to be processed in the future before used.

```
<class 'pandas.core.frame.DataFrame'>
RangeIndex: 10886 entries, 0 to 10885
Data columns (total 12 columns):
 #   Column      Non-Null Count   Dtype
---  ------      --------------   -----
 0   datetime    10886 non-null   object
 1   season      10886 non-null   int64
 2   holiday     10886 non-null   int64
 3   workingday  10886 non-null   int64
 4   weather     10886 non-null   int64
 5   temp        10886 non-null   float64
 6   atemp       10886 non-null   float64
 7   humidity    10886 non-null   int64
 8   windspeed   10886 non-null   float64
 9   casual      10886 non-null   int64
 10  registered  10886 non-null   int64
 11  count       10886 non-null   int64
dtypes: float64(3), int64(8), object(1)
memory usage: 1020.7+ KB
```

Fig. 6-2   Data descriptions

The describe( ) function directly gives some basic statistics of the sample data, including mean, standard deviation, maximum, minimum, quantile, etc. See Fig. 6-3 for the result of the function operation.

The isnull( ) function is used to determine whether there are missing values in the data. The commonly used methods include isnull( ). any( ) and isnull( ). sum, with the following code:

```
1.  data.isnull().values.any()
2.  data.isnull().sum()
```

①   The dataset in this unit is that for shared bicycles in the Washington area.

	season	holiday	workingday	weather	temp	atemp	humidity	windspeed	casual	registered
count	10886.000000	10886.000000	10886.000000	10886.000000	10886.0000	10886.000000	10886.000000	10886.000000	10886.000000	10886.000000
mean	2.506614	0.028569	0.680875	1.418427	20.23086	23.655084	61.886460	12.799395	36.021955	155.552177
std	1.116174	0.166599	0.466159	0.633839	7.79159	8.474601	19.245033	8.164537	49.960477	151.039033
min	1.000000	0.000000	0.000000	1.000000	0.82000	0.760000	0.000000	0.000000	0.000000	0.000000
25%	2.000000	0.000000	0.000000	1.000000	13.94000	16.665000	47.000000	7.001500	4.000000	36.000000
50%	3.000000	0.000000	1.000000	1.000000	20.50000	24.240000	62.000000	12.998000	17.000000	118.000000
75%	4.000000	0.000000	1.000000	2.000000	26.24000	31.060000	77.000000	16.997900	49.000000	222.000000
max	4.000000	1.000000	1.000000	4.000000	41.00000	45.455000	100.000000	56.996900	367.000000	886.000000

Fig. 6-3　Data statistics of shared bikes

After running the code, it can be seen that there is no missing data in the shared bike dataset, as shown in Fig. 6-4.

4. Meanings of data fields

The data. info ( ) function can be used to output the data of shared bikes, which is the data after data cleansing and data conversion. The meanings of the corresponding fields of data are as follows:

```
#检测是否有null值www.baidu
data.isnull().sum()
executed in 12ms, finished 14:53:41 2021-02-20

datetime      0
season        0
holiday       0
workingday    0
weather       0
temp          0
atemp         0
humidity      0
windspeed     0
casual        0
registered    0
count         0
dtype: int64
```

Fig. 6-4　Statistics of null data

datatime: Time and date.

season: Seasons. 1: Spring; 2: Summer; 3: Autumn; 4: Winter.

Holiday: Is it a holiday? 1: Holiday; 0: Noholiday.

Workingday: Is it a workingday? 1: Working day; 0: Noworking day.

Weather: Weather category. 1: Sunny day; 2: Cloudy; 3: Light rain or snow; 4: Adverse weather.

Temp: Actual temperature.

Atemp: Apparent temperature.

Humidity: Relative humidity/%.

Windspeed: Wind speed (km/h).

Casual: Random user, unregistered user.

Registered: Registered user.

Count: Total number of users.

6.2　Visualization Analysis of Seasonal Data

Season represents the season in the data. Based on basic common sense, seasonal data still has a certain impact on the usage of bikes. How to analyze and count the usage data of bikes according to seasons should be processed as follows:

Firstly, merge the duplicate data in the season. Use the unique( ) function to remove duplicate data values from the season and return the unique value in the season in list format. There are four values returned, 1,2,3,4, of type int64. Use unique( ) to count four different seasons.

Use values_counts( ) to count how many records there are in each season and summarize the total number of bike rentals for the four seasons. The code is as follows:

```
data.season.value_counts()
```

Use the groupby( ) function to group the season data, then summarize and count the number of rentals for season counts, and finally obtain the total number of rentals for each season.

Present data visually. Use the bar( ) function of plt to plot a bar chart and visually display the number of bike rentals for each season. The code is as follows:

```
1.  #sns.barplot(x = 'season',y = 'count',data = data)
2.  #s_list = data.season.unique()
3.  s_list = ['1','2','3','4']
4.  s_c = data.groupby('season')['count'].sum()
5.  plt.bar(range(len(s_list)),s_c,color = 'rgb',tick_label = s_list)
6.  plt.show()
```

After running the code, it can be seen that the number of bike rentals is summer > spring > autumn > winter, as shown in Fig. 6-5.

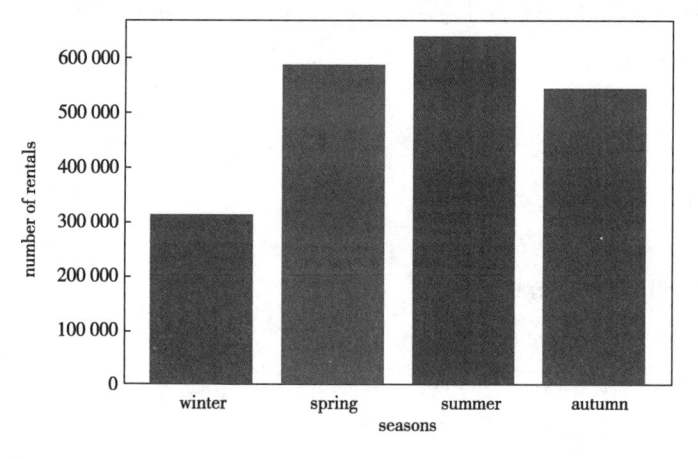

Fig. 6-5　Rental data calculated by grouping four seasons

6.3　Visualization Analysis of Weather and Holiday Data

The processing of weather data is similar to that of seasonal data, which involves grouping and

then statistical calculations. The code is as follows:

```
1.  s_list = data.season.unique()
2.  s_c = data.groupby('weather')['count'].sum()
3.  plt.bar(range(len(s_list)),s_c,color = 'rgb',tick_label = s_list)
4.  plt.show()
```

After running the code, it can be seen that the rental count of shared bikes is 1 sunny day > 2 cloudy day > 3 light rain and snow > 4 severe weather, as shown in Fig. 6-6.

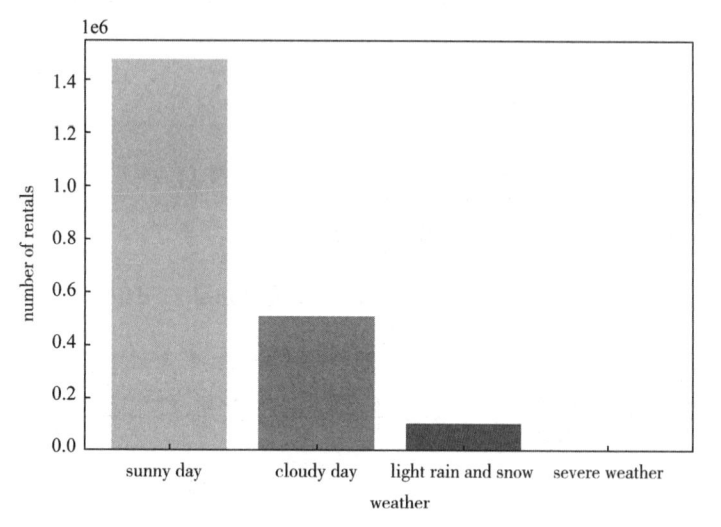

Fig. 6-6   Relationship between weather and rental data

Holiday features ('holiday') are divided into two types: 'holiday' and 'noholiday', while workingday features ('workingday') are divided into two types: 'workingday' and 'noworkingday'. These two features also require visual output comparison. The code is as follows:

```
1.  # comparison of information between workingdays and holidays
2.  d_list = ['holiday','noholiday','workingday','noworkingday']
3.  d_s = data.groupby('holiday')['count'].sum()
4.  ar = np.array(d_s)
5.  print(ar)
6.  ##print(ar.shape)
7.  d_2 = data.groupby('workingday')['count'].sum()
8.  d_s = [data.groupby('holiday')['count'].sum()] + [data.groupby('
    work-ingday')['count'].sum()]
9.  print(d_s)
10. print(np.array(d_s).shape)
```

```
11. #plt.bar(range(len(d_list)),d_s,color='rgb',tick_label=d_list)
12. #plt.show()
```

6.4   Processing of Time Data

From the above analysis, it can be seen that datetime is an important feature, but its type is object, and it needs to be splitted into two parts: date and time. The code is as follows:

```
1. # Time field processing
2. temp=pd.DatetimeIndex(data['datetime'])
3. data['date']=temp.date
4. data['time']=temp.time
5. data.head()
```

After the code runs, it can be seen that two new features, "date" and "time", have been generated, as shown in Fig. 6-7.

	datetime	season	holiday	workingday	weather	temp	atemp	humidity	windspeed	casual	registered	count	date	time
0	2011/1/1 0:00	1	0	0	1	9.84	14.395	81	0.0	3	13	16	2011-01-01	00:00:00
1	2011/1/1 1:00	1	0	0	1	9.02	13.635	80	0.0	8	32	40	2011-01-01	01:00:00
2	2011/1/1 2:00	1	0	0	1	9.02	13.635	80	0.0	5	27	32	2011-01-01	02:00:00
3	2011/1/1 3:00	1	0	0	1	9.84	14.395	75	0.0	3	10	13	2011-01-01	03:00:00
4	2011/1/1 4:00	1	0	0	1	9.84	14.395	75	0.0	0	1	1	2011-01-01	04:00:00

Fig. 6-7   Splitting results

Refine the time to hours and use the hour field as a more concise feature. The code is as follows:

```
1. # set the hour field
2. data['hour']=pd.to_datetime(data.time,format="% H:% M:% S")
3. data['hour']=pd.Index(data['hour']).hour
4. data
```

After running the code, a new "hour" feature can be generated, as shown in Fig. 6-8.

Finally, remove the original time field from the data subject to the code as follows:

```
1. #remove old data features
2. dataRel=data.drop(['datetime','count','date','time',
    'dayofweek'],axis=1)
3. dataRel.head()
```

	datetime	season	holiday	workingday	weather	temp	atemp	humidity	windspeed	casual	registered	count	date	time	hour
0	2011/1/1 0:00	1	0	0	1	9.84	14.395	81	0.0000	3	13	16	2011-01-01	00:00:00	0
1	2011/1/1 1:00	1	0	0	1	9.02	13.635	80	0.0000	8	32	40	2011-01-01	01:00:00	1
2	2011/1/1 2:00	1	0	0	1	9.02	13.635	80	0.0000	5	27	32	2011-01-01	02:00:00	2
3	2011/1/1 3:00	1	0	0	1	9.84	14.395	75	0.0000	3	10	13	2011-01-01	03:00:00	3
4	2011/1/1 4:00	1	0	0	1	9.84	14.395	75	0.0000	0	1	1	2011-01-01	04:00:00	4
5	2011/1/1 5:00	1	0	0	2	9.84	12.880	75	6.0032	0	1	1	2011-01-01	05:00:00	5
6	2011/1/1 6:00	1	0	0	1	9.02	13.635	80	0.0000	2	0	2	2011-01-01	06:00:00	6
7	2011/1/1 7:00	1	0	0	1	8.20	12.880	86	0.0000	1	2	3	2011-01-01	07:00:00	7
8	2011/1/1 8:00	1	0	0	1	9.84	14.395	75	0.0000	1	7	8	2011-01-01	08:00:00	8
9	2011/1/1 9:00	1	0	0	1	13.12	17.425	76	0.0000	8	6	14	2011-01-01	09:00:00	9
10	2011/1/1 10:00	1	0	0	1	15.58	19.695	76	16.9979	12	24	36	2011-01-01	10:00:00	10
11	2011/1/1 11:00	1	0	0	1	14.76	16.665	81	19.0012	26	30	56	2011-01-01	11:00:00	11
12	2011/1/1 12:00	1	0	0	1	17.22	21.210	77	19.0012	29	55	84	2011-01-01	12:00:00	12

Fig. 6-8　Results after time refinement

After the code is run, the results from time field refinement can be obtained, as shown in Fig. 6-9.

	season	holiday	workingday	weather	temp	atemp	humidity	windspeed	casual	registered	hour	dataDays	Saturday	Sunday
0	1	0	0	1	9.84	14.395	81	0.0	3	13	0	0.0	1	0
1	1	0	0	1	9.02	13.635	80	0.0	8	32	1	0.0	1	0
2	1	0	0	1	9.02	13.635	80	0.0	5	27	2	0.0	1	0
3	1	0	0	1	9.84	14.395	75	0.0	3	10	3	0.0	1	0
4	1	0	0	1	9.84	14.395	75	0.0	0	1	4	0.0	1	0

Fig. 6-9　Time field refinement

Finally, calculate the rental count of shared bikes by hour and visualize it with the code as follows:

```
1.  # time/hours
2.  sns.pointplot(data['hour'],data['count'])
3.  plt.title('the impact of hours on the number of shared bikes')
4.  plt.show()
```

After the code runs, a line chart is generated for each time period of the day and the count of shared bikes. The horizontal hour represents "hours", and the ordinate count represents "rental count of shared bikes". As can be seen, there are two peaks in the rental count around 8 o'clock and 17 o'clock, as shown in Fig. 6-10.

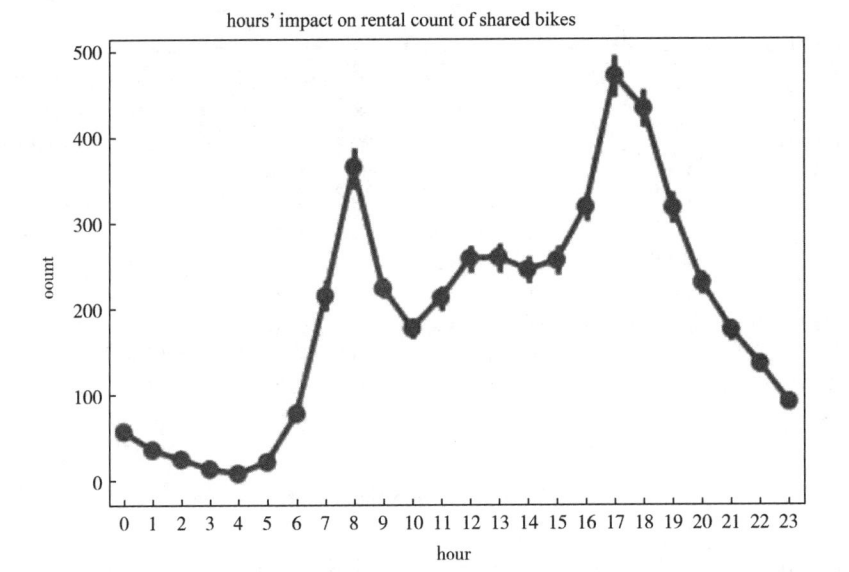

Fig. 6-10 Relationship between hours and rental count of shared bikes

6.5 Characterization Vectors

1. Separate discrete features from continuous features

We need to use sklearn for modeling. Moreover, there are methods/functions to convert pandas dataframe directly to dict in Python. In addition, it is necessary to distinguish between discrete and continuous features here, by placing continuous features in dict named X_dictCon subject to the code as follows:

```
1.  from sklearn.feature_extraction import DictVectorizer
2.  # place continuous feature values into a dict
3.  featureConCols = ['temp','atemp','humidity','windspeed',
    'dataDays', 'hour']
4.  dataFeatureCon = dataRel[featureConCols]
5.  dataFeatureCon = dataFeatureCon.fillna('NA')
6.  X_dictCon = dataFeatureCon.T.to_dict().values()
```

Place discrete features in the dict named X_dictCat subject to the code as follows:

```
1.  featureCatCols = ['season','holiday','workingday','weather',
    'Saturday','Sunday']
2.  dataFeatureCat = dataRel[featureCatCols]
```

```
3.  dataFeatureCat = dataFeatureCat.fillna('NA')
4.  X_dictCat = dataFeatureCat.T.to_dict().values()
```

Use DictVectorizer fit_transform ( ) to vectorize X_dictCat and X_dictCon subject to the following code:

```
1.  vec = DictVectorizer(sparse = False)
2.  X_vec_cat = vec.fit_transform(X_dictCat)
3.  X_vec_con = vec.fit_transform(X_dictCon)
4.  dataFeatureCon.head()
```

2. Standardized continuous eigenvalues

Next, it is necessary to standardize the continuous eigenvalues so that the mean is 0 and the variance is 1 after processing. Such data, if put into the model, is beneficial for both the convergence of model train and the accuracy of the model. The code is as follows:

```
1.  from sklearn import preprocessing
2.  # standardize continuous values
3.  scaler = preprocessing.StandardScaler().fit(X_vec_con)
4.  X_vec_con = scaler.transform(X_vec_con)
5.  X_vec_con
```

3. Category feature coding

Use one-hot coding to process category features. The code is as follows:

```
1.  from sklearn import preprocessing
2.  enc = preprocessing.OneHotEncoder()
3.  enc.fit(X_vec_cat)
4.  X_vec_cat = enc.transform(X_vec_cat).toarray()
5.  X_vec_cat
```

Features concatenation can combine discrete and continuous features together. The first six columns of the final features are the standardized continuous features, followed by the encoded discrete features. The code is as follows:

```
1.  import numpy as np
2.  X_vec = np.concatenate((X_vec_con, X_vec_cat), axis = 1)
3.  X_vec
```

Labels also need to be processed. Perform floating-point processing on the labels, and finally get the floating point value of the result. The code is as follows:

```
1.  # vectorize Y
2.  Y_vec_reg = dataRel['registered'].values.astype(float)
3.  Y_vec_cas = dataRel['casual'].values.astype(float)
4.  Y_vec_reg
```

After running the program, we can get the registered and casual values after floated

6.6 Data Preprocessing

Firstly, merge the train and test sets, and use append to merge the two fields train_std and test subject to the code as follows:

```
1.  train_std_y = np.log(train_std.pop('count'))
2.  # merge train_std and test for easy modification
3.  full = train_std.append(test, ignore_index = True)
4.  print ('merged dataset: ', full. shape)
```

Then select such features as year, month, weekday, hour, season, holiday, weather, temp, humidity and windspeed from the merged feature set. The code is as follows:

```
1.  full['datetime'] = full['datetime'].astype(str)
2.  full['datetime'] = full['datetime'].apply(lambdax:datetime.strp
    -time(x,'% Y - % m - % d% H:% M:% S'))
3.  # extract the number of years, months, hours, and Weekdays, and add
    them to the data in the form of columns
4.  full['year'] = full['datetime'].apply(lambdax:x.year)
5.  full['month'] = full['datetime'].apply(lambdax:x.month)
6.  full['hour'] = full['datetime'].apply(lambdax:x.hour)
7.  full['weekday'] = full['datetime'].apply(lambdax:x.strftime('% a'))
8.  # select the desired features
9.  columns = ['year','month','weekday','hour','season','holiday',
10.            'weather','temp','humidity','windspeed']
11. full_feature = full[columns]
12. full_feature.head()
```

Finally, use get_dummies( ) to convert multi-category data into multiple binary categories via one-hot subject to the code as follows:

```
1.  #convert multi-category data into multiple binary categories via
    one-hot
2.  dummies_year =pd.get_dummies(full_feature['year'],prefix='year')
3.  dummies_month =pd.get_dummies(full_feature['month'],prefix='
    month')
4.  dummies_weekday =pd.get_dummies(full_feature['weekday'],prefix
    ='weekday')
5.  dummies_hour =pd.get_dummies(full_feature['hour'],prefix='hour')
6.  dummies_season =pd.get_dummies(full_feature['season'],prefix=
    'season')
7.  dummies_holiday =pd.get_dummies(full_feature['holiday'],prefix
    ='holiday')
8.  dummies_weather =pd.get_dummies(full_feature['weather'],prefix
    ='weather')
9.  full_feature = pd.concat([full_feature,dummies_year,dummies_
    month,dummies_weekday,
10. dummies_hour, dummies_season,dummies_holiday, dummies_weather],
    axis=1)
11. # Delete original columns
12. dropFeatures = ['season','weather','year','month','weekday',
    'hour','holiday']
13. full_feature =full_feature.drop(dropFeatures,axis=1)
14. full_feature.head()
```

6.7 Modelling

Set up train and test sets, split train and test samples, and use the sample and subtractByKey
methods to split the dataset into train and test sets. The code is as follows:

```
1.  # use the sample and subtractByKey methods to split training
    samples-test samples
2.  data_with_idx = data.zipWithIndex().map(lambdapoint_index:
    (point_index[1], point_index[0]))
3.  test =data_with_idx.sample(False,0.2,42)
4.  train =data_with_idx.subtractByKey(test)
5.  train_data =train.map(lambdaindex_point:index_point[1])
```

```
6.  test_data = test.map (lambdaindex_point:index_point[1])
7.  train_size = train_data.count ()
8.  test_size = test_data.count ()
9.  print ("Training data size:% d"% train_size)
10. print ("Test data size:% d"% test_size)
11. print ("Total data size:% d"% num_data)
```

When a linear regression model is adopted to predict shared bike counts, the five variables such as city, time, temperature, weather conditions and wind speed have a significant impact on the counts, so it is used as a feature variable to fit the model. The code is as follows:

```
1.  # define metrics for evaluation models
2.  def evaluate (train, test, iterations, step, regParam, regType,
    intercept):
3.      model = LinearRegressionWithSGD. train (train, iterations,
    step, regParam = regParam, regType = regType, intercept =
    intercept)
4.      tp = test.map (lambdap: (p.label,model.predict (p.features)))
5.      rmsle = np.sqrt (tp.map (lambdalp:squared_log_error (lp[0],
    lp[1])).mean ())
6.      return rmsle
```

Adjust the parameters and select the optimal solution. Adjustable parameters include Iterations, step, L2/L1 regularization coefficient and Intercept subject to the code as follows:

```
1.  # example: Adjust the parameter - Iterations
2.  params = [1, 5, 10, 20, 50, 100, 200, 300]
3.  metrics = [evaluate (train_data, test_data, param,0.01, 0.0,'12',
    False) for param in params]
4.  print (params)
5.  print (metrics)
6.  plt.plot (params,metrics)
7.  fig = plt.gcf ()
8.  plt.xlabel ('log')
```

After running the code, the MSE error of the output model is 1 208. 955 627 135 937 7, and the R2 value on the training set is − 0. 483 487 258 551 061 45.

Unit 7

Logistic Regression and Decision Boundary

In linear regression, samples are used to learn the mapping f from x to y through supervised learning. Because y is a continuous value, it is a regression problem. If the variable y to be predicted is a discrete value, the logistic regression algorithm needs to be used to solve the classification problem.

There are many examples of logistic regression in real life. Moreover, there are a variety of classification problems: an online transaction website determines whether a deal is fraudulent, judges whether a tumor is benign or malignant, and confirm whether a spam occurs.

In these examples, a binary variable is predicted, either 0 or 1; a deal is either fraudulent or not; a tumor is malignant or not; an email is either a spam or not. The two classes that the dependent variables may belong to are called negative class and positive class, respectively. You can use 0 to represent negative classes and 1 to represent positive classes. The above classification problem is only limited to two classes: 0 or 1. Afterwards, we will discuss the multi-classification problem, which means that the variable y can take multiple values, such as 0,1,2,3.

7.1 Machine Learning Classification

Machine learning is generally divided into supervised learning, unsupervised learning and reinforcement learning, as shown in Fig. 7-1.

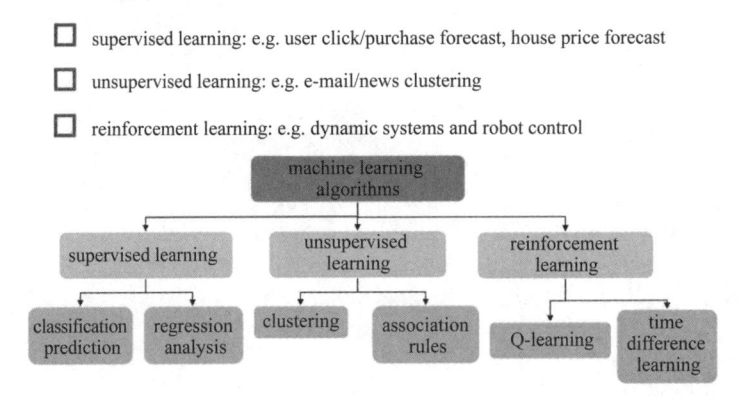

Fig. 7-1　Machine learning classification

1. Supervised learning

Under supervised learning, the input data is called "train data", and each group of train data has a clear identification or result, such as "spam" and "non-spam" in the anti-spam system; for handwritten digit recognition, "1", "2", "3", "4", etc. are shown in Fig. 7-2.

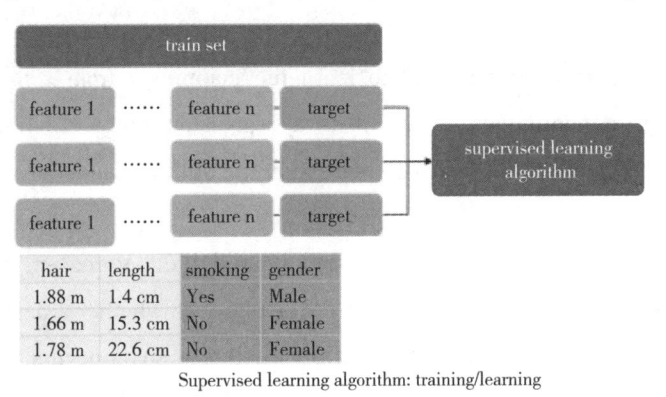

Fig. 7-2　Supervised learning

When building a prediction model, supervised learning establishes a learning process, compares the prediction results with the actual results of "train data", and constantly adjusts the prediction model until the prediction results of the model reach expected accuracy. For supervised learning, the common application scenarios include classification problems and regression problems. Common algorithms include logistic regression and back-propagation neural network.

Semi-supervised learning is a special case of supervised learning. In the semi-supervised learning mode, the input data is partially identified and partially not. This learning model can be used for prediction, but the model first needs to learn the internal structure of the data in order to organize the data for prediction in a reasonable manner. Application scenarios include classification and regression, and algorithms include some extensions of commonly used supervised learning

algorithms. These algorithms first try to model the unlabeled data, and then predict the identified data on this basis, such as graph inference algorithm or Laplacian SVM.

2. Unsupervised learning

In unsupervised learning, data is not specifically identified. The learning model is to infer some internal structures of data. Common application scenarios include learning association rules and clustering. Common algorithms include Apriori algorithm and k-means algorithm, as shown in Fig. 7-3.

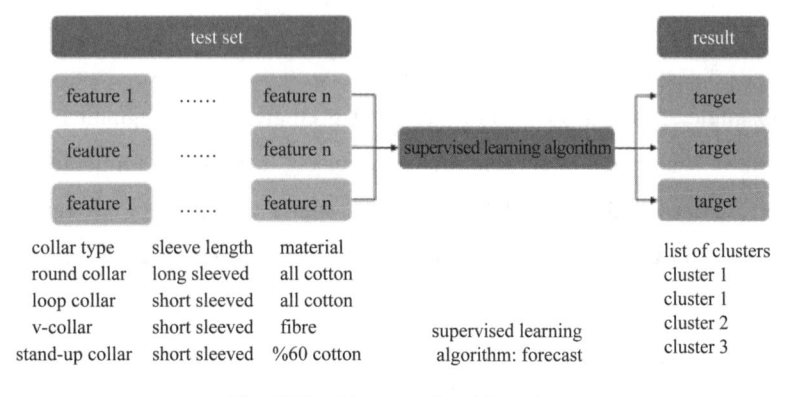

Fig. 7-3   Unsupervised learning

3. Reinforcement learning

In this learning mode, the input data is used as feedback to the model. Unlike the supervised learning model, the input data is only used as a way to check the correctness of the model. Under reinforcement learning, the input data is directly fed back to the model, and the model must make adjustments immediately. Common application scenarios include dynamic systems and robot control.

7.2   Logistic Regression

1. "Linear regression + threshold" to determine whether it is a malignant tumor

Now there is a classification task that requires determining whether a tumor is benign based on its size. See Fig. 7-4 for the training set, with the horizontal axis representing the size of the tumor, the vertical axis representing whether the tumor is benign, and × representing the patient sample point. Note that the vertical axis has only two values, 1 (for malignant tumors) and 0 (for benign tumors). Through previous learning, it has been known that the above dataset needs to be processed with linear regression. In fact, a straight line is used to fit these data.

We can first use linear regression to fit it, and then set a threshold of 0.5. The value below the threshold is 0 (benign tumor), and the value above the threshold is 1 (malignant tumor), which can be expressed as follows:

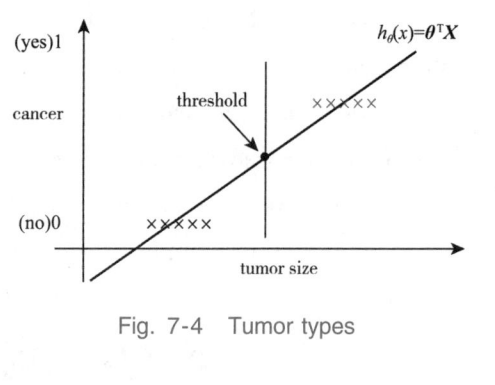

Fig. 7-4　Tumor types

If $h_\theta(x) \geqslant 0.5$, predicted $y = 1$

If $h_\theta(x) < 0.5$, predicted $y = 0$

Where, $h_\theta(x)$ represents the logistic regression model for tumor identification; y represents the predicted result.

The above example seems to solve the problem of prediction on malignant tumors very well, but the biggest problem of this model is its sensitivity to noise (insufficient robustness). If two training samples are added, following the idea of "linear regression + threshold", we can get a straight line via linear regression, and then set the threshold to 0.5, as shown in Fig. 7-5.

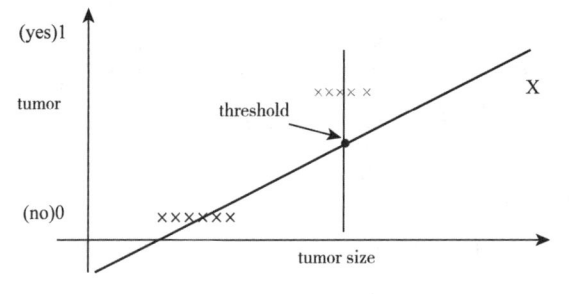

Fig. 7-5　Tumor types with training samples added

At this point, two false positives will occur, and it is not appropriate to use 0.5 as the threshold to predict whether it is a malignant tumor.

2. Use sigmoid to achieve logistic regression

In the above example, probability can be used as the basis for determining malignant tumors. The probability is valued between 0 and 1, and a function needs to be used to convert $h_\theta(x)$ value to 0 to 1, and then use this value as the basis for determining whether it is a malignant tumor.

The sigmoid function can be used to transform a real number value within $(-\infty, +\infty)$ to an interval of 0 to 1, increasing monotonically. The domain of definition is $(-\infty, +\infty)$, the value domain is $(0,1)$, and the sigmoid function is

$$y = \frac{1}{1 + e^{-x}}$$

See Fig. 7-6 for the function graph.

Where, e^x is an exponential function. The general expression of exponential function is $y = a^x$ ($a > 0$ and $a \neq 1$), and its function graph is shown in Fig. 7-7.

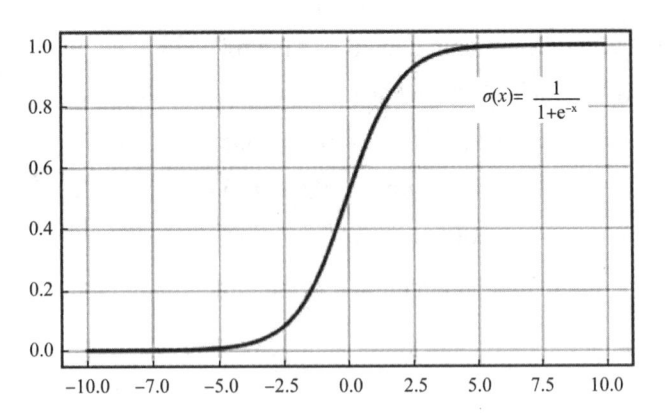

Fig. 7-6   Sigmoid function graph

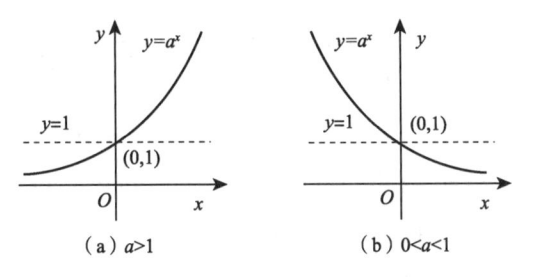

Fig. 7-7   Function graph of exponential function

Use $h_\theta(x)$ to replace z and the value of linear regression can be transformed to 0-1, as follows:

$$g(\boldsymbol{\theta}^{\mathrm{T}}\boldsymbol{X}) = \frac{1}{1+e^{-\boldsymbol{\theta}^{\mathrm{T}}\boldsymbol{x}}}$$

$$g(z) = \frac{1}{1+e^{-z}}$$

This function has a special point, where $h_\theta(x) = 0.5$; at this time, the value of $\boldsymbol{\theta}^{\mathrm{T}}\boldsymbol{X}$ is 0. If the value of $\boldsymbol{\theta}^{\mathrm{T}}\boldsymbol{X}$ is greater than 0, the probability is greater than 0.5; If the value of $\boldsymbol{\theta}^{\mathrm{T}}\boldsymbol{X}$ is smaller than 0, the probability is less than 0.5.

7.3   Decision Boundary

1. Linear judgment boundary

Fig. 7-8 is a typical binary classification problem where a straight line can be found to separate the two types of points. How to solve this problem? First do a linear regression $h_\theta(x) = \theta_0 + \theta_1 x_1 + \theta_2 x_2$, and when $\theta_0 + \theta_1 x_1 + \theta_2 x_2 = 0$, a straight line $-3 + x_1 + x_2 = 0$ can be obtained, as shown in Fig. 7-8.

How to use this straight line to separate two types of points? Wrap $h_\theta(x)$ with a sigmoid function, $h_\theta(x) = g(\theta_0 + \theta_1 x_1 + \theta_2 x_2)$, where the probability of the points above the line $(-3 + x_1 + x_2 > 0)$ is greater than 0.5 (positive sample), while the probability of the points below the line $(-3 + x_1 + x_2 < 0)$ is smaller than 0.5 (negative sample).

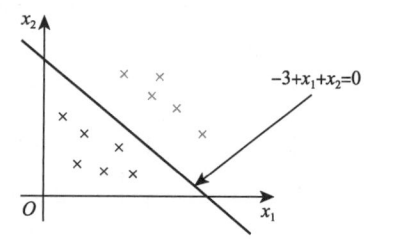

Fig. 7-8　Linear judgment boundary

This line $-3 + x_1 + x_2 = 0$ is the judgment boundary to be found, and it is linear, so it is called a linear judgment boundary.

2. Nonlinear judgment boundary

In practical use, it is often encountered that the sample is linearly indivisible. In this case, to distinguish sample points, a polynomial is needed. We can define a polynomial $h_\theta(x) = g(\theta_0 + \theta_1 x_1 + \theta_2 x_2 + \theta_3 x_1^2 + \theta_4 x_2^2)$, if a set $\theta = (-1, 0, 0, 1, 1)$ is taken, we can get $g(x_1^2 + x_2^2 - 1)$, which is a circle when $x_1^2 + x_2^2 - 1 = 0$. After the sample uses the sigmoid function, the probability is less than 0.5 within the circle, but greater than 0.5 for points outside the circle. $x_1^2 + x_2^2 - 1 = 0$ is just a nonlinear judgment boundary, as shown in Fig. 7-9.

$$h_\theta(x) = g(\theta_0 + \theta_1 x_1 + \theta_2 x_2 + \theta_3 x_1^2 + \theta_4 x_2^2)$$

Fig. 7-9　Nonlinear judgment boundary

From the above example, we can see that logistic regression is to find the decision boundary. If it is a high-dimensional polynomial, complex decision boundaries can be obtained.

As shown in Fig. 7-10, the horizontal axis represents one feature of the sample, while the vertical axis represents another feature of the sample. Different colors are used to represent different types of samples, γ represents the regularization coefficient. Fig. 7-10(a) shows that the decision boundary is suppressed by regularization. Fig. 7-10(b) does not use regularization, which can achieve a complex decision boundary.

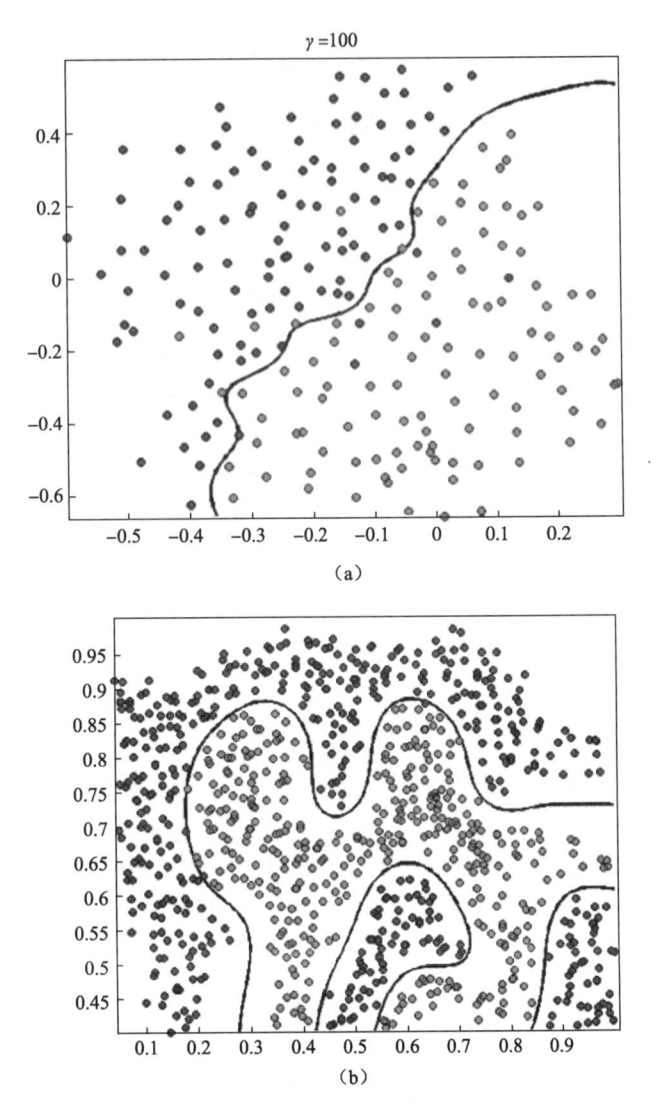

Fig. 7-10 Logistic regression seeking judgment boundary

7.4 Logistic Regression Loss Function

In logistic regression, one group θ can be trained from each group of sample points, each group θ can determine a decision boundary, a function needs to be defined to measure whether these decision boundaries are good or bad. In linear regression, MSE is used as the loss function. If logistic regression uses the loss function of linear regression, the curve shown in Fig. 7-11 can be available. This curve is a non-convex function, so logistic regression cannot use MSE as the loss function.

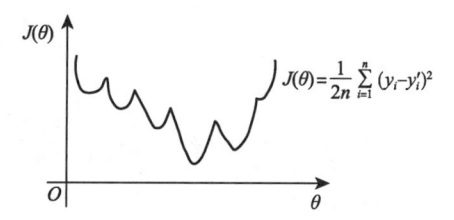

Fig. 7-11   Loss function of logistic regression using linear regression

In logistic regression, logarithmic function is usually used as loss function. See Fig. 7-12 for the function graph of the logarithmic function.

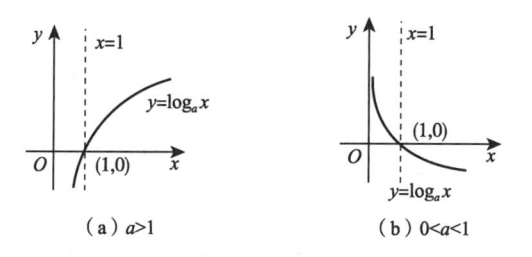

(a) $a>1$ (b) $0<a<1$

Fig. 7-12   The function graph of logarithmic function

The loss function of logistic regression is

$$\text{cost}(h_\theta(x), y) = \begin{cases} -\log(h_\theta(x)), & y=1 \\ -\log(1-h_\theta(x)), & y=0 \end{cases}$$

When $y=1$, $h_\theta(x)$ is a probability, where $h_\theta(x)=0.01$, which is when the model predicts negative samples in the case of positive samples, and then $-\log(h_\theta(x))$ is a large value, indicating a significant error, as shown in Fig. 7-13(a).

When $y=0$, $h_\theta(x)$ is a probability, where $h_\theta(x)=0.99$, which is when the model predicts a positive sample in the case of negative samples, and then $-\log(1-h_\theta(x))$ is a large value, indicating a significant error, as shown in Fig. 7-13(b).

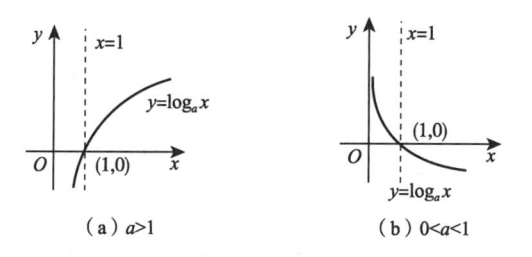

(a) $y=1$ (b) $y=0$

Fig. 7-13   Loss function of logistic regression

Generally, we can combine the loss function of the two cases to get the loss function - cross entropy loss function $J(\theta)$ of logistic regression via the formula as follows:

$$J(\theta) = \frac{1}{m} \sum_{i=1}^{m} \text{cost}(h_\theta(x), y)$$

$$J(\theta) = -\frac{1}{m} \Big[\sum_{i=1}^{m} y^i \log(h_\theta(x^i)) + (1 - y^i)\log(1 - h_\theta(x^i)) \Big]$$

In the above formula, the samples 1st to mth all have been calculated and the regularization term L2 is added to get the logistic regression loss function $J(\theta)$ with L2. The formula is as follows:

$$J(\theta) = -\frac{1}{m} \Big[\sum_{i=1}^{m} y^i \log(h_\theta(x^i)) + (1 - y^i)\log(1 - h_\theta(x^i)) \Big] + \frac{\lambda}{2m} \sum_{j=1}^{n} \theta_j^2$$

Apply gradient descent to loss function continuously, and seek partial derivative to obtain gradient descent function of logistic regression and determine the gradient value θ_j. The formula is as follows:

$$\theta_j := \theta_j - \alpha \sum_{i=1}^{m} (h_\theta(x^i) - y^i) x_j^i$$

$$\theta_j := \theta_j - \alpha \frac{\partial}{\partial \theta_j} J_\theta$$

Where, $h_\theta(x^i)$ represents the predicted value of the model; y^i represents the actual value; α represents the learning rate; $\theta_j:$ represents the new gradient value determined.

7.5   Code Implementation of Linear Decision Boundary

1. Load grade data

Currently there are test scores for two courses (Chinese and mathematics), with the x-axis showing the Chinese scores and the y-axis showing the mathematics scores. Now, it is necessary to determine whether the students' scores are qualified.

In Jupyter Notebook, import the minimize and PolynomialFeatures modules via the following code:

```
1.  import pandas as pd
2.  import numpy as np
3.  import matplotlib as mpl
4.  import matplotlib.pyplot as plt
5.  from scipy.optimize import minimize
6.  from sklearn.preprocessing import PolynomialFeatures
7.  pd.set_option('display.notebook_repr_html',False)
8.  pd.set_option('display.max_columns',None)
9.  pd.set_option('display.max_rows',150)
10. pd.set_option('display.max_seq_items',None)
11. #% config InlineBackend.figure_formats = {'pdf',}
```

```
12.  % matplotlib inline
13.
14.  import seaborn as sns
15.  sns.set_context('notebook')
16.  sns.set_style('white')
17.  #display Chinese
18.  plt.rcParams['font.sans-serif'] = ['SimHei']
19.  plt.rcParams['axes.unicode_minus'] = False
```

Read data, define the function loaddata( ) to read files, and define the function plotData( ) to plot sample points. Read the data code as follows:

```
1.  def loaddata(file,delimeter):
2.      data = np.loadtxt(file,delimiter = delimeter)
3.      print('Dimensions:',data.shape)
4.      print(data[1:6,:])
5.      return(data)
6.  def plotData(data,label_x,label_y,label_pos,label_neg,axes = None):
7.      # Obtain subscripts for positive and negative samples (i.e.
    which ones are positive and which ones are negative)
8.      neg = data[:,2] = =0
9.      pos = data[:,2] = =1
10.
11.     if axes = =None:
12.         axes = plt.gca()
13.     axes.scatter(data[pos][:, 0], data[pos][:, 1], marker = ' +', c =
    'k', s =60, linewidth =2, label = label_pos)
14.     axes.scatter(data[neg][:,0],data[neg][:,1],c = 'y',s = 60,
    label = label_neg)
15.     axes.set_xlabel(label_x)
16.     axes.set_ylabel(label_y)
17.     axes.legend(frameon = True,fancybox = True);
18. data = loaddata('data1.txt',',')
19. X = np.c_[np.ones((data.shape[0],1)),data[:,0:2]]
20. y = np.c_[data[:,2]]
21. plotData(data, 'Chinese score', 'mathematics score', 'passed', 'not
    passed')
```

Run the code and it will output a scatter plot of Chinese and math ematics scores, as shown in Fig. 7-14.

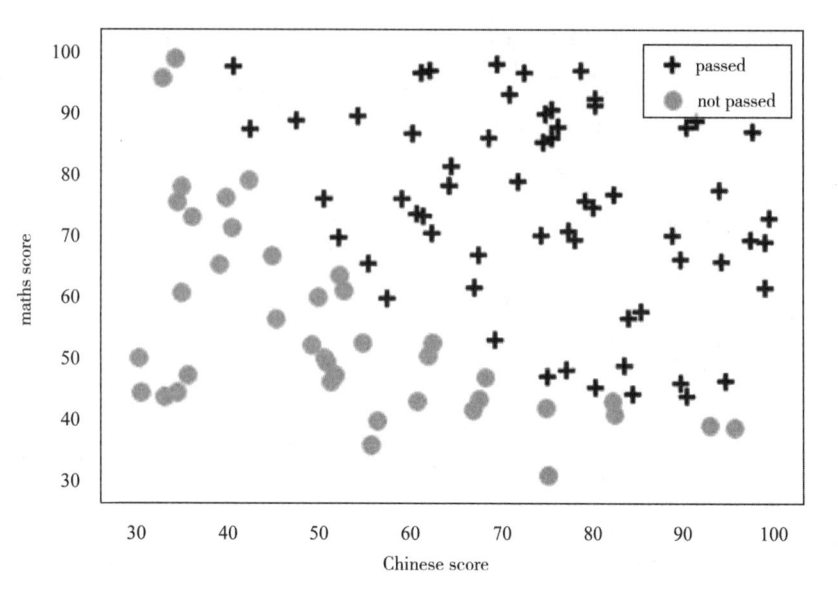

Fig. 7-14   Grade scatter plot

2. Model definition

Define logistic regression, define loss function, define gradient descent, and call loss function and gradient descent via the codes as follows:

```
1.  # define sigmoid function
2.  def sigmoid(z):
3.      return(1 / (1 + np.exp(-z)))
4.  # define loss function
5.  def costFunction(theta, X, y):
6.      m = y.size
7.      h = sigmoid(X.dot(theta))
8.
9.      J = -1* (1/m)* (np.log(h).T.dot(y) + np.log(1-h).T.dot(1-y))
10.
11.     if np.isnan(J[0]):
12.         return(np.inf)
13.     return(J[0])
14. # solve gradient
15. def gradient(theta, X, y):
16.     m = y.size
```

```
17.    h = sigmoid (X.dot (theta.reshape ( -1,1)))
18.
19.    grad = (1/m)*  X.T.dot (h - y)
20.
21.    return (grad.flatten ())
22. initial_theta = np.zeros (X.shape [1])
23. cost = costFunction (initial_theta, X, y)
24. grad = gradient (initial_theta, X, y)
25. print ('Cost: \n', cost)
26. print ('Grad: \n', grad)
```

3. Model training

Minimize the loss function, define and call the prediction function, and plot the decision boundary. The code is as follows:

```
1.    # Directly call the minimize function of the minimized loss
      function in scipy
2.    res = minimize (costFunction, initial_theta, args = (X,y), method
      = None, jac = gradient, options = {'maxiter':400})
3.    res
4.    def predict (theta, X, threshold = 0.5):
5.        p = sigmoid (X.dot (theta.T)) > = threshold
6.        return (p.astype ('int'))
7.
8.    #students who scored 45 points in the first course and 85 points in
      the second course
9.    sigmoid (np.array ([1, 45, 85]).dot (res.x.T))
10.
11. p = predict (res.x, X)
12. print ('Train accuracy {}% '.format (100 *  sum (p = = y.ravel ())/
      p.size))
13. plt.scatter (45, 85, s = 60, c = 'r', marker = 'v', label = '(45, 85)')
14. #plotData (data, 'Exam 1 score', 'Exam 2 score', 'Pass', 'Failed')
15. plotData (data, 'Chinese score ', 'mathematics score ', 'passed ',
      'not passed')
16. x1_min, x1_max = X[:,1].min (), X[:,1].max (),
17. x2_min, x2_max = X[:,2].min (), X[:,2].max (),
```

```
18. xx1, xx2 =np.meshgrid(np.linspace(x1_min, x1_max), np.linspace
    (x2_min, x2_max))
19. h =sigmoid(np.c_[np.ones((xx1.ravel().shape[0], 1)), xx1.ravel
    (),xx2.ravel()].dot(res.x))
20. h =h.reshape(xx1.shape)
21. plt.contour(xx1, xx2, h, [0.6], linewidths =1, colors ='b');
```

7.6   Code Implementation of Nonlinear Decision Boundary

Firstly, use PolynomialFeatures to generate multiple features, and then use regularization Lambda =0, Lambda =1 to plot a nonlinear decision boundary. The code is as follows:

```
1.  # read data
2.  data2 =loaddata('data2.txt', ',')
3.  # split data
4.  y =np.c_[data2[:,2]]
5.  X =data2[:,0:2]
6.  # plot a scatter chart
7.  plotData (data2, 'Chinese score', 'mathematics score', 'y =passed',
    'y =not passed')
8.  # feature mapping via the code as follows: perform feature mapping
    and generate polynomial features, with the maximum degree up to 6
9.  poly =PolynomialFeatures(6)
10. XX =poly.fit_transform(data2[:,0:2])
11. XX.shape
12. # define the loss function via the code as follows
13. def costFunctionReg(theta, reg, * args):
14.     m =y.size
15.     h = sigmoid(XX.dot(theta))
16.
17.     J = -1* (1/m)* (np.log(h).T.dot(y) +np.log(1 -h).T.dot(1 -
        y)) + (reg/(2* m))* np.sum(np.square(theta[1:]))
18.
19.     if np.isnan(J[0]):
20.         return(np.inf)
```

```
21.      return(J[0])
22. # define the gradient descent function via the code as follows
23. def gradientReg(theta, reg, * args):
24.     m = y.size
25.     h = sigmoid(XX.dot(theta.reshape(-1,1)))
26.
27.     grad = (1/m) * XX.T.dot(h - y) + (reg/m) * np.r_[[[0]],theta
    [1:]. reshape(-1,1)]
28.
29.     return(grad.flatten())
30. # train the model via the code as follows
31. initial_theta = np.zeros(XX.shape[1])
32. costFunctionReg(initial_theta, 1, XX, y)
33.
34. fig, axes = plt.subplots(1,3, sharey = True, figsize = (17,5))
35.
36. # what happens to the decision boundary when the regularization
    coefficient Lambda is too large or too small
37. # Lambda = 0: no regularization exists and overfitting occurs
38. # Lambda = 1: this is the correct open mode
39. # Lambda = 100: over regularization, so that no decision boundary
    has been fitted
40.
41. for i, C in enumerate([0, 1, 100]):
42.     # optimal costFunctionReg
43.     res2 = minimize(costFunctionReg, initial_theta, args = (C, XX, y),
    method = None, jac = gradientReg, options = {'maxiter':3000})
44.
45.     # accuracy rate
46.     accuracy = 100* sum(predict(res2.x, XX) = =y.ravel())/y.size
47.
48.     # X, y hash plot
49.     plotData(data2, 'Microchip Test 1', 'Microchip Test 2', 'y =1',
    'y =0', axes.flatten()[i])
50.
51.     # plot the decision boundary
```

```
52.      x1_min, x1_max = X[:,0].min(), X[:,0].max(),
53.      x2_min, x2_max = X[:,1].min(), X[:,1].max(),
54.        xx1, xx2 = np.meshgrid(np.linspace(x1_min, x1_max), np.
      linspace(x2_min, x2_max))
55.      h = sigmoid(poly.fit_transform(np.c_[xx1.ravel(), xx2.ravel
      ()]).dot(res2.x))
56.      h = h.reshape(xx1.shape)
57.        axes.flatten()[i].contour(xx1,xx2,h,[0.5],linewidths = 1,
      colors = 'g');
58.        axes.flatten()[i].set_title('Train accuracy {}% with Lambda
      = {}'.format(np.round(accuracy, decimals = 2), C))
```

Run the code and it will output the nonlinear decision boundary. In Fig. 7-15 (a), no regularization coefficient is used, resulting in a train accuracy of 91.53%; in Fig. 7-15 (b), the regularization coefficient is set to 1, resulting in a train accuracy of 83.05%. As can be seen from Fig. 7-15, overfitting occurs in Fig. 7-15 (a).

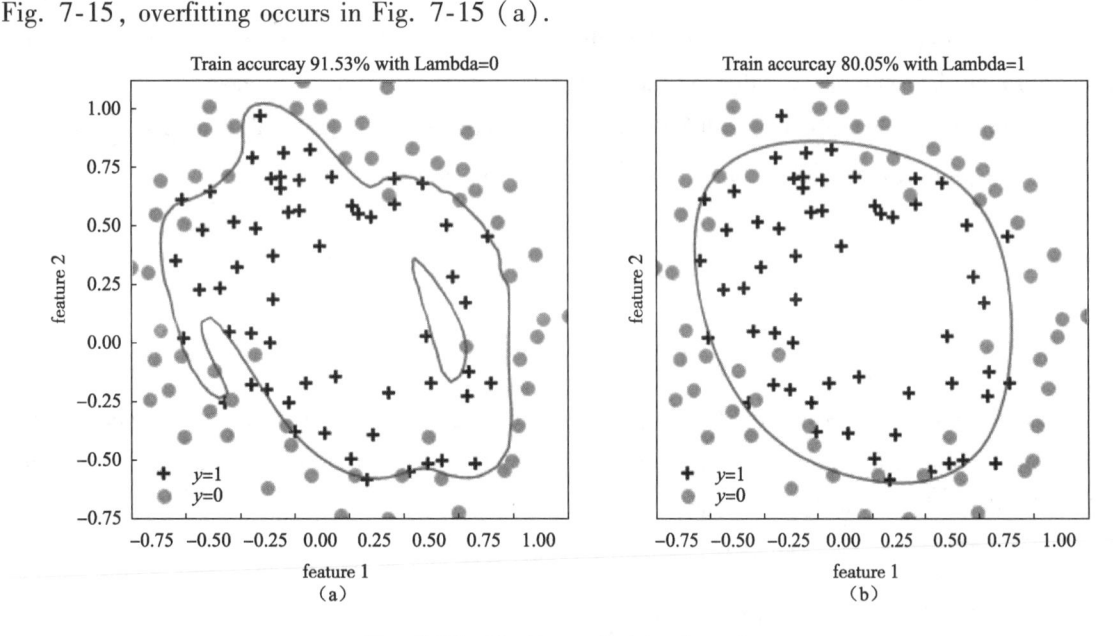

Fig. 7-15   Nonlinear decision boundary

Classification of Heart Disease Patients via Logistic Regression

Heart disease is the number one killer of human health. If we can extract relevant body measurement indicators from the human body and analyze the impact of different features on heart disease through data mining, it will play a crucial role in preventing heart disease.

The dataset in this unit contains 303 pieces of physical examination data of heart disease patients in a certain region of the United States, with a total of 14 features.

8.1 Data Loading and Display

1. Import data load packages

Import numpy, pandas and matplotlib packages into Jupyter Notebook to load the heart disease patient dataset and analyze the data; use read_csv to open the CSV file and use the info( ) function to view the data. The code is as follows:

```
1.   import numpy as np
2.   import pandas as pd
3.   import seaborn as sns
4.   import matplotlib.pyplot as plt
5.   # display Chinese
6.   plt.rcParams['font.sans-serif'] = ['SimHei']
7.   plt.rcParams['axes.unicode_minus'] = False
8.   df_data = pd.read_csv("../data/heart.csv")
9.   df_data.head()
```

After running the code, it will output the heart disease data in heart. csv, with a total of 14 features and 303 samples, as shown in Fig. 8-1.

	age	sex	cp	trestbps	chol	fbs	restecg	thalach	exang	oldpeak	slope	ca	thal	target
0	63	1	3	145	233	1	0	150	0	2.3	0	0	1	1
1	37	1	2	130	250	0	1	187	0	3.5	0	0	2	1
2	41	0	1	130	204	0	0	172	0	1.4	2	0	2	1
3	56	1	1	120	236	0	1	178	0	0.8	2	0	2	1
4	57	0	0	120	354	0	1	163	1	0.6	2	0	2	1

Fig. 8-1   Display results of heart disease data

See Table 8-1 for the meaning of the fields in the heart disease dataset.

Table 8-1   Meaning of fields in the heart disease dataset

Field Name	Type	Descriptions
age	STRING	Age
sex	STRING	Gender, with values of either female or male
cp	STRING	Types of chest pain, from severe to mild, are typical, atypical, non-anginal, and asymptomatic
trestbps	STRING	Blood pressure
chol	STRING	Cholesterol
fbs	STRING	Fasting blood glucose. If the blood glucose content is greater than 120 mg/dL, the value is true; otherwise, the value is false
restecg	STRING	Whether there are T waves in the electrocardiogram results, from mild to severe, norm and hyp
thalach	STRING	Maximum heartbeats
exang	STRING	Does the patient have angina pectoris? "True" indicates angina pectoris; "false" indicates no angina pectoris
oldpeak	STRING	Motion ST Depression relative to resting, i. e. ST segment pressure value
slope	STRING	Inclination of the electrocardiogram ST Segment, with values including down, flat, and up
ca	STRING	Number of main blood vessels found in fluoroscopic examination
thal	STRING	Types of diseases, from mild to severe, are normal, fix, and rev
target	STRING	Get sick or not? buff indicates health, while sick indicates illness

8.2   Data Analysis

1. Analysis of data relationship coefficients

Use seaborne's heatmap( ) to plot a relationship coefficient graph, with the following code:

```
1.  plt.figure(figsize = (10,10))
2.  sns.heatmap(df_data.corr(),cmap = "YlGnBu",annot = True)
```

After the code is run, it will generate a heat map of the correlation coefficients of the heart disease dataset features, as shown in Fig. 8-2. As can be seen from the results:

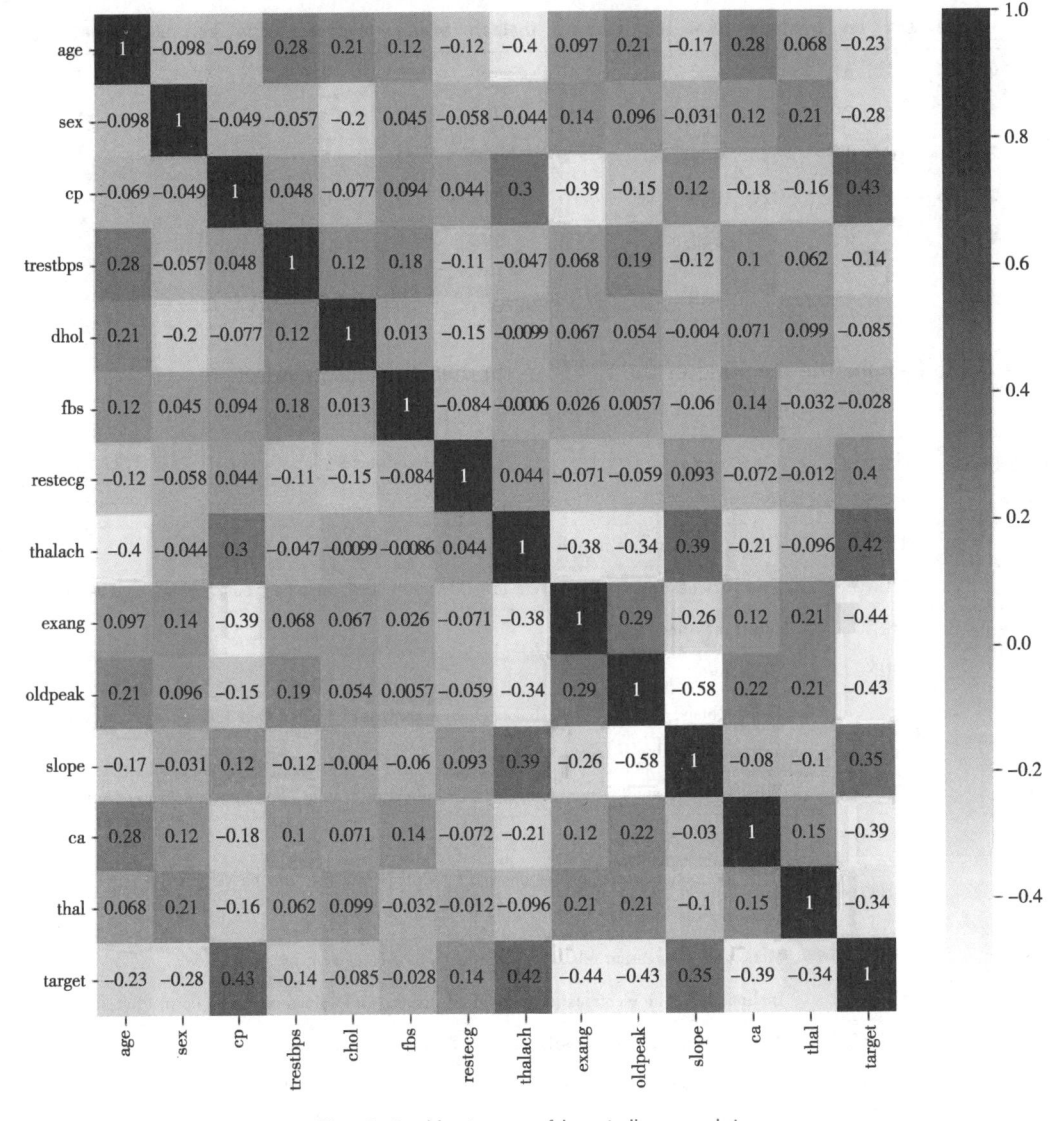

Fig. 8-2　Heat map of heart disease data

① There is a strong correlation between heart disease and several features such as cp (chest pain type), thalach (maximum heart rate), exang (presence or absence of angina), and oldpeak (ST segment pressure value), with the correlation coefficient exceeding 0.4.

② There is a certain correlation between heart disease and such features as age, sex, slope (slope of ST segment in electrocardiogram), ca (number of main blood vessels), and tal (type of disease).

③ The correlation is weak between heart disease and chol (cholesterol) and fbs (fasting blood

glucose).

2. Draw an age-blood pressure scatter plot

As can be seen from the heat map, there is a correlation between age-blood pressure and heart disease. The seaborne's scatterplot ( ) can be used to draw the scatter plot between the three, with the code as follows:

```
1.  df_heart_target = df_data[['age','thalach','target']]
2.  markers = {'heart disease patients':'s', 'health': 'x'}
3.  sns.scatterplot(x = 'age', y = 'thalach', hue = 'target', style =
    'target ',data = df_heart_target)
4.  plt.legend(title = 'Age - Blood Pressure - Heart Disease', labels
    = ['heart disease patient', 'health'])
5.  plt.show()
```

After running the code, as can be seen, there are more heart disease patients between the ages of 35 and 50, and their blood pressure is on the high side. Therefore, blood pressure is an important feature in determining whether to have heart disease or not. See Fig. 8-3 for the scatter plot.

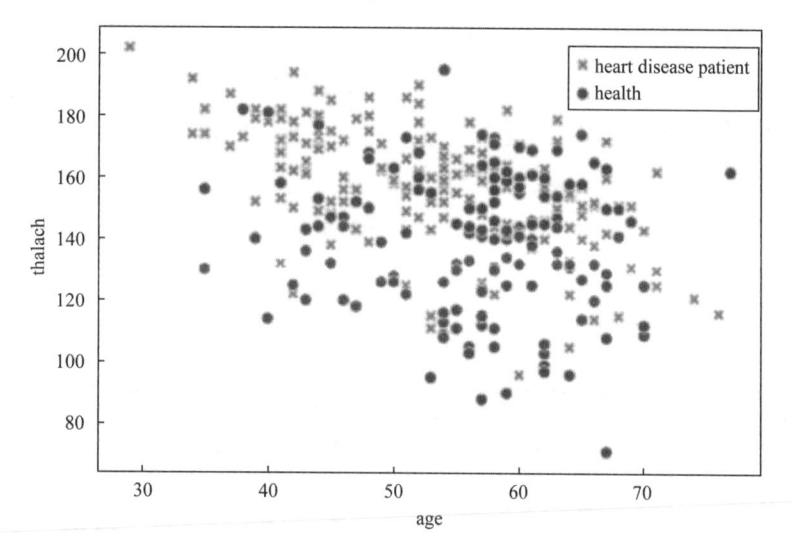

Fig. 8-3 Age-blood pressure scatter plot

8.3 Feature Data Processing

1. Processing of type data

The cp, restecg, slope, and thal features are non-continuous classification data of the type data. The get_dummies( ) encoding method can be used to processes non-continuous classification data,

with the following code:

```
1.  df_data.cp.unique()
2.  df_data.cp.value_counts()
3.
4.  df_data.restecg.unique()
5.  df_data.restecg.value_counts()
6.
7.  df_data.slope.unique()
8.  df_data.slope.value_counts()
9.
10. df = df_data[['cp','thal','slope']]
11. df.apply(pd.value_counts)
12.
13. a = pd.get_dummies(df_data['cp'],prefix = 'cp')
14. b = pd.get_dummies(df_data['thal'],prefix = 'thal')
15. c = pd.get_dummies(df_data['slope'],prefix = 'slope')
16  df_heart_all = [df_data,a,b,c]
17. df_heart = pd.concat(df_heart_all,axis =1)
18. df_heart = df_data.drop(columns = ['cp','thal','slope'])
19. df_heart.head()
```

After running the code, continue to perform one-hot encoding on the three types of features in the dataset: cp, thal, and slope, and generate new feature fields with the results as shown in Fig. 8-4.

	age	sex	trestbps	chol	fbs	restecg	thalach	exang	oldpeak	ca	...	cp_1	cp_2	cp_3	thal_0	thal_1	thal_2	thal_3	slope_0	slope_1	slope_2
0	63	1	145	233	1	0	150	0	2.3	0	...	0	0	1	0	1	0	0	1	0	0
1	37	1	130	250	0	1	187	0	3.5	0	...	0	1	0	0	0	1	0	1	0	0
2	41	0	130	204	0	0	172	0	1.4	0	...	1	0	0	0	0	1	0	0	0	1
3	56	1	120	236	0	1	178	0	0.8	0	...	1	0	0	0	0	1	0	0	0	1
4	57	0	120	354	0	1	163	1	0.6	0	...	0	0	0	0	0	1	0	0	0	1

5 rows × 22 columns

Fig. 8-4　Processing results from "cp, thal, and slope" classification data

2. Dataset split

Firstly, remove the label, and then divide the data into two parts by splitting the components in accordance with the ratio 8 : 2 between the training set and the prediction set. The training set data is used for the training of the logistic regression binary model, and the prediction set data is used for the

prediction on the training model. The code is as follows:

```
1.  X =df_data.drop(['target'],axis =1)
2.  y =df_data.target.values
3.  y =y.reshape(-1,1)
4.  X.shape,y.shape
5.
6.  from sklearn.model_selection import train_test_split
7.  from sklearn.preprocessing import MinMaxScaler
8.
9.  X_train,X_test,y_train,y_test =train_test_split(X,y,test_size =
    0.2)
```

3. Normalization operation

Use Min-Max Scaling normalization to range the numerical value of each feature from 0 to 1 and then we can remove the influence of dimensionality on the results. In sklearn, use preprocessing. MinMaxScaler to implement this function, and the code is as follows:

```
1.  scaler =MinMaxScaler()
2.  X_train =scaler.fit_transform(X_train)
3.  X_test =scaler.fit_transform(X_test)
```

8.4   Logistic Regression Modeling

1. sigmoid function

The output result of logistic regression is to determine the dichotomy, which can be used to solve the dichotomy problem in practice. The key to logistic regression is to understand the sigmoid function, also known as Logistic function, whose expression is as follows:

$$h(x) = \frac{1}{1 + e^{-\theta^T X}}$$

The output of the Logistic function is a number between 0 and 1. Consider the output value of this function as a probability of belonging to Class 1. If the data is in Class 1, make the value of this function close to 1; if this data is in Class 0, let the value with 1 subtracted by this function close to 1.

The sigmoid function is used to implement dichotomy problems. Use it to work out a probability based on the calculation result and compare it with the threshold to determine whether it belongs to a certain class. The code is as follows:

```
1.  def sigmoid(z):
2.      y_hat =1/(1 +np.exp(-z))
3.      return y_hat
```

2. Loss function

The sigmoid function represents the probability of taking 1 from 0 to 1. Therefore, the loss function can be defined as:

When $y = 0$, $\text{cost} = -\log(1 - h(x))$;

When $y = 1$, $\text{cost} = -\log(h(x))$.

Let the loss function correspond to the law of the 0-1 distribution and set $p = h(x)$; the loss function is to take the logarithm based on the 0-1 distribution and then take the negative number. The loss function requires that the closer the predicted result is to the real result, the smaller the function value, so a negative sign will be added before it. When $y = 0$, the probability of $1 - p$ will be relatively high, and if a negative sign is added before it, the cost value will be very small; when $y = 1$, the probability of p will be relatively high, and if a negative sign is added before it, the cost value will be very small. As for taking logarithm, it is related to the maximum likelihood function. Taking logarithm does not affect the monotonicity of the original function, but will amplify the difference between probabilities and better distinguish the categories of each sample. The loss function code of logistic regression is implemented as follows:

```
1.  def loss_function(X,y,w,b):
2.      y_hat = sigmoid(np.dot(X,w) +b)
3.      loss = - (y* np.log(y_hat) + (1 -y)* np.log(1 -y_hat))
4.      cost =np.sum(loss)/X.shape[0]
5.      return cost
```

3. Define gradient descent function

The gradient descent function of logistic regression is the same as that of linear regression and it is macro differential; then multiply the calculated derivative by the learning rate, and update w and b iteratively. By seeking partial derivative, a gradient can be obtained. The gradient descent function has the code as follows:

```
1.  l_history =np.zeros(iter)
2.  w_history =np.zeros((iter,w.shape[0],w.shape[1]))
3.  b_history =np.zeros(iter)
4.  for i in range(iter):
5.      y_hat = sigmoid(np.dot(X,w) +b)
6.      loss = - (y* np.log(y_hat) + (1 -y)* np.log(1 -y_hat))
```

```
7.        derivative_w = np.dot(X.T,((y_hat - y)))/X.shape[0]
8.        derivative_b = np.sum(y_hat - y)/X.shape[0]
9.        w = w - lr* derivative_w
10.       b = b - lr* derivative_b
11.       l_history[i] = loss_function(X,y,w,b)
12.       print("round",i + 1," Current round training and loss:",l_
          history[i])
13.       w_history[i] = w
14.       b_history[i] = b
15. return l_history,w_history,b_history
```

4. Predict classification results

Define the results from the classification via the predict( ) function. Here a threshold is defined with the probability of 0.5. The samples above the threshold can be predicted as 1, while the samples below the threshold can be predicted as 0. The code is as follows:

```
1.   def predict(X,w,b):
2.        z = np.dot(X,w) + b
3.        y_hat = sigmoid(z)
4.        y_pred = np.zeros((y_hat.shape[0],1))
5.        for i in range(y_hat.shape[0]):
6.            if y_hat[i,0] < 0.5:
7.                y_pred[i,0] = 0
8.            else:
9.                y_pred[i,0] = 1
10.       return y_pred
```

5. Define models

Define the logistic_regression( ) function train model, initialize the parameters to start training, complete 500 rounds of training, and output the final loss of 0.331 373 with the accuracy of 87.19%, subject to the code as follows:

```
1.   def logistic_regression(X,y,w,b,lr,iter):
2.        l_history,w_history,b_history = gradient_descent(X,y,w,b,lr,
          iter)
3.        print("training final loss:",l_history[-1]) #print final loss
4.        y_pred = predict(X,w_history[-1],b_history[-1]) # make a
          prediction
```

```
5.      traning_acc = 100 - np.mean(np.abs(y_pred - y_train)) * 100
        # calculation accuracy
6.       print ("logistic regression training accuracy: {:. 2f}% ".
        format (traning_acc)) #print accuracy
7.        return l_history, w_history, b_history # return training
        history
8.   # initialization parameters
9.   dimension = X.shape[1]
10.  weight = np.full((dimension,1),0.1) # weight vector, usually 1D, but
        here a 2D tensor is actually created
11.  bias = 0 # bias value
12.  #initialize hyperparameters
13.  alpha = 1 # learning rate
14.  iterations = 500 # number of iterations
15.  # train the machine with logistic regression function
16.  loss_history, weight_history, bias_history = logistic_regression (X_
        train,y_train,weight,bias,alpha,iterations)
```

8.5 Model Testing

After the model training is completed, the model shall be tested with the test set to output the model acc and plot the loss curve. The code is as follows:

```
1.  y_pred = predict(X_test,weight_history[-1],bias_history[-1])
2.  testing_acc = 100 - np.mean(np.abs(y_pred - y_test)) * 100
3.  print ("logistic regression test accuracy: {:. 2f}% ". format
        (testing_acc))
4.  loss_history_test = np.zeros(iterations) # initialize historical
        losses
5.  for i in range(iterations): # calculate the test set loss caused by
        different parameters during training
6.  loss_history_test[i] = loss_function(X_test,y_test,
7.                          weight_history[i],bias_history [i])
8.  index = np.arange(0,iterations,1)
9.  plt.plot(index, loss_history,c = 'blue',linestyle = 'solid')
```

```
10. plt.plot(index, loss_history_test,c = 'red',linestyle = 'dashed')
11. plt.legend(["Training Loss", "Test Loss"])
12. plt.xlabel("Number of Iteration")
13. plt.ylabel("Cost")
14. plt.show() # Display loss curves for both training and test sets
    simultaneously
```

Run the code and it will output the loss curves for the training and test sets, as shown in Fig. 8-5. The horizontal axis Number of Iteration represents the number of training iterations, and the vertical axis Cost represents the loss. As can be seen from Fig. 8-5, after iterated to 80-100 times, the Training Loss (training set loss) will further decrease and become smaller. However, the Test Loss (test set loss) does not decrease, but instead shows an upward trend, which is obvious overfitting. Therefore, the optimal number of iterations should be kept between 80 and 100.

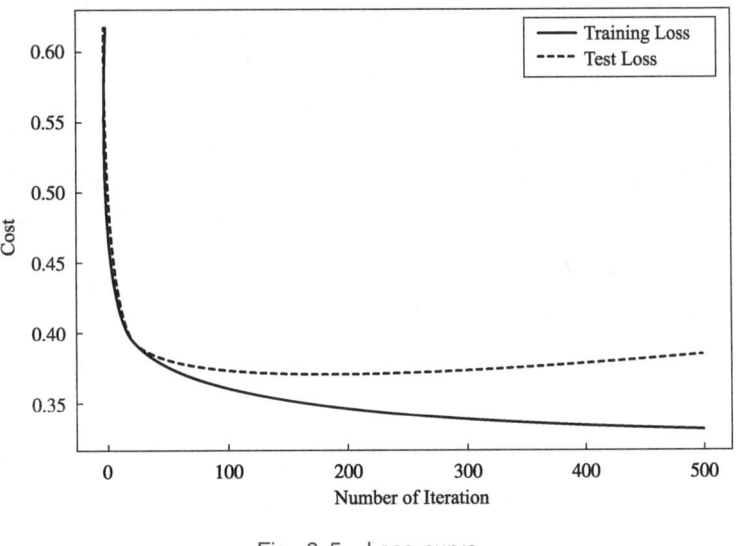

Fig. 8-5　Loss curve

8.6　Use sklearn to Realize Logistic Regression

1. Use sklearn to establish a logistic regression model

In practical work, the previous methods will not be used for modeling, but instead models are directly called from the sklearn library. The following code directly calls the LogisticRegression() function to achieve logistic regression, and then uses the score() function to calculate the accuracy rate and the training set score subject to the code as follows:

```
1.  from sklearn.linear_model import LogisticRegression # import
    logistic regression model
2.  lr = LogisticRegression () # lr, which represents the logistic
    regression model
3.  lr.fit(X_train,y_train) # fit, which is equivalent to a gradient
    descent
4.  print ("sklearn logistic regression test accuracy {:.2f}%".
    format (lr.score (X_test, y_test) * 100))
```

Calculate the accuracy of classification and the test set score, subject to the code as follows:

```
1.  lr.score(X_test,y_test)
2.  from sklearn.metrics import accuracy_score
3.  y_pre_lr = lr.predict(X_test)
4.  accuracy_score(y_test,y_pre_lr)
```

2. Usage of logisticRegression( ) function in sklearn

(1) Basic syntax

LogisticRegression(C = 1. 0, class_weight = None, dual = False, fit_intercept = True, intercept_ scaling = 1, max_iter = 100, multi_class = 'ovr', n_jobs = 1, penalty = 'L2', random_state = None, solver = 'liblinear', tol = 0. 0001, verbose = 0, warm_start = False)

(2) Main parameter descriptions

penalty: Penalty term, str type, optional parameters L1 and L2, and L2 by default. Used to specify the specifications used in penalty terms.

dual: Dual or primitive method, bool type, False by default. The dual method is only used to solve the L2 penalty terms of linear multi-core (liblinear). When the sample size > sample features, dual is usually set to False.

tol: The standard to stop solving, float type, $1e - 4$ by default. When the solution reaches a certain point, it is considered that the optimal solution has been found.

c: Reciprocal of the regularization coefficient λ, float type, 1. 0 by default. It must be a positive floating-point number. Like SVM, smaller values represent stronger regularization.

fit_iIntercept: Whether there is an intercept or deviation, bool type, True by default.

intercept_scaling: Useful only when the regularization term is "liblinear" and the fit_intercept is set to True. Float type, 1 by default.

class_weight: Used to indicate various types of weights in the classification model, which can be a dictionary or a 'balanced' string. No entry by default, which means that no weight is considered, namely "None".

8.7   Use Grid Search for Hyperparameter Tuning

1. Initialize model parameter range

The parameters in machine learning are variables learned from training data, such as weights and biases in the network. And hyperparameters are used to determine some parameters of the model. Different hyperparameters result in different models. Hyperparameters generally need to be determined based on experience, and these parameters cannot be directly learned from data and need to be set before model training.

For example, in the LogisticRegression model, C, penalty, class_weight and other parameters; in deep learning, the learning rate, number of iterations, number of layers, and number of neurons in each layer are all hyperparameters.

For the LogisticRegression model, first determine the range of hyperparameters based on experience, and then determine the optimal value by searching for all points within the search range. Here select such three hyperparameters as C, penalty, and class_weight, subject to the code as follows:

```
1.  parm_grid = [
2.      {
3.          'C':[0.01,0.1,1,10,100],
4.          'penalty':['L2','L1',],
5.          'class_weight':['balanced',None] #if the parameter is
        'balanced', the weight will be automatically adjusted according to
        the frequency of the input sample
6.      }
7.  ]
```

2. Use grid search parameters

Grid search: A parameter tuning method. Exhaustive search: Among all candidate parameters, by iterating through each possibility, the best performing parameter is the final result. The principle is like finding the maximum value in an array. Why is it called grid search? Taking two parameters model as an example, the parameter a has 3 possibilities and the parameter b has 4 possibilities. List all the possibilities represented as a 3 ×4 form, where each cell is a grid, and the loop process is like traversing and searching within each grid, so it is called network search. You can use the GridSearchCV function provided by sklearn to achieve grid search.

（1）Basic syntax of GridSearchCV

sklearn. model_selection. GridSearchCV(estimator,param_grid,scoring,refit,cv,n_jobs)

（2）Main parameter descriptions

estimator: Select the classifier to use and import other parameters besides the one that needs to be determined as the best.

param_grid: It needs the value of the parameter to be optimized, namely a dictionary or list.

scoring: Model evaluation criteria. The evaluation criteria vary depending on the selected model. The default value is "None", indicating the error estimation function using the estimator.

n_jobs: The number of processes used in model training, 1 by default. If the value is −1, all CPUs are used for the operation; if the value is 1, parallel operations are not performed for convenient debugging.

refit: "True" by default; the program will use the best parameter obtained from the cross validation of the training set as the final optimal model parameter for performance evaluation. After searching for parameters, train all datasets once again with the best parameter results.

cv: Cross validation parameter, "None" by default, using triple folded cross validation.

3. Use of GrideSearchCV

Use the fit method of GridSearchCV to train the model, and then use. best_estimator_to output the optimal parameters with the following code:

```
1.  from sklearn.model_selection import GridSearchCV
2.  gride_search =GridSearchCV(lr,parm_grid,cv =10,n_jobs = -1)
3.  % % time
4.  gride_search.fit(X_train,y_train)
5.  gride_search.best_estimator_
```

Run the code and it will output the optimal parameters of the current model, $C = 0.01$, class_weight = 'balanced' and penalty = 'L2'.

8.8 Measurement Indicators for Logistic Regression Model

1. Confusion matrix

The confusion matrix is a situation analysis table that summarizes the classification model and prediction results in machine learning, and summarizes the records in the dataset in the form of a matrix according to the two criteria of real category and category judgment predicted by the classification model. Among them, the rows of the matrix represent the true values while the columns of the matrix represent the predicted values. Let's take the dichotomy as an example to see the

representation of the confusion matrix, as shown in Fig. 8-6, where a, b, c, d represent the number of samples.

confusion matrix		predicted value	
		P	N
true value	P	a	b
	N	c	d

Fig. 8-6   Confusion matrix

The confusion matrix contains four categories: TP, FN, FP and TN, as shown in Fig. 8-7. The meanings of the four categories are as follows:

TP (true positive): It can predict a positive class as a positive one; 0 for true and then 0 for prediction.

FN (false negative): It can predict a positive class as a negative one; 0 for true and then 1 for prediction.

FP (false positive): It can predict a negative class as a positive one; 1 for true and then 0 for prediction.

TN (true negative): It can predict a negative class as a negative one; 1 for true and then 1 for prediction.

2. Indicators for confusion matrix evaluation

(1) Correctness/Accuracy

Correctness refers to the proportion or quantity of correctly classified samples. The accuracy is defined as the percentage of correctly predicted results in the total samples, as shown in Fig. 8-7.

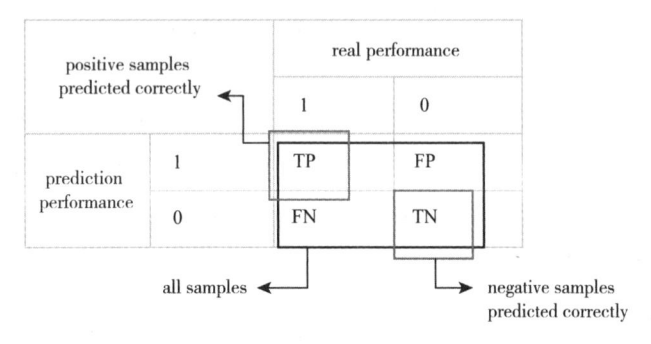

Fig. 8-7   Accuracy chart

The formula is as follows:

$$\text{Accuracy} = \frac{\text{TP} + \text{TN}}{\text{TP} + \text{FP} + \text{TN} + \text{FN}}$$

Although accuracy can determine the overall correctness, it cannot be used as a good indicator to measure the results in case of imbalanced samples. For example, in total samples, if positive samples account for 90% and negative samples account for 10%, the samples are severely imbalanced. For this situation, it is only necessary to predict all samples as positive samples to obtain a high accuracy of 90%, but in reality, this classification is ineffective. This indicates that due to the problem of imbalanced samples, the high accuracy results obtained contain a lot of

moisture. If the samples are imbalanced, the accuracy will fail. Consequently, two other indicators have been derived: accuracy and recall rate.

(2) Precision

Precision (accuracy/precision/precision ratio) is an evaluation of the "TP (true positive rate)" prediction. That is, in the data predicted positive (denominator TP + FP), how much is actually correct (numerator TP). In this example, it refers to the proportion of accurately predicted cases of heart disease, representing the proportion of the actual positive cases in the predicted cases predicted positive (TP + FP), as shown in Fig. 8-8.

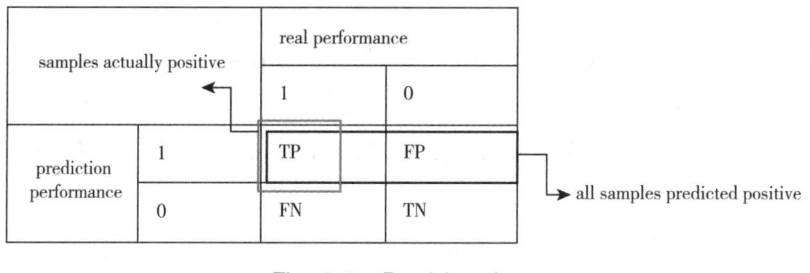

Fig. 8-8 Precision chart

The formula is as follows:

$$\text{Precision} = \frac{\text{TP}}{\text{TP} + \text{FP}}$$

(3) Recall

Recall also known as recall rate, is specific to the original sample and refers to the probability of being predicted as a positive sample in the actual positive samples. The recall rate is concerned with the guarantee of predicting positive cases, which is a problem of selecting positive cases from positive cases, as shown in Fig. 8-9.

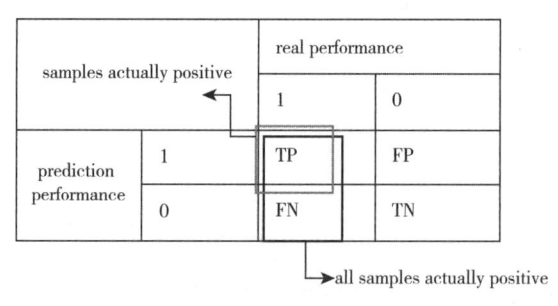

Fig. 8-9 Recall chart

The formula is as follows:

$$\text{Recall} = \frac{\text{TP}}{\text{TP} + \text{FN}}$$

TP/ (TP + FN) refers to the actual positive P (denominator TP + FN), where the proportion of

those predicted correct (numerator TP).

Application scenario of recall rate: Taking the default rate of online loans as an example, we are more concerned about bad users and cannot miss any bad users relative to good users. If too many bad users are treated as good users, the potential default amount that may occur in the future will far exceed the loan interest amount repaid by good users, resulting in serious losses. The higher the recall rate, the higher the probability of actual bad users being predicted.

Ideally, the higher the accuracy and recall, the better. However, in fact, these two are contradictory in some cases; when the accuracy is high, the recall rate is low; when the accuracy is low, the recall rate is high instead.

For example, there are 10 000 people who have a certain type of cancer predicted, as shown in Table 8-2.

Table 8-2 Hypothetical samples

True value/prediction result	Sample label	
	0	1
0	9990	0
1	10	0

$$Accuracy = \frac{9990}{9990 + 10} = 99.9\%$$

$$Precision = 0\%$$

$$Recall = 0\%$$

The incidence rate of this cancer itself is only 0.1%, even if we directly predict that everyone is healthy without training the model, this prediction accuracy can reach 99.9%, but the accuracy and recall are 0%, so this model is invalid.

(4) F1 score

In dichotomy, accuracy and recall are usually used to evaluate the analytical performance of dichotomy models. However, there is a conflict between these two indicators. When the accuracy is high, the recall rate is low, and vice versa. Moreover, it is difficult to compare between models when these two indicators conflict. For example, there are two models A and B. The recall rate of Model A is higher than that of Model B, but the accuracy of Model B is higher than that of Model A. Which of the two models A and B has the better overall performance? As shown in Table 8-3.

Table 8-3 Performance of models A and B

Model	Accuracy	Recall rate
A	80%	90%
B	90%	80%

At this point, accuracy and recall will be combined into one indicator, namely, F1 score. Especially when a simple method is needed to compare two classifiers, F1 score is the harmonic mean of accuracy and recall.

3. Generate a confusion matrix

We can import the metrics library in sklearn to generate a confusion matrix. Firstly, obtain the measurement indicators of the classification model, with the code as follows:

```
1.  gride_search.best_estimator_
2.  gride_search.best_params_
3.  gride_search.best_score_
4.  lr = gride_search.best_estimator_
5.  lr.score(X_test,y_test)
6.  lr.score(X_train,y_train)
```

Then use the lr prediction y_pre_lr value from online search via the code as follows:

```
1.  from sklearn.metrics import f1_score
2.  y_pre_lr = lr.predict(X_test)
3.  f1_score(y_test,y_pre_lr)
```

Finally, import the metrics package of sklearn, and work out the confusion matrix via the code as follows:

```
1.  from sklearn.metrics import classification_report
2.  print(classification_report(y_test,y_pre_lr))
```

Run the code and it will output the values of precision, recall, and f1, as shown in Fig. 8-10.

	precision	recall	f1-score	support
0	0.85	0.65	0.74	26
1	0.78	0.91	0.84	35
accuracy			0.80	61
macro avg	0.82	0.78	0.79	61
weighted avg	0.81	0.80	0.80	61

Fig. 8-10　Model metrics output results

4. Plot a confusion matrix

When using a confusion matrix, we usually need to plot its image, so that we can observe the results more intuitively. This function can be achieved by defining a function via the code as follows:

```
1.  from sklearn.metrics import confusion_matrix
2.  cnf_matrix = confusion_matrix(y_test,y_pre_lr)
3.  cnf_matrix
```

```
4.
5.   def plot(matrix):
6.       f,ax =plt.subplots()
7.       print(matrix)
8.       sns.heatmap(matrix,annot =True,cmap ="Blues",ax =ax)
9.       ax.set_title ("confusion matrix for heart disease prediction")
10.      ax.set_xlabel ("actual value 0/1")
11.      ax.set_ylabel ("predicted value 0/1")
12.
13. plot(cnf_matrix)
```

Run the code and draw the confusion matrix for the classification results of the current model, as shown in Fig. 8-11. As can be seen from the figure, if the non-diagonal elements in the confusion matrix are all 0, a nearly perfect classifier will be obtained. The data in the figure means as follows:

00 (TN) represents 22 people who have not actually had heart disease; the model predicts that 22 people also have not suffered from heart disease.

01 (FP) represents 5 people who have not suffered from heart disease; the model predicts that 5 people have had heart disease.

10 (FP) represents 4 people who have suffered from heart disease; the model predicts that 4 people have not had heart disease.

12 (TP) represents 30 people who have suffered from heart disease; the model predicts that 30 people have had heart disease.

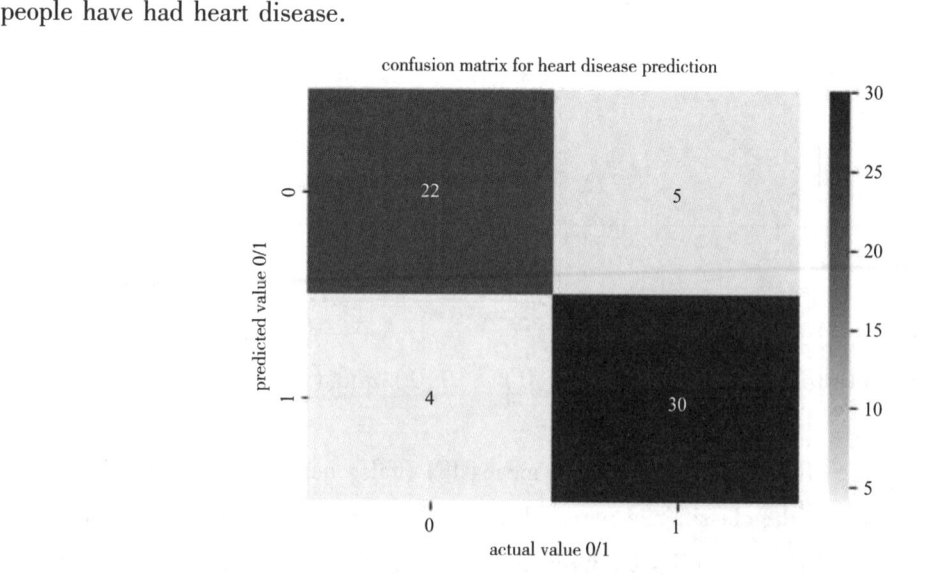

Fig. 8-11   Running results

8.9	Logistic Regression Evaluation Indicators

In logistic regression, the two indexes (ROC and AUC) are usually used to determine whether the model is optimal.

1. ROC

In the above example, the threshold value set by logistic regression is 0.5. In practical projects, the threshold value needs to be set according to requirements. For example, in some virus detection experiments, it is required to be able to recognize all true positives and tolerate a certain degree of false positives. At this point, the threshold 0.1 is the best choice. If false positives cannot be tolerated, it is required that the identified samples should be true positives, and the threshold can be set to 0.9.

In the process of threshold changes, TPR (true positive rate) and FPR (false positive rate) can be calculated respectively. Take FPR as the horizontal axis and TPR as the vertical axis. Connect all points and we can get the ROC curve, as shown in Fig. 8-12.

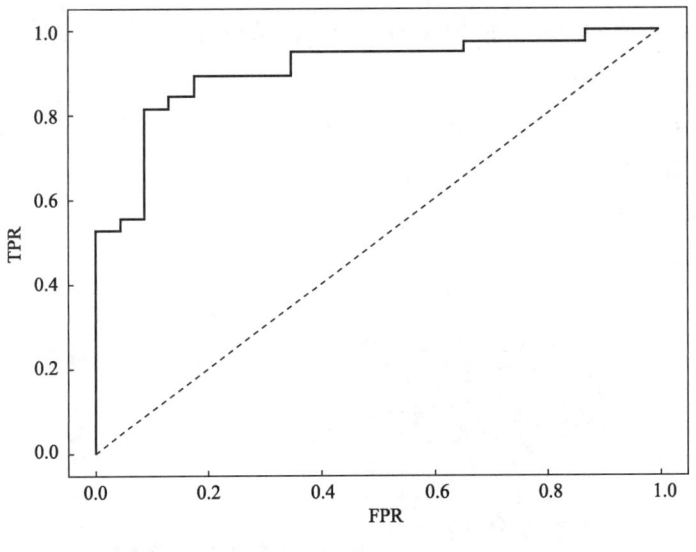

Fig. 8-12   ROC curve

Let's first look at the four points $(0,1)$, $(1,0)$, $(0,0)$ and $(1,1)$ and the diagonal in Fig. 8-12.

$(0, 1)$ point: i. e. FPR $=0$, TPR $=1$, which means FN (false negative) $=0$, and FP (false positive) $=0$, indicating that the classifier is perfect because it correctly classifies all samples.

$(1, 0)$ point: i. e. FPR $=1$, TPR $=0$, indicating that this classifier is the worst because it successfully avoids all correct answers.

$(0, 0)$ point: i. e. FPR $=$ TPR $=0$, FP $=$ TP $=0$, at which point the classifier predicts all

samples as negative samples.

(1, 1) point: The classifier predicts all samples as positive samples.

The points on the diagonal represent that the classifier predicts half of the samples as positive and the other half as negative. Therefore, the closer the ROC curve is to the upper left corner, the better the performance of the classifier will be. The code for ploting ROC curve is as follows:

```
1.  from sklearn.metrics import roc_curve
2.  fprs,tprs,thresholds = roc_curve(y_test,deision_scores)
3.
4.  def plot_roc_curve(fprs,tprs):
5.      plt.figure(figsize = (8,6),dpi =80)
6.      plt.plot(fprs,tprs)
7.      plt.plot([0,1],linestyle = '- -')
8.      plt.xticks(fontsize =13)
9.      plt.yticks(fontsize =13)
10.     plt.ylabel('TP rate',fontsize =15)
11.     plt.xlabel('FP rate',fontsize =15)
12.     plt.title('ROC curve',fontsize =17)
13.     plt.show()
14.
15. plot_roc_curve(fprs,tprs)
```

2. AUC

The AUC value is the area covered by the ROC curve. Obviously, the larger the AUC, the better the classification performance of the classifier. AUC is a numerical value. When it can't be told which classifier is better just from the ROC curve, you can use AUC for judgement.

When $AUC = 1$, it is a perfect classifier, and when using this prediction model, no matter what threshold is set, a perfect prediction can be achieved. In most prediction scenarios, there is no perfect classifier.

When $0.5 < AUC < 1$, it is better than random guess. If the threshold is properly set for this classifier (model), it can be valuable in prediction.

When $AUC = 0.5$, like random guess (for example, throwing copper coins), the model has no predictive value.

When $AUC < 0.5$, it is worse than random guess; but as long as one always goes against prediction, it is better than random guess. The area of AUC can be calculated via such code as follows:

```
1.  # AUC - ROC curve area
2.  from sklearn.metrics import roc_auc_score
3.  roc_auc_score(y_test,deision_scores)
```

Unit 9

k-means and DBSCAN Clustering Algorithms

For the previously classification algorithms, the datasets all have labels during model training. For example, in the fruit classification algorithm, what kind of a certain fruit is already known in the training set. This classification algorithm here is called supervised learning. The clustering algorithm also needs to determine the category of an object, but unlike previous classification problems, there is no category defined here in advance, namely, the dataset is unlabeled. The clustering algorithm itself needs to find a way to separate a batch of samples into multiple classes, ensuring that the samples in each class are similar, while the samples of different classes are different. Here, the type is called "cluster", and the clustering algorithm is a typical unsupervised learning algorithm.

There are two commonly used clustering algorithms: *k*-means and DBSCAN. Clustering algorithms are often applied as follows: user portrait, advertising recommendations, data segmentation, traffic recommendations for search engines, and malicious traffic identification; business push, news clustering, filtering and sorting based on location information; image segmentation, dimensionality reduction, recognition; outlier detection; abnormal credit card consumption; exploration on gene fragments with the same function, as shown in Fig. 9-1 and Fig. 9-2.

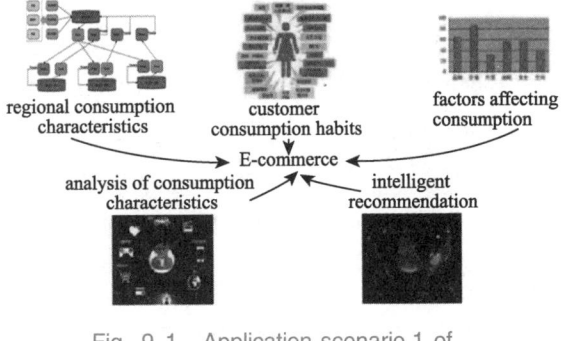

Fig. 9-1 Application scenario 1 of clustering algorithm

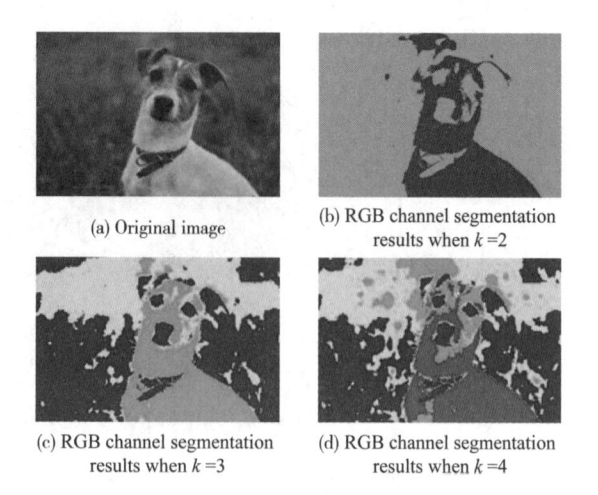

(a) Original image

(b) RGB channel segmentation results when $k=2$

(c) RGB channel segmentation results when $k=3$

(d) RGB channel segmentation results when $k=4$

Fig. 9-2   Application scenario 2 of clustering algorithm

9.1 *k*-means Algorithm Process

The *k*-means algorithm is a very simple algorithm. It belongs to unsupervised classification, which measures the similarity between samples in a certain way and iteratively updates the cluster center. When the cluster center no longer moves or the difference in movement is less than the threshold, then the samples are divided into different categories. Five steps are needed to achieve image classification via the *k*-means algorithm:

① Select a cluster center at random.

② Use the selected measurement method to classify all sample points based on the current cluster center.

③ Calculate the mean of the sample points of the same category in the current class as the cluster center for the next iteration.

④ Calculate the difference between the current cluster center and the cluster center of the next iteration.

⑤ If the difference is less than the given iteration threshold, the iteration ends; on the contrary, continue with the next iteration.

Firstly, it is necessary to select the appropriate "cluster" value, namely, the value of k, for example, specify $k=3$, as shown in Fig. 9-3.

Calculate the distance from the sample to the centroid of the "cluster", select the sample data at random based on the value of the "cluster", or specify an array (n_clusters, n_features) to initialize the centroids of k clusters, and calculate the distance from each sample point to each centroid

(Euclidean distance or cosine similarity).

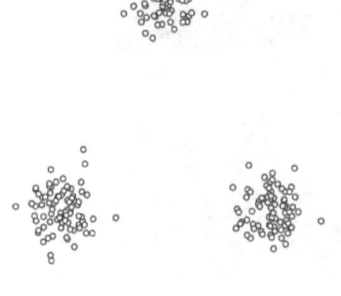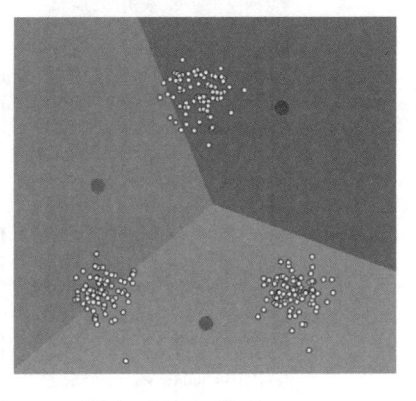

Fig. 9-3　Sample cluster with $k=3$ specified

Redivide the clusters according to the distance from the sample to the centroid. In the formula $\min\sum_{i=1}^{k}\sum_{x \in C_i}\mathrm{dist}(C_i, x)^2$ for redividing the clusters according to the optimization objective, dist $(C_i, x)^2$ represents the distance from the point x in the sample to the centroids of all clusters, and min represents finding the minimum value from all distances. Divide the samples into the corresponding clusters and repeat the above steps, as shown in Fig. 9-4.

Calculate the centroid repeatedly and update the cluster, as shown in Fig. 9-5.

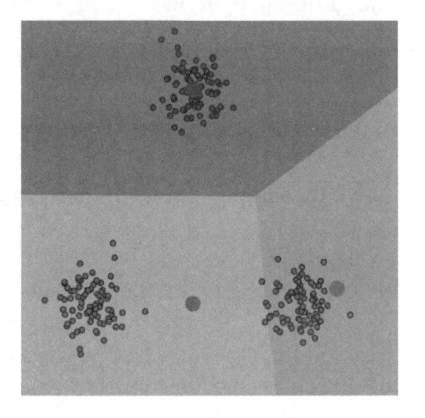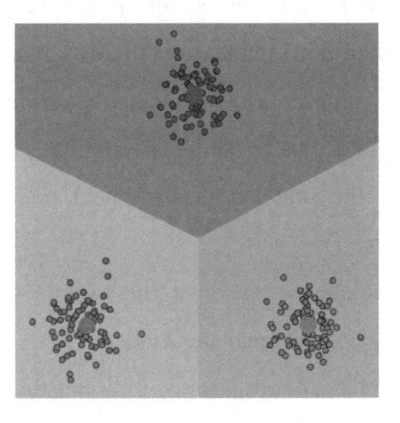

Fig. 9-4　Redivide clusters　　　　　　Fig. 9-5　Update clusters

9.2　Implementation of k-means Algorithm

1. Generate dataset

Import the package in Jupyter Notebook and set the number of random seeds. Define five point blok_centers as the centroids of the five clusters in the clustering algorithm, and define the variance

std corresponding to the cluster subject to the code as follows:

```
1.  from sklearn.datasets import make_blobs
2.  #generate the center points of clusters, five clusters in all
3.  blok_centers =np.array(
4.      [[0.2,2.3],
5.      [-1.5,2.3],
6.      [-2.8,1.8],
7.      [-2.8,2.8],
8.      [-2.8,1.3]])
9.  #define variance and distance between center points
10. blok_std =np.array([0.4,0.3,0.1,0.1,0.1])
```

2. Clusters and centroids in clustering

Usually, k is used to represent the number of clusters ultimately obtained. In the code, n_clusters is used to specify the number of clusters, which needs to be specified in advance.

The centroid is the center of a cluster, namely the center of all sample points in a cluster, which usually takes the mean of the dimensions of the vector.

3. make_blobsk( ) function

Use make_blobsk( ) to generate X and y coordinates, where four parameters need to be specified, n_samples = 2000 is the number of generated sample points; centers = blok_centers specify the centroid of the cluster; cluster_std = blok_std specifies the variance generating sample points; random_state = 7 specifies the number of random seeds, and then it uses shape to detect the shape of X and y. The code is as follows:

```
1.  X,y =make_blobs(n_samples =2000,centers =blok_centers,
2.                  cluster_std =blok_std,random_state =7)
3.  X.shape,y.shape
```

Run the code and it will output the results: ((2000, 2), (2000,)).

4. Visualization of generated datasets

Use the plot scatter method to display the original distribution of data subject to the code as follows:

```
1.  def plot_clusters(X,y =None):
2.      plt.figure(figsize = (8,4))
3.      plt.scatter(X[:,0],X[:,1],c =y,s =1)
4.      plt.xlabel("x1")
5.      plt.ylabel("x2")
```

```
6.      plt.show()
7.  plot_clusters(X)
```

See Fig. 9-6 for the code running results.

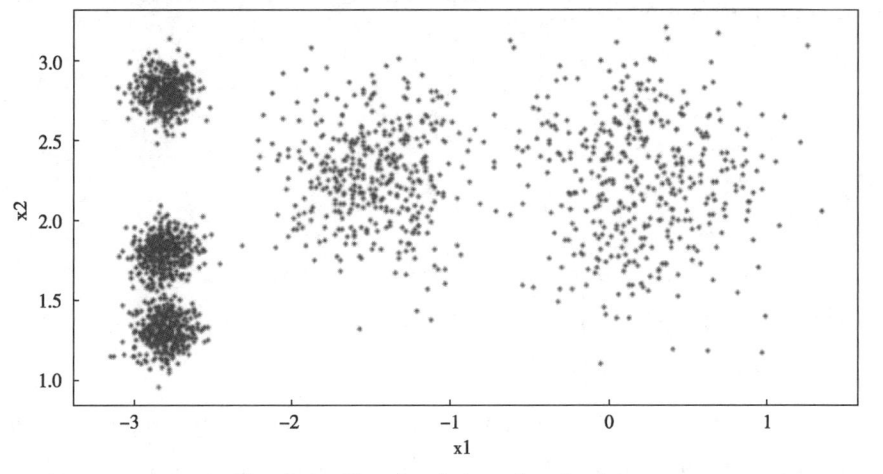

Fig. 9-6   Results of data visualization

5. Establish a k-means model and predict the results

(1) Establishing a k-means model

Use the sklearn library to implement the k-means algorithm, and then use fit_predict to cluster the samples in X, so as to get the trained model k-means, and return the predicted clustering result y_pred subject to the code as follows:

```
1.  from sklearn.cluster import KMeans
2.  k = 5
3.  kmeans = KMeans(n_clusters = k, random_state = 42)
4.  y_pred = kmeans.fit_predict(X)
5.  y_pred, y_pred.shape
```

Run the code and it will output: (array ([4, 1, 0, ···, 3, 0, 1]), (2000,)), where the first output is the clustering result and the second output is the shape of y_pre.

(2) Output cluster labels and cluster centroids

Use labels_ to output the serial number of the cluster while use the cluster_centers_ to output the centroid of the cluster, with the following code:

```
       kmeans.labels_, kmeans.cluster_centers_
```

Run the code and it will output the result as shown in Fig. 9-7. The output contents include the labels (categories) to which each sample belongs and the centroid coordinates of the

```
(array([4, 1, 0, ..., 3, 0, 1]),
array([[ 0.20876306,  2.25551336],
       [-2.80389616,  1.80117999],
       [-1.46679593,  2.28585348],
       [-2.79290307,  2.79641063],
       [-2.80037642,  1.30082566]]))
```

Fig. 9-7   Cluster labels and centroid of clusters

five clusters.

(3) *k*-means function

We can use the *k*-means function to establish a *k*-means clustering model. The Kmeans function has multiple parameters subject to such syntax as follows:

```
KMeans(n_clusters = 8, init = 'k - means + + ', n_init = 10 , max_iter = 300,
tol = 0.0001 , verbose = 0 , random_state = None , copy_x = True , algorithm = '
auto')
```

Parameter descriptions:

n_clusters: The int type value of the number of clusters, the number of clusters to be formed, and the number of centroids to be generated, 8 by default.

init: Initialization method, which can specify an array (n_clusters, n_features) to initialize the cluster. The default value default = 'kmeans' can intelligently select the initial cluster center for the *k*-means cluster to accelerate convergence.

n_init: The int type value, the number of times the *k*-means algorithm is run with different centroid seeds, 10 by default.

max_iter: The int type value, maximum number of iterations, 300 by default, which is the maximum number of iterations for a single run of the *k*-means algorithm.

random_state: The int type value, the number of random seeds, which determines the random number generation of the centroid initialization.

(4) *k*-means method function

fit (X): Train the *k*-means model and return the trained *k*-means model.

fit_predict (X): Calculate the cluster center and return the cluster index for each sample.
fit_transform (X): Calculate the distance from the sample point to the centroid of the cluster, and return an ndarray (n_samples,), n_samples refers to the number of samples.

6. Predicted actual value

Generate the sample point X_new at random, and use *k*-means to predict the results, with the following code:

```
1.  X_new = np.array([[0,2],[3,2],[ -3,3],[ -3,2.5]])
2.  kmeans.predict(X_new)
```

Return the result: array ([0, 0, 3, 3]).

The result is an ndarray (4, 1), where the value represents which cluster each point belongs to. Use the transform ( ) function to output the distance between samples and the centroid of the cluster, subject to the code as follows:

```
kmeans.transform(X_new)
```

See Fig. 9-8 for the output results.

```
array([[0.32995317, 2.81093633, 1.49439034, 2.9042344 , 2.88633901],
       [2.80290755, 5.80730058, 4.4759332 , 5.84739223, 5.84236351],
       [3.29399768, 1.21475352, 1.69136631, 0.29040966, 1.71086031],
       [3.21806371, 0.72581411, 1.54808703, 0.36159148, 1.21567622]])
```

Fig. 9-8 Distance from sample to cluster centroid

The output result is an ndarray（n_samples, n_clusters）, with four values output for each sample point, representing the distance between the sample point and the centroids of the four clusters. The smallest of the four values can be found to determine which cluster this sample point belongs to.

7. Visualize the predicted results

Define functions to visualize and display all data. The code is as follows:

```
1.  # display all data
2.  def plot_data(X):
3.      plt.plot(X[:,0],X[:,1],'k.',markersize =2)
```

Display the center point of the cluster, with the following code:

```
1.  # display the center point of the cluster
2.  def plot_centroids(centroids,crircle_color ='w',cross_color ='k'):
3.      plt.scatter (centroids [:,0 ],centroids [:,1 ],marker ='o',s =30,
    linewidths =8,color =crircle_color,zorder =10,alpha =0.9)
4.      plt.scatter (centroids [:,0 ],centroids [:,1 ],marker ='x',s =50,
    linewidths =8,color =crircle_color,zorder =11,alpha =1)
```

Plot the decision boundary with the following code:

```
1.  # plot the decision boundary
2.  def plot_decision_boundaries(clusterer,X,resolution =1000,show_
    cen-troids =True,show_xlabels =True,show_ylabels =True):
3.      # obtain the coordinate checkerboard
4.      mins =X.min(axis =0) -0.1
5.      maxs =X.max(axis =0) +0.1
6.      xx,yy =np.meshgrid(np.linspace(mins [0],maxs [0],resolution),
7.                  np.linspace(mins [1],maxs [1],resolution))
8.      Z =clusterer.predict(np.c_[xx.ravel(),yy.ravel()])
```

```
9.      Z = Z.reshape(xx.shape)
10.     print(xx.shape,yy.shape,Z.shape)
11.     #plot the contour line
12.     plt.contourf(Z,extent = (mins[0], maxs[0], mins[1], maxs[1]),
   cmap = 'Pastel2')
13.     plt.contour(Z,extent = (mins[0],maxs[0],mins[1],maxs[1]),
   line - widths =1,colors = 'k')
14.
15.     plot_data(X)
16.     plot_centroids(clusterer.cluster_centers_)
17.     plt.xlabel("x1")
18.     plt.ylabel("x2")
19.     plt.tick_params(labelbottom = 'off')
20.     plt.tick_params(labelleft = 'off')
```

Call the function with the following code:

```
1.  plt.figure(figsize = (15,5))
2.  plot_decision_boundaries(kmeans,X)
3.  plt.show()
```

After running the code, it can be seen that the data is divided into five clusters, as shown in Fig. 9-9.

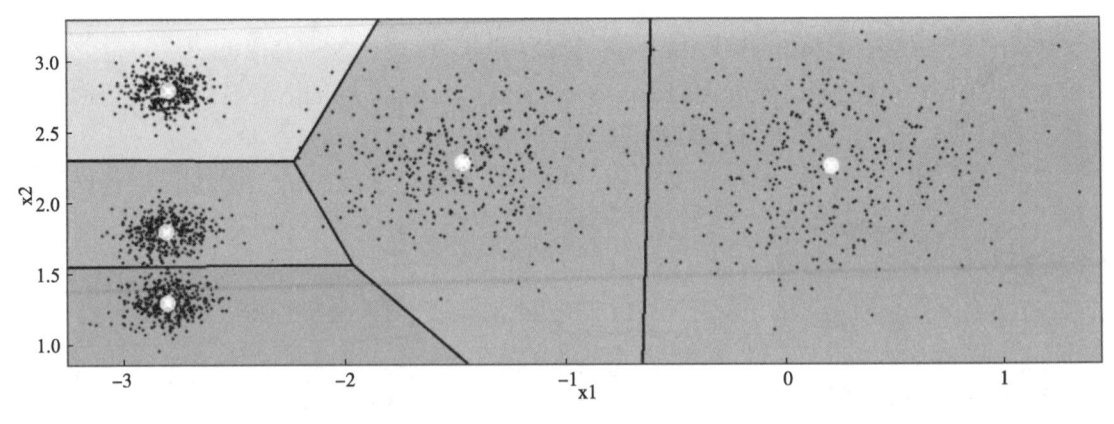

Fig. 9-9   Visualization of decision boundary

The above code only displays the final result after five update iterations. You can write a code to display the process of five "cluster" updates, as follows:

```
1.  plt.figure(figsize = (15,15))
2.   # initialization point and centroid of original data in the
     first figure
3.  plt.subplot(421)
4.  plot_data(X)
5.  plot_centroids(kmeans_iter1. cluster_centers_, crircle_color
     = ' r ', cross_color = 'k')
6.  plt. title ('original cluster centroid ')
7.
8.  # divide clusters by centroid for the first time
9.  plt.subplot(422)
10. plot_decision_boundaries(kmeans_iter1,X,show_xlabels = False,
     show_ ylabels = False)
11. plt. title ('first clustering ')
12.
13. # recalculate the centroid based on the divided clusters
14. plt.subplot(423)
15. plot_decision_boundaries(kmeans_iter1,X,show_xlabels = False,
     show_ ylabels = False)
16. plot_centroids(kmeans_iter2. cluster_centers_, crircle_color
     = ' r ', cross_color = 'k')
17. plt. title ('first updated cluster centroid')
18. plt.subplot(424)
19. plot_decision_boundaries(kmeans_iter2,X,show_xlabels = False,
     show_ ylabels = False)
20. plt. title ('after updating the cluster, divide the cluster for the
     second time')
21.
22. # update cluster centroids for the second time based on the divided
     clusters
23. plt.subplot(425)
24. plot_decision_boundaries(kmeans_iter2,X,show_xlabels = False,
     show_ ylabels = False)
25. plot_centroids(kmeans_iter3. cluster_centers_, crircle_color
     = ' r ', cross_color = 'k')
26. plt. title ('second update of centroid')
```

```
27.
28. # third clustering
29. plt.subplot(426)
30. plot_decision_boundaries(kmeans_iter3,X,show_xlabels=False,
      show_ylabels=False)
31. plt.title('third clustering')
32.
33. # recalculate the centroid based on the divided clusters
34. plt.subplot(427)
35. plot_decision_boundaries(kmeans_iter4,X,show_xlabels=False,
      show_ylabels=False)
36. plot_centroids(kmeans_iter5.cluster_centers_,crircle_color
      ='r',cross_color='k')
37. plt.title('fourth update of centroid')
38.
39. # fifth clustering
40. plt.subplot(428)
41. plot_decision_boundaries(kmeans_iter5,X,show_xlabels=False,
      show_ylabels=False)
42. plt.title('fifth clustering')?
43.
44.
45. plt.show()
```

Run the code and it will output the results from each clustering and visualize the results. From the results, we can see the process, in which the *k*-means algorithm updates the centroid of the cluster five times and redivides the "cluster", as shown in Fig. 9-10.

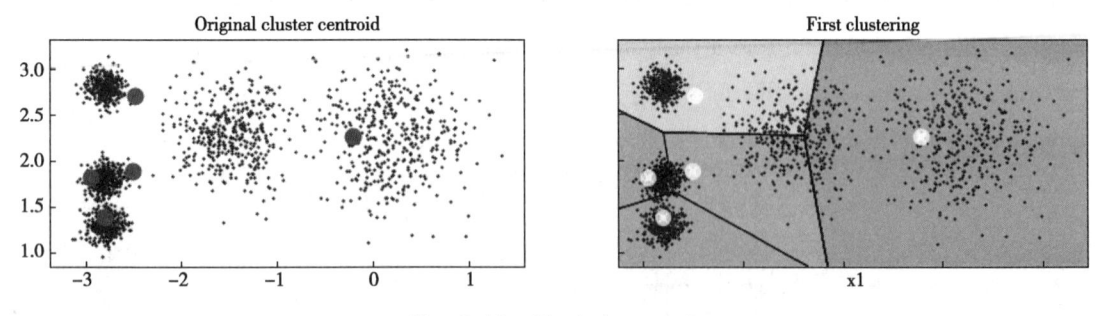

Fig. 9-10 Clustering results

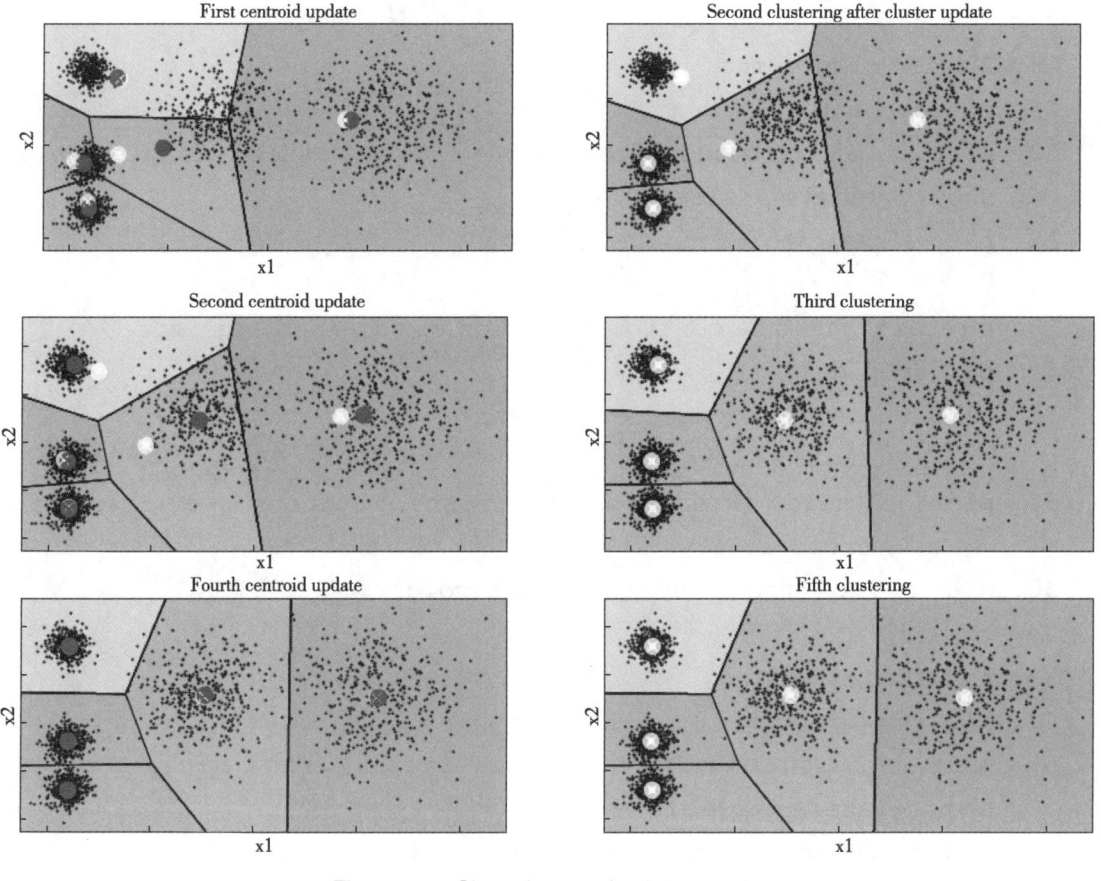

Fig. 9-10 Clustering results (continued)

9.3 Instability of k-means Algorithm

When using the k-means algorithm, the same algorithm produces different results for different random seed numbers because it highly depends on the position of the first centroid (randomly selected). This algorithm can find a local minimum, and subsequent algorithms will be limited to a certain range. As shown in the following code, running the k-means algorithm twice produces different results due to different settings of random_ state. The code is as follows:

```
1.  c1 =KMeans(n_clusters =5,init = 'random',n_init =1,random_state =10)
2.  c2 =KMeans(n_clusters =5,init = 'random',n_init =1,random_state =20)
3.  def plot_clusterer_comparsion(c1,c2,X):
4.      c1.fit(X)
5.      c2.fit(X)
```

```
6.
7.      plt.figure(figsize = (15,5))
8.      plt.subplot(121)
9.      plot_decision_boundaries(c1,X)
10.     plt.subplot(122)
11.     plot_decision_boundaries(c2,X)
12.
13. plot_clusterer_comparsion(c1,c2,X)
```

See Fig. 9-11 for the effect after the program is run. From the figure, it can be seen that the initialization position of the cluster will affect the clustering results.

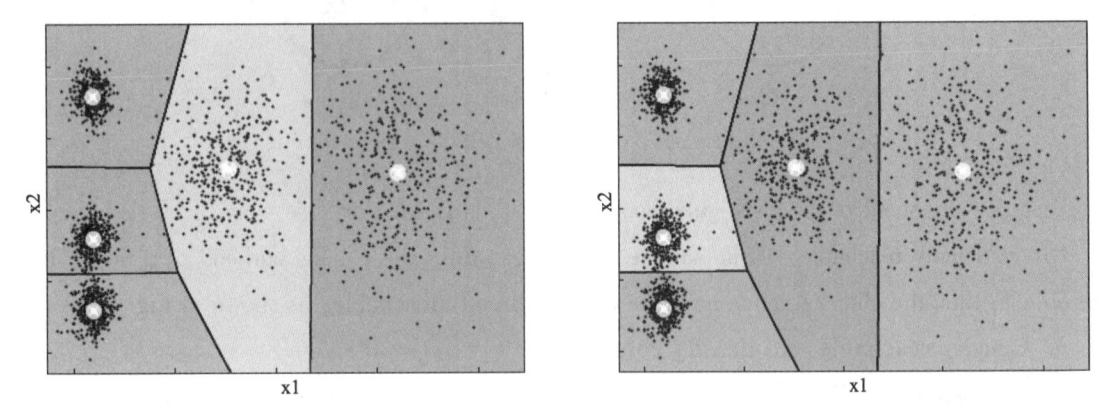

Fig. 9-11   Results from k-means algorithm running

9.4   DBSCAN Clustering Algorithm

1. Introduction to DBSCAN clustering algorithm

The k-means algorithm is simple and straightforward, but it is somewhat deficient, for example, we should first determine the k value each time, and the k value will directly affect the clustering results. The clustering algorithm is greatly influenced by the initial value, as it requires calculating the distance from each sample point to the centroids of all "clusters". The complexity of the algorithm is linearly related to the sample size, and it is difficult to discover clusters of any shape.

There is another clustering algorithm, also known as DBSCAN clustering algorithm (DBSCAN algorithm for short), which is based on density clustering algorithm. The so-called density clustering algorithm is to cluster based on the compactness of the samples.

2. Basic concepts of DBSCAN algorithm

The DBSCAN algorithm contains many basic terms, such as r-neighborhood, core object, etc.

The specific meaning is as follows:

r-neighborhood means that the area within a given object radius of r is called the r-neighborhood of this object. As shown in Fig. 9-12 (a), the circle formed within the area covered by the radius r of the object p is its r-neighborhood.

Core object: If the number of sample points in the r-neighborhood of a given object is greater than or equal to MinPoints (MinPts for short), then this object is called a core object. As shown in Fig. 9-12 (b), if the point of MinPoints is set to 2, then there are four points in the e-neighborhood of the object p, greater than MinPoints, then the object p is the core object.

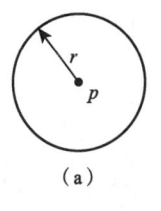 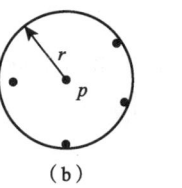 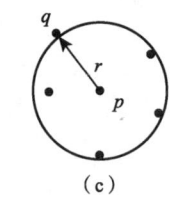

(a)　　　　　(b)　　　　　(c)

Fig. 9-12　Basic concepts of DBSCAN algorithm

Distance threshold of the ε-neighborhood: The radius is set as r.

Direct density reachable: If the sample point q is within the r-neighborhood of p and p is the core object, then the object p-q can have the density directly reachable, as shown in Fig. 9-12 (c).

3. Density reachable and density connected

Density reachable: If a sequence of points q_0, q_1, $\cdots$, q_k is directly density reachable for any q_0-q_k, then the density from q_0 to q_k is called density reachable. This is actually a "propagation" of direct density reachable, as shown in Fig. 9-13. If q-p is directly density reachable and m-q is directly density reachable, then m-p density can be reached.

Density connected: If starting from a certain core point p, both the point q and point k are density reachable, then the point q and point k are said to be density connected, as shown in Fig. 9-14. q-o is density reachable, p-o is density reachable, and q-p is density connected. In the DBSCAN algorithm, certain samples can be considered as a class (also known as a cluster), namely the set of samples with the highest density connected.

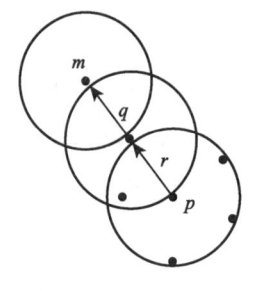 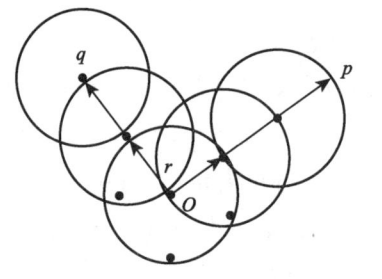

Fig. 9-13　Density reachable　　　　　Fig. 9-14　Maximum density connected

As shown in Fig. 9-15, if MinPts =5 is set, the points on the arrow are all core objects because there are at least 5 samples in their ε-neighborhood. The samples outside the single arrow line are not core objects. All samples with direct density access to the core object are located within a cirde centered on the starting point of the single arrow line. If not within the cirde, they cannot have density directly reachable. The core objects connected by arrow lines in the figure form a sample sequence with density reachable. Within the ε-neighborhood of these density reachable sample sequences, all samples are density connected to each other.

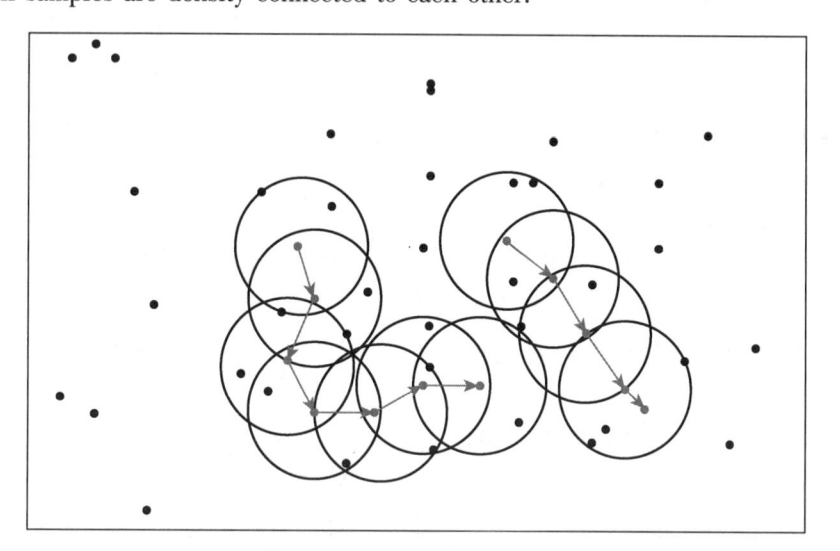

Fig. 9-15   Density connected

4. Boundary points and noise points

The boundary point belongs to a non-core point of a certain class and cannot develop referrals, as shown in Fig. 9-16 where the points B and C are boundary points.

Noise points are points that do not belong to any class/cluster and are density unreachable from any core point, like point N as shown in Fig. 9-16.

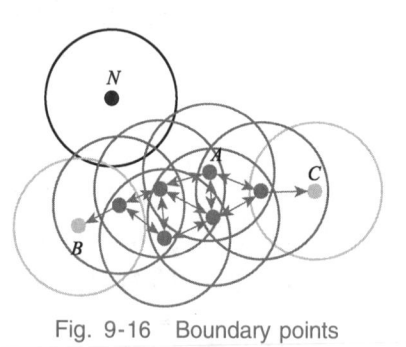

Fig. 9-16   Boundary points

9.5 | Implementation of DBSCAN Clustering Algorithm

The DBSCAN algorithm requires three parameters to be set, namely D input dataset, neighborhood radius specified by the parameter ε and MinPts density threshold (the minimum number of sample points in the neighborhood). Firstly, set the parameter ε based on the K distance. Find out the mutation point and set the K distance. Given the dataset $P = \{p(i); i = 0, 1, \cdots, n\}$,

calculate the distances between the point $p(i)$ and all points in the subset S of the set D, and sort the distances in an ascending order. Finally, to determine MinPts, it is generally necessary to take a smaller value first and try repeatedly. The specific code is as follows:

```
1.   np.unique(dbscan.labels_)
2.   # adjust different eps and min_samples and different results can be
     achieved
3.   dbscan2 = DBSCAN(eps = 0.2, min_samples = 5)
4.   dbscan2.fit(X)
5.   # define functions
6.   def plot_dbscan(dbscan, X, size, show_xlabels = True, show_ylabels =
     True):
7.       # define 1000 sets classified
8.       core_mask = np.zeros_like(dbscan.labels_, dtype = bool)
9.       # set the core point to True
10.      core_mask[dbscan.core_sample_indices_] = = True
11.      # find outliers
12.      anomalies_maske = dbscan.labels_ = = -1
13.      # point not an outlier or a core point
14.      non_core_mask = ~ (core_mask |anomalies_maske)
15.      # get the coordinates of the core point
16.      cores = dbscan.components_
17.      # get outliers
18.      anomalies = X[anomalies_maske]
19.      non_cores = X[non_core_mask]
20.      # plot feature points
21.      plt.scatter(cores[:,0], cores[:,1],
22.       c = dbscan.labels_[core_mask], marker = 'o', s = size, cmap =
         'Paired')
23.      # plot
24.      plt.scatter(cores[:, 0], cores[:, 1], marker = '* ', s = 20, c =
     dbscan.labels_[core_mask])
25.       plt.scatter(anomalies[:,0], anomalies[:,1], c = 'r', marker =
     "x", s = 100)
26.       plt.scatter(non_cores[:,0], non_cores[:,1], c = dbscan.labels_
     [non_core_mask], marker = ".")
```

```
27.     plt.title("DBSCAN clustering algorithm eps = {:.2f}, min_samples =
        {}".format (dbscan.eps, dbscan.min_samples), fontsize =14)
28. # visualization of data results
29. plt.figure(figsize = (10,5))
30. plt.subplot(121)
31. plot_dbscan(dbscan,X,size =100)
32.
33. plt.subplot(122)
34. plot_dbscan(dbscan2,X,size =600)
35. plt.show()
```

After running the code, it can be seen that the same data sample, if eps, namely the distance thresholds for ε-neighborhood is set differently, can have different results. In Fig. 9-17 (a), eps is set to 0.08 and the DBSCAN algorithm has divided the data into four "clusters"; in Fig. 9-17 (b), eps is set to 0.20 and the DBSCAN algorithm has divided the data into two "clusters".

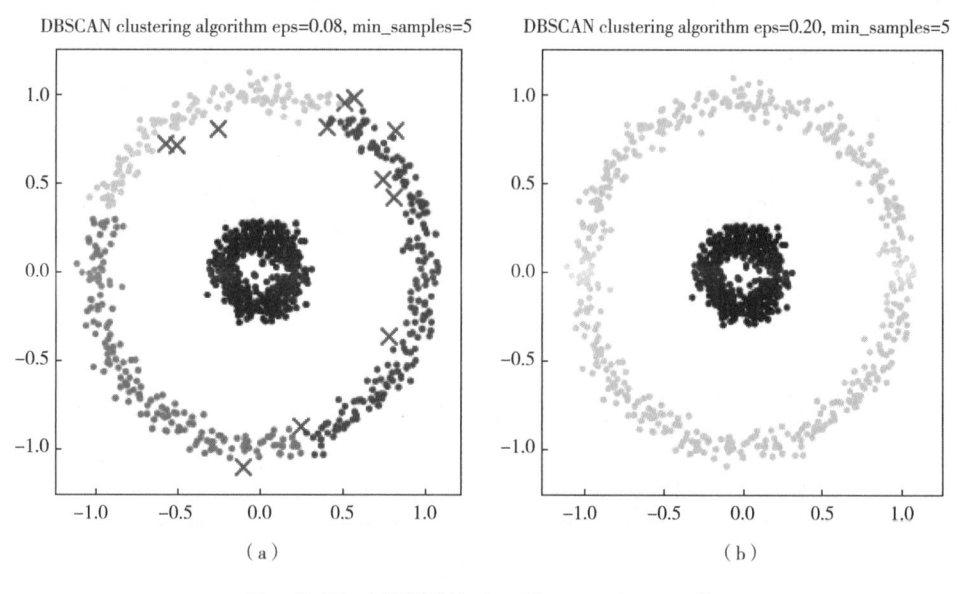

Fig. 9-17   DBSCAN algorithm running results

9.6 Using the *k*-means Algorithm to Complete Image Segmentation

Image segmentation just uses such features as grayscale, color, texture, and shape to divide an image into several regions not overlapped, and to make these features appear similar within the same

region; there are significant differences between different regions. Then, the regions with unique properties can be extracted from the segmented image for different research purposes.

Whether it is a grayscale image or an RGB color image, it is actually a matrix containing grayscale values, so the data format of the image determines that it is very easy and concrete to use the k-means clustering algorithm in the direction of image segmentation.

1. Import packages and read images in Jupyter Notebook

The code is as follows:

```
1.  #ladybug.png
2.  from matplotlib.image import imread
3.  #image = imread('ladybug.png')
4.  image = imread('street.png')
5.  #image = imread('street.jpg')
6.  image.shape
```

2. Define the k-means algorithm

8 clusters are specified with a random seed count of 42. The code is as follows:

```
kmeans = KMeans(n_clusters =8,random_state =42).fit(X)
```

Output the information about the centroid of the cluster. The code is as follows:

```
1.  kmeans =KMeans(n_clusters =8,random_state =42).fit(X)
2.  kmeans.cluster_centers_
```

Run the code and it has output the centroid information of 8 clusters. See Fig. 9-18 for the running results.

```
array([[0.11479929, 0.131998  , 0.07966766, 1.        ],
       [0.8284333 , 0.8657306 , 0.8369371 , 1.        ],
       [0.34023207, 0.36337155, 0.3184119 , 1.        ],
       [0.21859105, 0.23419523, 0.21826884, 1.        ],
       [0.11070266, 0.6104674 , 0.5286546 , 1.        ],
       [0.5076874 , 0.51638305, 0.46688914, 1.        ],
       [0.5988326 , 0.6959467 , 0.65107864, 1.        ],
       [0.9320028 , 0.95879555, 0.95750713, 1.        ]], dtype=float32)
```

Fig. 9-18   Centroid information of 8 clusters

3. Calculate the clustered image

Use k-means to get the number of clusters (10, 8, 6, 4, 2) and use the current cluster center to get the clustered image. The code is as follows:

```
1.  segmented _img = kmeans. cluster _centers _[kmeans. labels _].
    reshape (311, 500, 4)
```

```
2.  segmented_imgs = []
3.  n_colors = (10,8,6,4,2)
4.  for n_cluster in n_colors:
5.      kmeans = KMeans(n_clusters = n_cluster,random_state = 42).fit(X)
6.      segmented_img = kmeans.cluster_centers_[kmeans.labels_]
7.      segmented_imgs.append(segmented_img.reshape(image.shape))
```

4. Use matplotlib to display clustered images

The code is as follows:

```
1.  plt.figure(figsize = (20,10))
2.  plt.subplot(231)
3.  plt.imshow(image)
4.  plt.title('Original image')
5.
6.  for idx,n_clusters in enumerate(n_colors):
7.      plt.subplot(232 + idx)
8.      plt.imshow(segmented_imgs[idx])
9.      plt.title('{}colors'.format(n_clusters))
```

After running the code, it can be seen that 10, 8, 6, 4, 2 can be used as the number of clusters for image clustering segmentation and different clusters can be set to segment objects in the image. This method is usually used for image preprocessing. For example, when $K = 2$, the boundary and centerline of the highway can be extracted for subsequent processing. See Fig. 9-19 for the output results.

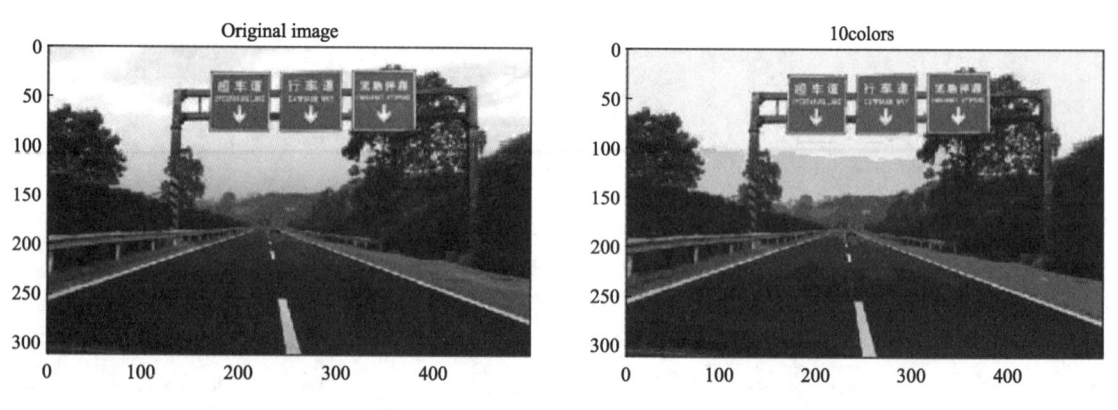

Fig. 9-19   Image clustering results

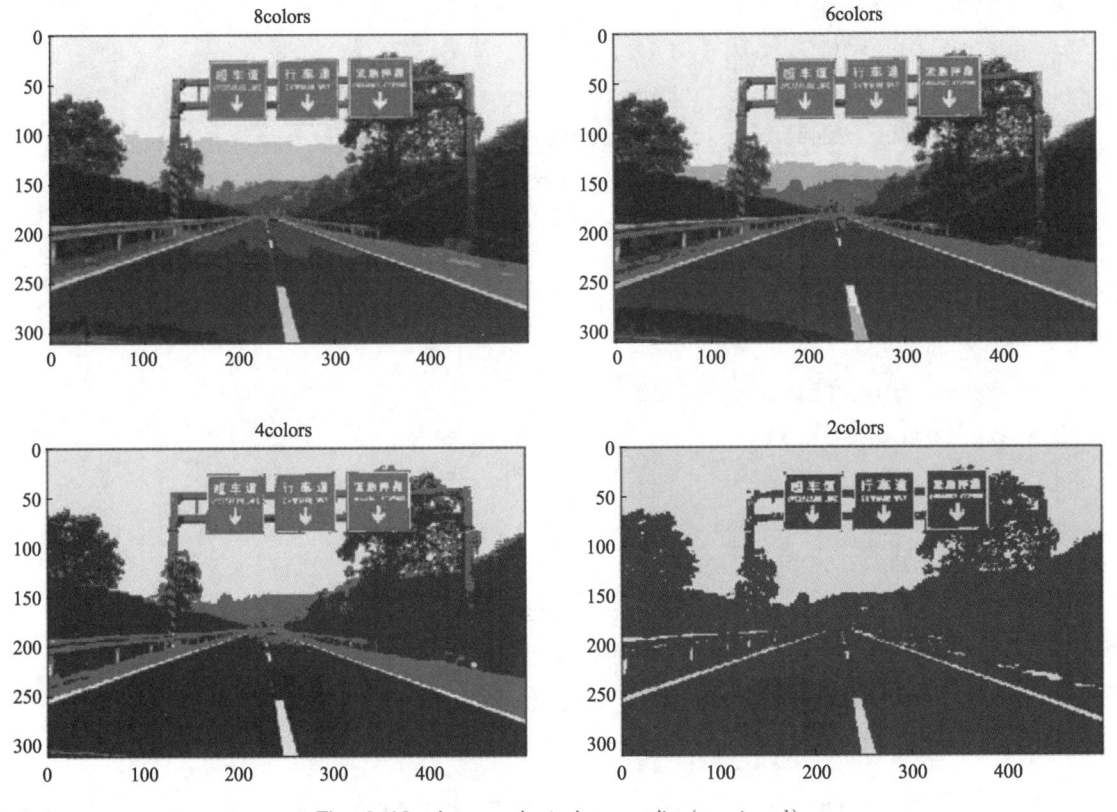

Fig. 9-19   Image clustering results（continued）

Unit 10

Use Decision Trees to Complete Iris Classification

Iris classification is a typical classification problem. Botanists have collected some measurement data of iris, including the length and width of petals and calyxes, measured in cm. These flowers have previously been identified by botanical experts as belonging to one of the three varieties, Setosa, Versicolor, or Virginica, as shown in Fig. 10-1. The variety to which each iris belongs can be determined based on the data area of the measurement.

Fig. 10-1 Three varieties of iris

Now the goal is to build a decision tree model and learn from the measured data of these known varieties of iris, so as to predict new varieties of iris. Specifically, there are known measurement data of iris and these data are labeled, so this is a supervised learning problem. In this problem, it is

necessary to predict one of multiple options (the variety of iris). This is a classification problem. The output (variety of iris) is called a class. Each iris in the dataset belongs to one of the three categories, so this is a trinary classification problem.

10.1   Decision Tree

1. Decision tree overview

The decision tree algorithm originated from the paper "Experiments in Induction" published by E. B. Hunt et al. in 1966, but it was Quinlan who really made the decision tree become the mainstream algorithm of machine learning. In 2011, Quinlan won the KDD Innovation Award, the highest prize in the field of data mining. In 1979, he proposed the ID3 algorithm, setting off the climax of the decision tree research. The most commonly used decision tree algorithm now is C4.5 proposed by Quinlan in 1993. (Why is it called C4.5? There is an anecdote: After Quinlan proposed ID3, it sparked a wave of decision tree research, and then ID4, ID5, and other names were taken up, so Quinlan had to name his ID3 update as C4.0, and C4.0 update as C4.5). The new version for commercial applications is C5.0link.

2. Basic concepts of decision trees

A decision tree is a tree that contains a root node, several internal nodes, and several leaf nodes; the leaf nodes correspond to the decision results, while each other node corresponds to an attribute test; the sample set contained in each node is divided into child nodes based on the results of attribute testing; the root node contains the complete set of samples, and the path from the root node to each leaf node corresponds to a decision test sequence. Decisions can be made based on the characteristics of the sample, as shown in Fig. 10-2, indicating that a sample makes category decisions based on the decision tree according to the characteristics.

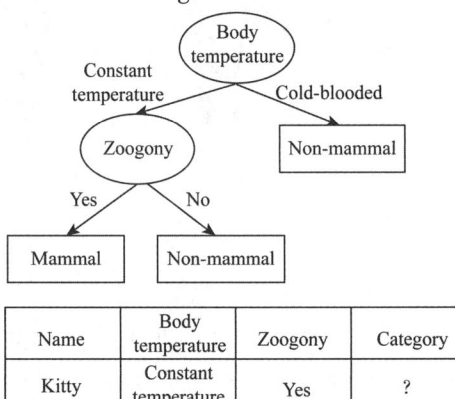

Name	Body temperature	Zoogony	Category
Kitty	Constant temperature	Yes	?

Fig. 10-2   Basic structure of decision making tree

10.2　Information Entropy

The key to decision tree learning is how to select the optimal classification attribute. The so-called optimal classification attribute, for dichotomy, is to try to make the divided samples belong to the same category, that is, the attribute with the highest "purity". So how to measure the purity of features requires the use of information entropy. Let's first see the definition of information entropy: If the proportion of the k th class of samples in the current sample set D is $p_k(k=1, 2, 3, 4, \cdots, |y|)$, and y is the total number of categories, then the information entropy of the sample set should be

$$\text{Ent}(D) = -\sum_{k=1}^{|y|} p_k \log_2 p_k$$

The smaller the value of Ent (D), the higher the purity of D. This formula determines a defect of the information gain, namely, the information gain has a preference for features with a large number of values that can be taken (i. e., the more values the attribute can take, the more information gain is partial to this attribute). Because the more values a feature can take, the higher the "purity", which means that Ent (D) will be very small. If the number of discrete features is equal to the number of samples, the Ent (D) value will be 0.

Here follows an example on how to calculate the information entropy of the dataset shown in Fig. 10-3.

Features (attributes)
⇩

S/N	Color	Root/Base	Knock Sound	Texture	Navel	Touch	Good Melon (label)
1	Dark green	Curled	Turbid	Clear	Depressed	Hard&smooth	Yes
2	Ebony	Curled	Dull	Clear	Depressed	Hard&smooth	Yes
3	Ebony	Curled	Turbid	Clear	Depressed	Hard&smooth	Yes
4	Dark green	Curled	Dull	Clear	Depressed	Hard&smooth	Yes
5	Light white	Curled	Turbid	Clear	Depressed	Hard&smooth	Yes
6	Dark green	Slightly curled	Turbid	Clear	Slightly depressed	Soft&sticky	Yes
7	Ebony	Slightly curled	Turbid	Slightly fuzzy	Slightly depressed	Soft&sticky	Yes
8	Ebony	Slightly curled	Turbid	Clear	Slightly depressed	Hard&smooth	Yes
9	Ebony	Slightly curled	Dull	Slightly fuzzy	Slightly depressed	Hard&smooth	No
10	Dark green	Stiff	Clear	Clear	Flat	Soft&sticky	No
11	Light white	Stiff	Clear	Fuzzy	Flat	Hard&smooth	No
12	Light white	Curled	Turbid	Slightly fuzzy	Flat	Soft&sticky	No
13	Dark green	Slightly curled	Turbid	Slightly fuzzy	Depressed	Hard&smooth	No
14	Light white	Slightly curled	Dull	Slightly fuzzy	Depressed	Hard&smooth	No
15	Ebony	Slightly curled	Turbid	Clear	Slightly depressed	Soft&sticky	No
16	Light white	Curled	Turbid	Fuzzy	Flat	Hard&smooth	No
17	Dark green	Curled	Dull	Slightly fuzzy	Slightly depressed	Hard&smooth	No

Fig. 10-3　Watermelon dataset

This dataset contains 17 samples, classified as binary. The proportion of positive examples (samples in category 1) is $p_1 = \frac{8}{17}$, and the proportion of negative examples (samples in category 0) is $p_2 = \frac{9}{17}$. According to the formula of information entropy, the information entropy of the dataset D can be calculated as

$$\text{Ent}(D) = -\left(\frac{8}{17}\log_2\frac{8}{17} + \frac{9}{17}\log_2\frac{9}{17}\right) = 0.998$$

10.3　Information Gain

Assuming that the discrete attribute a has V possible values $\{a^1, a^2, a^3, \cdots, a^v\}$, if the feature a is used to classify the dataset D, then V branch nodes will be generated, where the v th node contains the total number of samples in the dataset D that have a value of a^v on the feature a, denoted as D^v. Therefore, the information entropy can be calculated based on the formula for information entropy above. Considering that different branch nodes contain different numbers of samples, the weight $\frac{|D^v|}{|D|}$ can be given to the branch nodes, which means that the branch nodes with more samples will have a greater impact. Therefore, the "information gain" obtained by classifying the sample set D with the feature a can be calculated as follows:

$$\text{Gain}(D,a) = \text{Ent}(D) - \sum_{v=1}^{V} \frac{|D^v|}{|D|}\text{Ent}(D^v)$$

Generally speaking, a greater information gain indicates greater "purity improvement" obtained by using a certain feature to classify the dataset. So, information gain can be used to select attributes for decision tree classification, actually selecting the attribute with the highest information gain. The ID3 algorithm uses information gain to classify attributes.

10.4　Calculation of Information Gain

1. Calculate the information gain of color

Now, take the watermelon dataset shown in Fig. 10-3 as an example to calculate the information gain. From the dataset, it can be seen that the feature set is {color, root, tapping sound, texture, navel and tactile}. Let's calculate the information gain for each feature. Let's first see "color", which has three values: {dark green, dull black and light white}. If the "color" is used to classify the dataset D, we can obtain three subsets, namely D^1 (color = dark green), D^2 (color = dull

black), and D^3 (color = light white), as shown in Fig. 10-4.

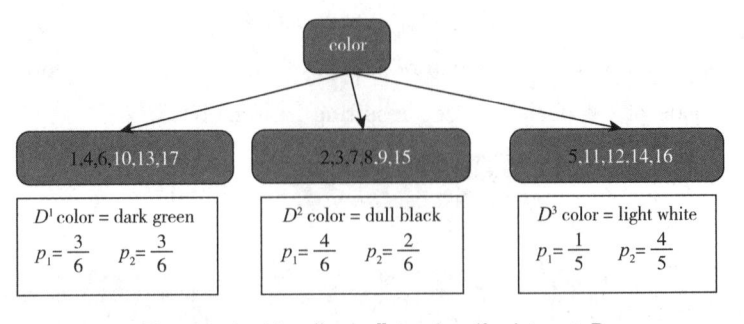

Fig. 10-4　Use "color" to classify dataset D

D^1 contains six samples $\{1, 4, 6, 10, 13, 17\}$, where the proportion of positive examples is $p_1 = \dfrac{3}{6}$ and the proportion of negative examples is $p_2 = 3$. D^2 contains six samples $\{2, 3, 7, 8, 9, 15\}$, of which the proportion of positive examples is $p_1 = \dfrac{4}{6}$ and the proportion of negative examples is $p_2 = \dfrac{2}{6}$. D^3 contains five samples $\{5, 11, 12, 14, 16\}$, of which the proportion of positive examples is $p_1 = \dfrac{1}{5}$ and the proportion of negative examples is $p_2 = \dfrac{4}{5}$. Therefore, the information entropy of the three branch nodes obtained after classified by "color" can be calculated as follows:

$$\text{Ent}(D^1) = -\left(\frac{3}{6}\log_2\frac{3}{6} + \frac{3}{6}\log_2\frac{3}{6}\right) = 1.00$$

$$\text{Ent}(D^2) = -\left(\frac{4}{6}\log_2\frac{4}{6} + \frac{2}{6}\log_2\frac{2}{6}\right) = 0.918$$

$$\text{Ent}(D^3) = -\left(\frac{1}{5}\log_2\frac{1}{5} + \frac{4}{5}\log_2\frac{4}{5}\right) = 0.722$$

Calculate the information gain of the features such as "color", "root" and "texture":

$$\text{Gain}(D, \text{color}) = \text{Ent}(D) - \sum_{v=1}^{3} \frac{|D^v|}{|D|}\text{Ent}(D^v)$$

$$= 0.998 - \left(\frac{6}{17}\times 1.000 + \frac{6}{17}\times 0.918 + \frac{5}{17}\times 0.772\right) = 0.109$$

2. Information gain of all features

Similarly, we can calculate the information gain of features and determine the root node of the decision tree:

$$\text{Gain}(D, \text{root}) = 0.143$$

$$\text{Gain}(D, \text{tapping sound}) = 0.141$$

$$\text{Gain}(D, \text{texture}) = 0.381$$

$$\text{Gain } (D, \text{ navel}) = 0.289$$

$$\text{Gain } (D, \text{ touch}) = 0.006$$

By comparison, it is found that the feature "texture" has the greatest information gain, so it is selected as the root node of the decision tree, resulting in Fig. 10-5.

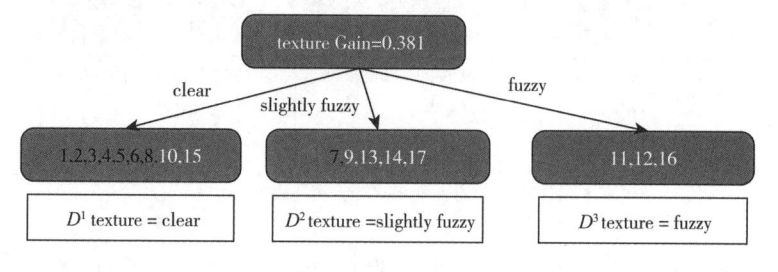

Fig. 10-5   Information gain of "texture"

3. Calculate information gain to determine the feature nodes of the second layer

Continue to classify each branch in Fig. 10-5, using the first branch node {"texture = clear"} in Fig. 10-5 as an example to classify this node. Assuming the sample set $D^1\{1, 2, 3, 4, 5, 6, 8, 10, 15\}$ for this node contains 9 samples and the feature set available is {color, root, tapping sound, navel, touch}. Therefore, based on D^1, the information gain of each feature can be calculated:

$$\text{Gain } (D^1, \text{ root}) = 0.458$$

$$\text{Gain } (D^1, \text{ tapping sound}) = 0.331$$

$$\text{Gain } (D^1, \text{ color}) = 0.043$$

$$\text{Gain } (D^1, \text{ navel}) = 0.458$$

$$\text{Gain } (D^1, \text{ touch}) = 0.458$$

By comparison, it is found that the three attributes of "root", "navel" and "touch" all have achieved the maximum information gain, and one of them can be randomly selected as the classification attribute. Here, "root" is selected, thus resulting in Fig. 10-6.

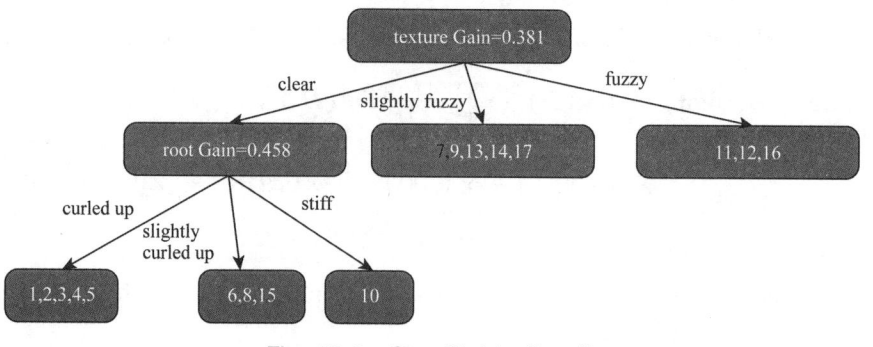

Fig. 10-6   Classified by "root"

4. Calculate information gain to determine the feature nodes of the third layer

Continue to recursively classify each branch node in Fig. 10-6. Taking the node {"root = curled

up"} in Fig. 10-6 as an example, let the sample set of this node be {1, 2, 3, 4, 5}, consisting of five samples. However, the labels of these five samples are all "good melon". Therefore, all samples contained in the current node belong to the same category and do not need to be classified. Mark the current node as a Class C (in this example, "good melon") leaf node and return recursively. Therefore, Fig. 10-6 has become Fig. 10-7.

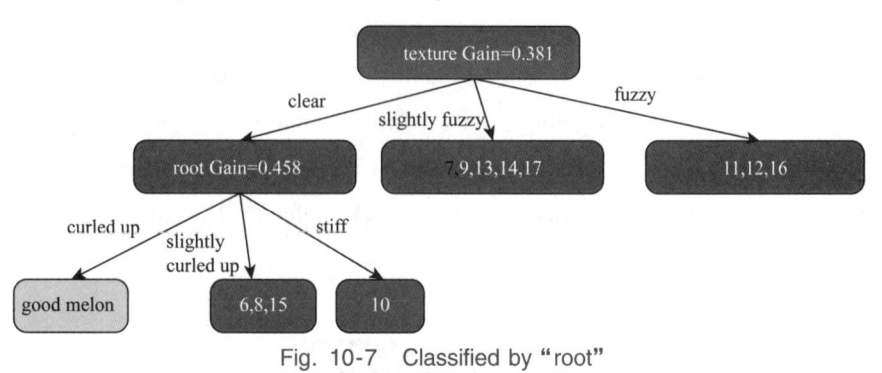

Fig. 10-7 Classified by "root"

Next, classify the node {"root = slightly curled up"} in Fig. 10-7, with a sample set of D^1 {6, 8, 15}, consisting of three samples. The feature set available is {color, tapping sound, navel, touch}, and the information gain of each feature can also be calculated subject to the process as follows:

$$Gain\ (D^1,\ touch) = 0.251$$
$$Gain\ (D^1,\ tapping\ sound) = 0$$
$$Gain\ (D^1,\ navel) = 0$$
$$Gain\ (D^1,\ color) = 0.251$$

From the calculation results, it can be seen that both the "color" and "touch" attributes have achieved the maximum information gain, and one attribute can be selected for classification. Here, "color" is selected for classification, resulting in Fig. 10-8.

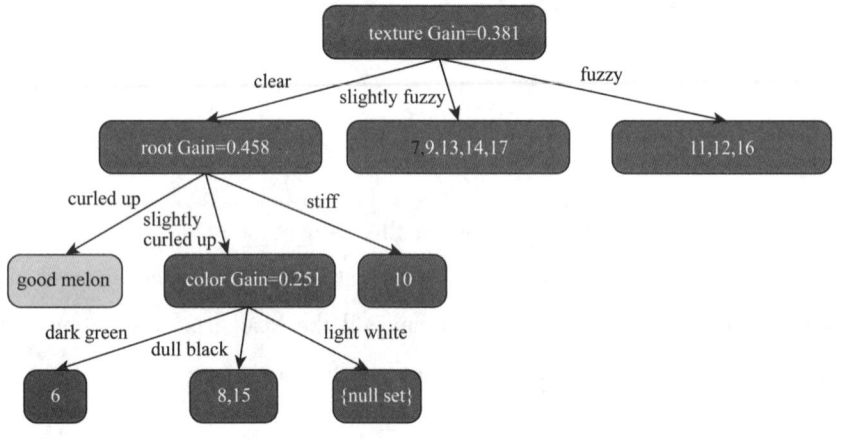

Fig. 10-8 Classified by "Color"

Continuing recursively and let's see the node "color = dark green", which only contains one sample and does not need to be further classified. Instead, mark the current node as a leaf node with the category of the current sample, namely, good melon. Return recursively. Then recursively classify the node "color = dull black", with no more repeated here. After the recursive depth has processed the "color = dull black" branch, return to process the "color = light white" node, because the sample set contained in the current node is empty and cannot be classified. The corresponding processing measures are as follows: set it as a leaf node, and set the category to the one with the most samples in its parent node (root = slightly curled). "root = slightly curled" contains three samples {6, 8, 15}, with 6, 8 being positive samples and 15 being negative samples, Therefore, the category of the "color = light white" node is positive (good melon). The final decision tree obtained is shown in Fig. 10-9.

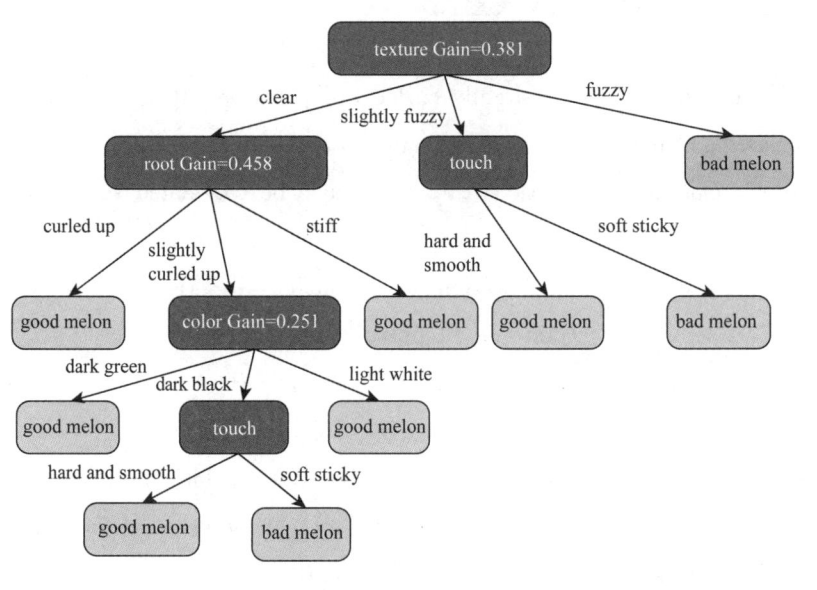

Fig. 10-9   Watermelon classification decision tree

10.5 Code Implementation for Iris Classification Using Decision Trees

1. Import the data load package and open the dataset

The iris dataset is used here. As a classic dataset in machine learning and statistics, it is included in the Scikit-learn datasets model. We can call the load_iris() function to load data, or download the dataset locally and open it. The code is as follows:

```
1.  import numpy as np
2.  import os
3.  % matplotlib inline
4.  import matplotlib
5.  import matplotlib.pyplot as plt
6.  plt.rcParams['axes.labelsize']=14
7.  plt.rcParams['xtick.labelsize']=12
8.  plt.rcParams['ytick.labelsize']=12
9.  # display Chinese
10. plt.rcParams['font.sans-serif'] = ['SimHei'] 11.plt.rcParams['
    axes.unicode_minus'] = False
12. import warnings
13. warnings.filterwarnings('ignore')
```

2. Load data

Use the load_iris method to open the dataset and use the head( ) function to view the first five pieces of data. At the same time, we can use the info( ) or head( ) function to view the data information via the code as follows:

```
1.  from sklearn.datasets import load_iris
2.  from sklearn.tree import DecisionTreeClassifier
3.  iris = load_iris()
4.  X = iris.data[:,2:]
5.  y = iris.target
```

Each row of data represents one aspect of the flower information, with four features and one label: calyx length 'sepal length (cm)', calyx width 'sepal width (cm)', petal length 'petal length (cm)', and petal width 'petal width (cm)'. The "Species" field indicates the type of flower, namely, its label.

3. Define decision trees

Here, a decision tree is defined, setting the max_depth value as 2; then use the fit method to train it. From the code execution effect, we can see the information of the generated decision tree tree_clf, with the code as follows:

```
1.  tree_clf = DecisionTreeClassifier(max_depth = 2)
2.  tree_clf.fit(X,y)
```

After the code runs, it will output the basic parameters of the defined decision tree. The gini

coefficient is used as the basis for generating the decision tree, with the maximum depth (max_depth) 2 and the minimum number of leaf nodes 2. See Fig. 10-10 for the results.

```
DecisionTreeClassifier(ccp_alpha=0.0, class_weight=None, criterion='gini',
                       max_depth=2, max_features=None, max_leaf_nodes=None,
                       min_impurity_decrease=0.0, min_impurity_split=None,
                       min_samples_leaf=1, min_samples_split=2,
                       min_weight_fraction_leaf=0.0, presort='deprecated',
                       random_state=None, splitter='best')
```

Fig. 10-10　DecisionTreeClassifier running results

4. Decision trees visualization

We can use the dot command-line tool in the graphviz package to convert this dot file to various formats, such as PDF or png. The following code can convert the dot file into a png file:

```
1.  from sklearn.tree import export_graphviz
2.  # modify the visible function parameters
3.  export_graphviz(
4.      tree_clf,
5.      out_file = 'iris_tree1.dot',
6.      # introduce the feature values
7.      feature_names = iris.feature_names[2:],
8.      # introduce the label name
9.      class_names = iris.target_names,
10.     rounded = True,
11.     filled = True
12. )
```

After the code runs, it will generate a dot file and execute $ dot-Tpng iris_tree1. dot-o iris_tree1. png command to convert a dot file into a png file and display it again. The code is as follows:

```
1.  from IPython.display import Image
2.  Image(filename = 'iris_tree1.png',width =400,height =400)
```

After the code runs, it will display the process of generating the decision tree in a graphical manner, and also display the results of gini coefficient calculation and set classification during each decision process, as shown in Fig. 10-11.

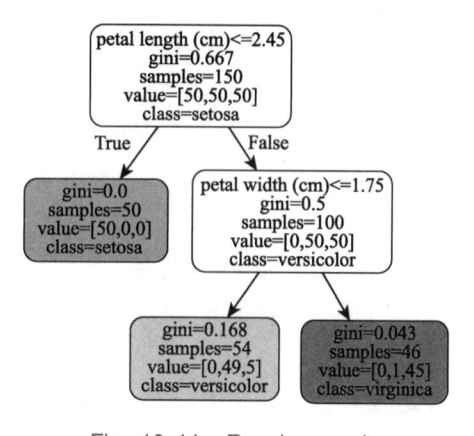

Fig. 10-11 Running results

10.6 Visualization of Iris Classification Effect

After classification, the decision tree can have its decision boundary plotted, through which the results of classification can be displayed more intuitively via the code as follows:

```
1.  from matplotlib.colors import ListedColormap
2.
3.  def plot_decision_boundary(clf, X, y, axes = [0, 7.5, 0, 3], iris =
    True, legend = False, plot_training = True):
4.      x1s = np.linspace(axes[0], axes[1], 100)
5.      x2s = np.linspace(axes[2], axes[3], 100)
6.      x1, x2 = np.meshgrid(x1s, x2s)
7.      X_new = np.c_[x1.ravel(), x2.ravel()]
8.      y_pred = clf.predict(X_new).reshape(x1.shape)
9.      custom_cmap = ListedColormap(['#fafab0','#9898ff','#a0faa0'])
10.     plt.contourf(x1, x2, y_pred, alpha = 0.3, cmap = custom_cmap)
11.     if not iris:
12.         custom_cmap2 = ListedColormap(['#7d7d58','#4c4c7f','#507d50'])
13.         plt.contour(x1, x2, y_pred, cmap = custom_cmap2, alpha = 0.8)
14.     if plot_training:
15.         plt.plot(X[:, 0][y = =0], X[:, 1][y = =0], "yo", label = "Iris -
    Setosa")
```

```
16.           plt.plot(X[:, 0][y = =1], X[:, 1][y = =1], "bs", label = "Iris -
   Versicolor")
17.           plt.plot(X[:, 0][y = =2], X[:, 1][y = =2], "g^", label = "Iris -
   Virginica")
18.       plt.axis(axes)
19.    if iris:
20.       plt.xlabel("Petal length", fontsize =14)
21.       plt.ylabel("Petal width", fontsize =14)
22.    else:
23.       plt.xlabel(r" $ x_1 $ ", fontsize =18)
24.       plt.ylabel(r" $ x_2 $ ", fontsize =18, rotation =0)
25.    if legend:
26. plt.legend(loc = "lower right", fontsize =14) 27.
28. plt.figure(figsize = (8, 4))
29. plot_decision_boundary(tree_clf, X, y)
30. plt.plot([2.45, 2.45], [0, 3], "k - ", linewidth =2)
31. plt.plot([2.45, 7.5], [1.75, 1.75], "k - - ", linewidth =2)
32. plt.plot([4.95, 4.95], [0, 1.75], "k:", linewidth =2)
33. plt.plot([4.85, 4.85], [1.75, 3], "k:", linewidth =2)
34. plt.text(1.40, 1.0, "Depth =0", fontsize =15)
35. plt.text(3.2, 1.80, "Depth =1", fontsize =13)
36. plt.text(4.05, 0.5, "(Depth =2)", fontsize =11)
37. plt.title('Decision Tree decision boundaries')
38.
39. plt.show()
```

After running the code, we can see the entire process of generating the decision tree, as shown in Fig. 10- 11. The horizontal axis represents the length of the petals while the vertical axis represents the width of the petals. In Fig. 10-12, the horizontal lines are classified according to the width of the petals, while the vertical lines are classified according to the length of the petals. Inclusively, the black solid line represents the first classification, the thick dashed line represents the second classification, and the thin dashed line represents the third classification. In this way, the iris data samples are divided into three categories.

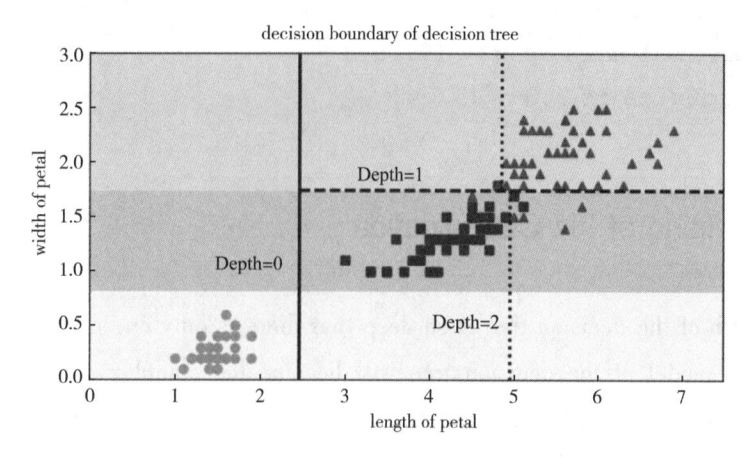

Fig. 10-12 Decision tree running results

10.7 | Probability Estimation for Iris Classification

Decision trees can be used to classify samples, but such a situation can also be encountered in algorithm use as follows: The model predicts the sample X, with a probability of 51% belonging to class A and 49% belonging to class B. At this time, the algorithm concludes that the sample X belongs to class A, but it is not so firmly confident in the results of this prediction. For certain scenarios in reality where it is better not to make mistakes than to have a try, the current prediction is not what is needed. So, when making algorithms, the model needs to output the probability that samples belong to each category.

In this example, the input data is a flower with the petal length and width of 5 cm and 1.5 cm respectively. The corresponding leaf node is a left node with the depth of 2, so the decision tree should output the following probability:

```
tree_clf.predict_proba([[5,1.5]])
```

Use the predict( ) function again:

```
tree_clf.predict([[5,1.5]])
```

The output of the code represents the probabilities that the flowers 5 cm in petal length and 1.5 cm in petal width are classified as three kinds of iris flowers respectively as follows:

```
Iris-Setosa is 0% (0/54);
Iris-Versicolor is 90.7% (49/54);
Iris-Virginica is 9.3% (5/54).
```

10.8 Overfitting of Iris Classification

When the depth of the decision tree is so deep that there is only one or very few objects left in the leaf nodes, the model of the decision tree will become too complex, which can easily cause overfitting and reduce generalization ability. One solution is to find a point (depth) to stop the decision tree from splitting, avoid making the tree grow too long, and avoid making the tree too detailed. We can set the maximum depth of the tree and the minimum size of the leaves to prevent overfitting. The following code sets min_samples_leaf = 4 to prevent overfitting in the decision tree, with the specific code as follows:

```
1.  tree_clf1 = DecisionTreeClassifier(random_state = 42)
2.  tree_clf2 = DecisionTreeClassifier(min_samples_leaf = 4, random_
    state = 42)
3.  tree_clf1.fit(X,y)
4.  tree_clf2.fit(X,y)
```

Plot the decision boundary output subject to the code as follows:

```
1.  plt.figure(figsize = (12,4))
2.  plt.subplot(121)
3.  plot_decision_boundary(tree_clf1,X,y,axes = [-1.5,2.5, -1,1.5])
4.
5.  plt.subplot(122)
6.  plot_decision_boundary(tree_clf2,X,y,axes = [-1.5,2.5, -1,1.5])
7.  plt.show()
```

After the code runs, visualize the decision boundary of the two data tests. As can be seen, Fig. 10-13 (a) shows the overfitting state of the decision tree; Fig. 10-13 (b) suppresses the occurrence of overfitting by setting min_samples_leaf = 4.

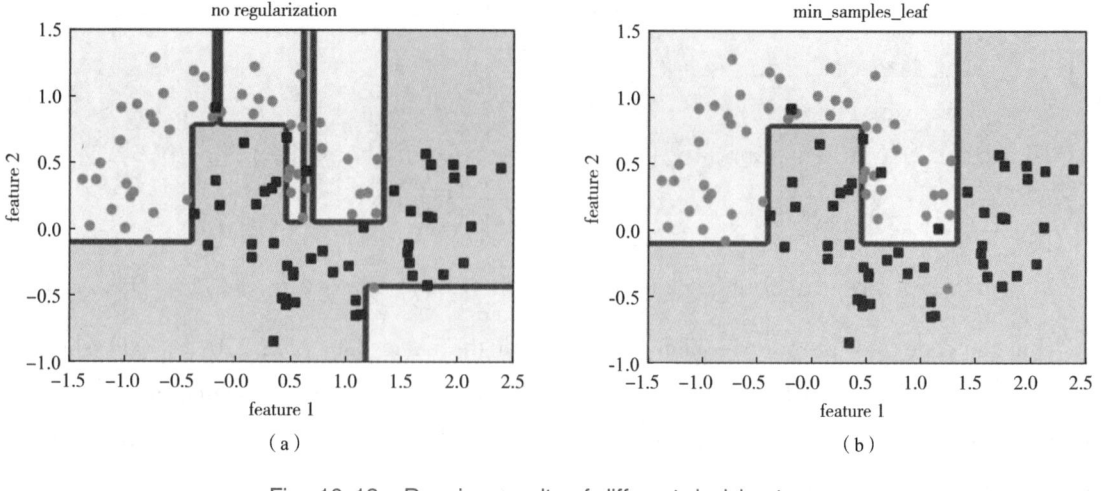

Fig. 10-13　Running results of different decision trees

10.9 Sensitivity of Decision Trees to Data

Decision trees are sensitive to data. The most obvious point is the division of the decision boundary. The decision boundary of the decision tree is quite regular and straight, perpendicular to the horizontal or vertical axis, which leads to the fact that the decision boundary plotted may not be the real one. Moreover, decision trees are very sensitive to data rotation and training details. The following code demonstrates the use of decision trees to classify data before and after data rotation, resulting in completely different results, as follows:

```
1.  np.random.seed(6)
2.  Xs = np.random.rand(100,2) - 0.5
3.  ys = (Xs[:,0] >0).astype(np.float32) * 2
4.
5.  angle = np.pi/4
6.  rotation_matrix = np.array([[np.cos(angle), - np.sin(angle)],
    [np.sin(angle),np.cos(angle)]])
7.  Xsr = Xs.dot(rotation_matrix)
8.
9.  tree_clf_s = DecisionTreeClassifier(random_state = 42)
10. tree_clf_sr = DecisionTreeClassifier(random_state = 42)
11. tree_clf_s.fit(Xs,ys)
12. tree_clf_sr.fit(Xsr,ys)
```

```
13.
14. plt.figure(figsize = (12,4))
15. plt.subplot(121)
16. plot_decision_boundary(tree_clf_s,Xs,ys,axes = [-1,1, -0.6,1])
17.
18. plt.subplot(122)
19. plot_decision_boundary(tree_clf_sr,Xsr,ys,axes = [-1,1, -0.6,1])
```

Run the code and use two decision trees to train the same data. Fig. 10-14 (a) shows the original data, while Fig. 10-14 (b) shows the rotated data. As can be seen, the shape of the decision boundary before and after rotation has changed, which proves that the decision boundary of the decision tree is highly sensitive to data rotation.

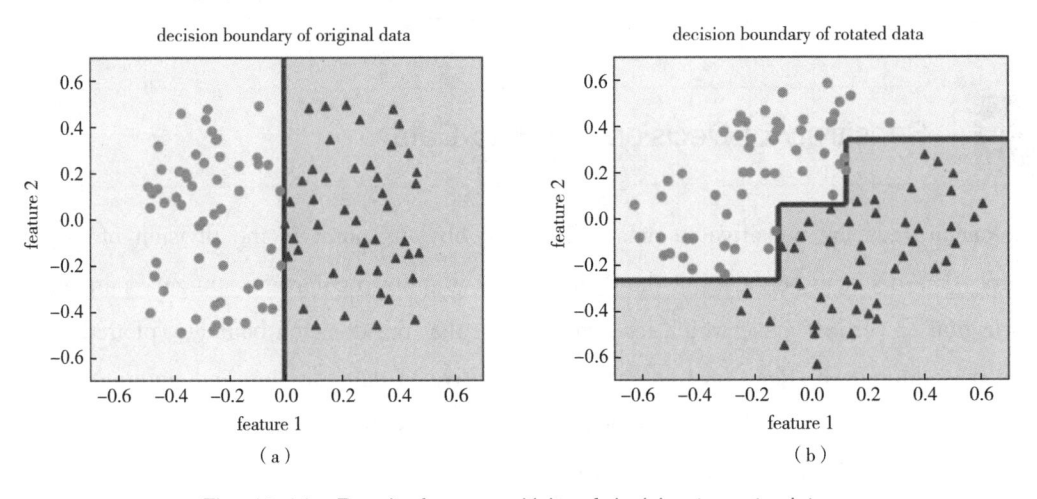

Fig. 10-14　Results from sensitivity of decision trees to data

Unit 11

Use Support Vector Machine for Iris Classification

SVM (support vector machine) is a supervised learning algorithm for regression and classification, but it is more often used for classification. SVM performs well in various settings and is generally considered the best "out-of-the-box" classifier. In this unit, SVM is used to classify iris flowers.

11.1 SVM Principles

The support vector machine is a dichotomy model. Its basic model is the linear classifier spaced largest and defined in the feature space. The maximum spacing sets it apart from the perceptron. SVM also includes kernel techniques, which makes it a virtually nonlinear classifier. The learning strategy of SVM is to maximize the space, which can be formalized as a problem of solving convex quadratic programming, and also equivalent to the minimization of regularized hinge loss function. The learning algorithm of SVM is an optimal algorithm for solving convex quadratic programming.

The basic idea of SVM learning is to solve the separation hyperplane that can correctly classify the training dataset and has the largest geometric spacing. As shown in Fig. 11-1, $w \cdot x + b = 0$ is the separation hyperplane. For linearly separable datasets, there are

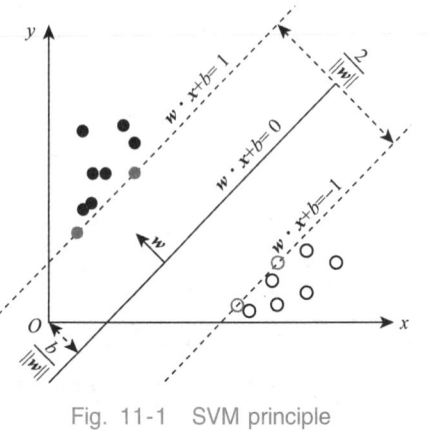

Fig. 11-1 SVM principle

infinite hyperplanes (i. e. perceptrons), but the separation hyperplane with the largest geometric spacing is unique.

 Linear Separable SVM

1. Basic concepts of linearly separable support vector machines

For linearly separable datasets, the goal of learning is to find a separating hyperplane in the feature space that can divide instances into different classes. The separating hyperplane can divide the feature space into two parts: positive class and negative class. The normal vector of the separating hyperplane points to the positive class on one side and the negative class on the other. Moreover, it is required that the sum of the distances of the two points closest to the separating hyperplane is the largest and this separating hyperplane is called the separating hyperplane with the largest separation space. The mathematical model of linearly separable support vector machines is as follows:

$$f(x) = \mathrm{sgn}(wx + b)$$

Where, $wx + \mathrm{b} = 0$ can be written as $w \cdot x + b = 0$ if represented with a vector (w, x represent the vector of weight w and feature x respectively), which is the separating hyperplane with the maximum space. The model to be solved is this separating hyperplane with the maximum space, as shown in Fig. 11-2. Inclusively, negative objects indicate negative sample objects while positive objects indicate positive sample objects; $w \cdot x + b = \pm 1$ refers to the equations of the upper and lower hyperplanes, and the points on the two hyperplanes are support vectors.

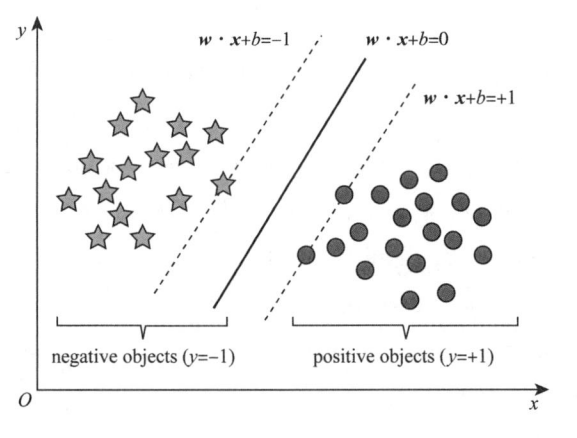

Fig. 11-2 Separating hyperplanes with the maximum space

2. Code implementation

There are two types of linear support vector machines: SVC (support vector classification), used for classification problems; SVR (support vector regression), used for regression problems.

The following example uses the iris dataset, which is classified based on the calyx length and width of iris. The two-dimensional features used only can facilitate visualization. The code is as follows:

```
1.  from sklearn.svm import SVC
2.  from sklearn import datasets
3.
4.  iris = datasets.load_iris()
5.  X = iris['data'][:,(2,3)]
6.  y = iris['target']
7.
8.  setosa_or_versicolor = (y == 0)|(y == 1)
9.  X = X[setosa_or_versicolor]
10. y = y[setosa_or_versicolor]
11. # kernel = " linear " (linear kernel function) gives a linear
    decision boundary. The separation boundary between the two types
    is a straight line
12. svm_clf = SVC(kernel = 'linear',C = float('inf'))
13. svm_clf.fit(X,y)
```

Firstly, generate an SVC vector machine, set the kernel function as a linear kernel function, let the classification boundary be a straight line and then visualize the effectiveness of SVC classification with the code as follows:

```
1.  # general model
2.  x0 = np.linspace(0, 5.5, 200)
3.  pred_1 = 5* x0 - 20
4.  pred_2 = x0 - 1.8
5.  pred_3 = 0.1 * x0 + 0.5
6.  def plot_svc_decision_boundary(svm_clf, xmin, xmax,sv = True):
7.      w = svm_clf.coef_[0]
8.      b = svm_clf.intercept_[0]
9.      print(w)
10.     x0 = np.linspace(xmin, xmax, 200)
11.     decision_boundary = - w[0]/w[1] * x0 - b/w[1]
12.     margin = 1/w[1]
13.     gutter_up = decision_boundary + margin
14.     gutter_down = decision_boundary - margin
15.     if sv:
```

```
16.            svs = svm_clf.support_vectors_
17.            plt.scatter(svs[:,0],svs[:,1],s =180,facecolors = '#FFAAAA')
18.       plt.plot(x0,decision_boundary,'k - ',linewidth =2)
19.       plt.plot(x0,gutter_up,'k - - ',linewidth =2)
20.       plt.plot(x0,gutter_down,'k - - ',linewidth =2)
21. plt.figure(figsize = (14,4))
22. plt.subplot(121)
23. plt.plot(X[:,0][y = =1],X[:,1][y = =1],'bs')
24. plt.plot(X[:,0][y = =0],X[:,1][y = =0],'ys')
25. plt.plot(x0,pred_1,'g - - ',linewidth =2)
26. plt.plot(x0,pred_2,'m - ',linewidth =2)
27. plt.plot(x0,pred_3,'r - ',linewidth =2)
28. plt.axis([0,5.5,0,2])
29.
30. plt.subplot(122)
31. plot_svc_decision_boundary(svm_clf, 0, 5.5)
32. plt.plot(X[:,0][y = =1],X[:,1][y = =1],'bs')
33. plt.plot(X[:,0][y = =0],X[:,1][y = =0],'ys')
34. plt.axis([0,5.5,0,2])
```

After running the code, the results as shown in Fig. 11-3. As can be seen from the visualization results, the two types of data have been completely separated with the boundary points marked, as shown in Fig. 11-3.

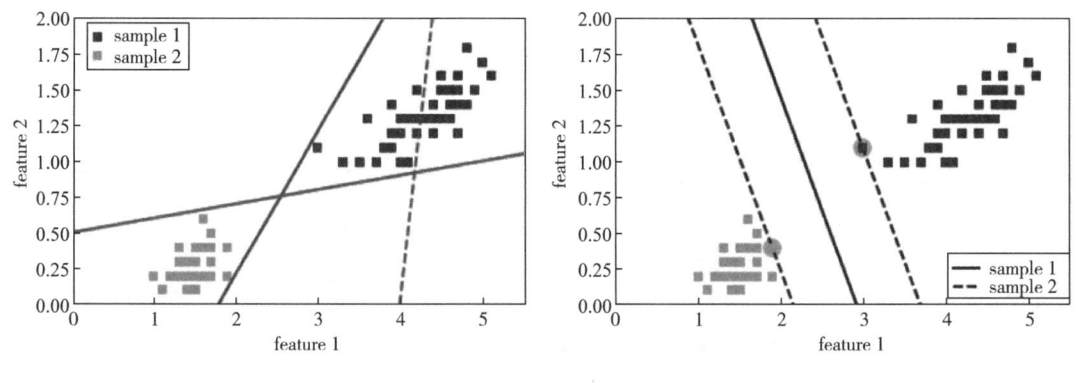

Fig. 11-3　Running results of SVC vector machine

3. Soft margin SVM

The role of support vector machines is to make the two types of data points as far apart as possible, while ensuring complete classification accuracy. In the iris classification shown in Fig. 11-4,

the horizontal axis indicates the petal length while the vertical axis the petal width. As shown in Fig. 11-4 (a), the decision boundary is not good enough because of the influence of outliers [the decision boundary in Fig. 11-4 (b) deviated]. How to solve this problem? The soft margin can be applied, in essence, to adjust the division conditions of the decision boundary and allow error overfitting.

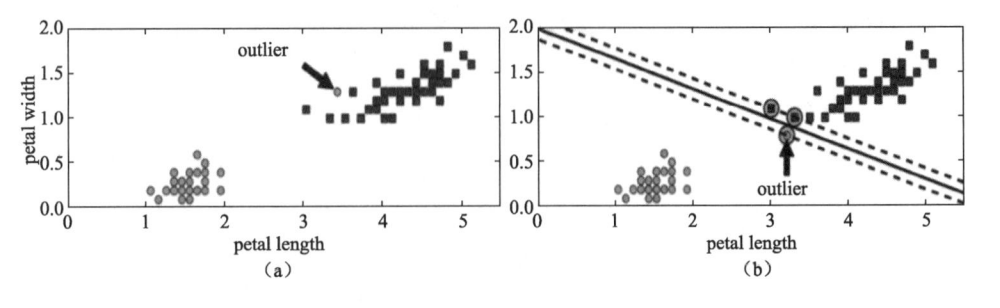

Fig. 11-4 Soft margin instance

The relaxation factor C has been added in the following example. When C tends to positive infinity, it means that the classification is strict, allowing no errors; when C tends to negative infinity, it means that there can be greater error tolerance. The code is as follows:

```
1.  iris = datasets.load_iris()
2.  X = iris["data"][:,(2,3)] # petal length, petal width
3.  y = (iris["target"] = =2).astype(np.float64) # Iris -Viginica
4.  svm_clf = Pipeline((
5.  ('std',StandardScaler()),
6.  ('linear_svc',LinearSVC(C =1))
7.  ))
8.  svm_clf.fit(X,y)
```

Set C = 1 and it will output the structure of the LinearSVC model, with the code running results as shown in Fig. 11-5.

```
Pipeline(memory=None,
         steps=[('std',
                 StandardScaler(copy=True, with_mean=True, with_std=True)),
                ('linear_svc',
                 LinearSVC(C=1, class_weight=None, dual=True,
                           fit_intercept=True, intercept_scaling=1,
                           loss='squared_hinge', max_iter=1000,
                           multi_class='ovr', penalty='12', random_state=None,
                           tol=0.0001, verbose=0))],
         verbose=False)
```

Fig. 11-5 LinearSVC running results

Different classification effects can be achieved by adjusting the value of C, with the following code:

```
1.  scaler = StandardScaler()
2.  svm_clf1 = LinearSVC(C = 1, random_state = 42)
3.  svm_clf2 = LinearSVC(C = 100, random_state = 42)
4.
5.  scaled_svm_clf1 = Pipeline((
6.      ('std', scaler),
7.      ('linear_svc', svm_clf1)
8.  ))
9.
10. scaled_svm_clf2 = Pipeline((
11.     ('std', scaler),
12.     ('linear_svc', svm_clf2)
13. ))
14. scaled_svm_clf1.fit(X, y)
15. scaled_svm_clf2.fit(X, y)
16.
17. b1 = svm_clf1.decision_function([ - scaler.mean_ / scaler.scale_])
18. b2 = svm_clf2.decision_function([ - scaler.mean_ / scaler.scale_])
19. w1 = svm_clf1.coef_[0] / scaler.scale_
20. w2 = svm_clf2.coef_[0] / scaler.scale_
21. svm_clf1.intercept_ = np.array([b1])
22. svm_clf2.intercept_ = np.array([b2])
23. svm_clf1.coef_ = np.array([w1])
24. svm_clf2.coef_ = np.array([w2])
```

Visualize the results with the following code：

```
1.  plt.figure(figsize = (14, 4.2))
2.  plt.subplot(121)
3.  plt.plot(X[:, 0][y = =1], X[:, 1][y = =1], "g^", label = "Iris -
    Virginica")
4.  plt.plot(X[:, 0][y = =0], X[:, 1][y = =0], "bs", label = "Iris -
    Versicolor")
5.  plot_svc_decision_boundary(svm_clf1, 4, 6, sv = False)
6.  plt.xlabel("petal length", fontsize =14)
7.  plt.ylabel("petal width", fontsize =14)
8.  plt.legend(loc = "upper left", fontsize =14)
```

```
9.  plt.title(" $ C={} $ ".format(svm_clf1.C), fontsize=16)
10. plt.axis([4, 6, 0.8, 2.8])
11.
12. plt.subplot(122)
13. plt.plot(X[:, 0][y==1], X[:, 1][y==1], "g^")
14. plt.plot(X[:, 0][y==0], X[:, 1][y==0], "bs")
15. plot_svc_decision_boundary(svm_clf2, 4, 6, sv=False)
16. plt.xlabel("petal length", fontsize=14)
17. plt.title(" $ C={} $ ".format(svm_clf2.C), fontsize=16)
18. plt.axis([4, 6, 0.8, 2.8])
```

Run the code and see Fig. 11-6 for the results. As can be seen from the visualization results, when a higher C value is used in Fig. 11-6 (b), the classifier will reduce misclassification, but ultimately there will be a smaller interval. Moreover, when a lower C value is used in Fig. 11-6 (a), the interval will be much larger, but many instances will eventually appear within the interval.

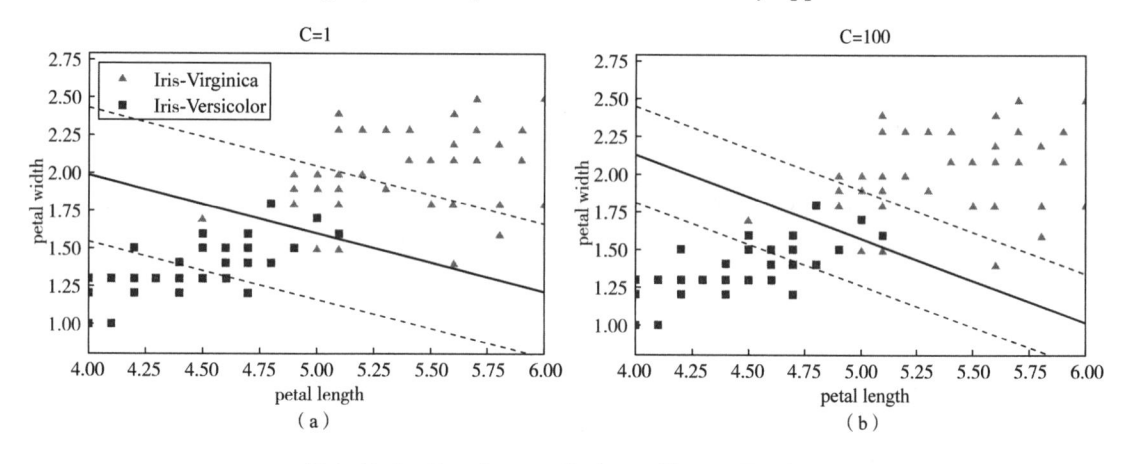

Fig. 11-6 Running results from different C values

11.3   Nonlinear SVM

1. Basic concepts of nonlinear support vector machines

For a set of datasets that are neither linearly separable nor approximately linearly separable, it is called linearly inseparable, as shown in Fig. 11-7.

In Fig. 11-7, circles and crosses are used to represent different label classifications. It is obvious that this dataset cannot be correctly classified through a straight line, but it can be correctly classified by an elliptical curve, as shown in Fig. 11-8.

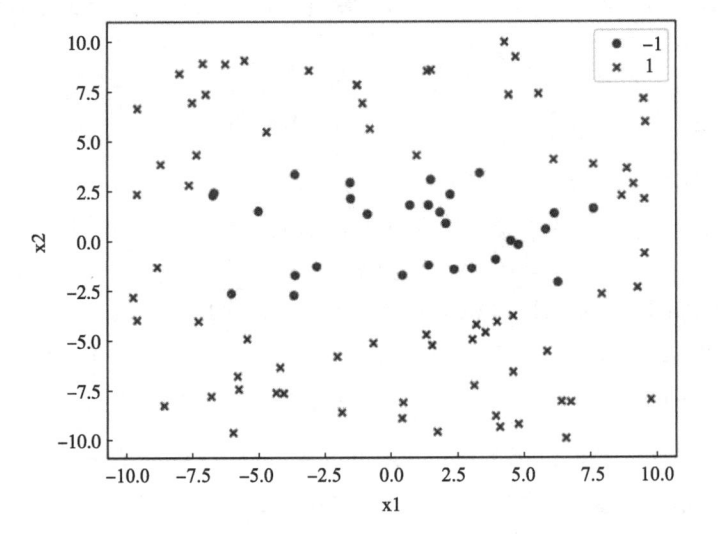

Fig. 11-7　Datasets linearly separable, approximately linearly separable and linearly inseparable

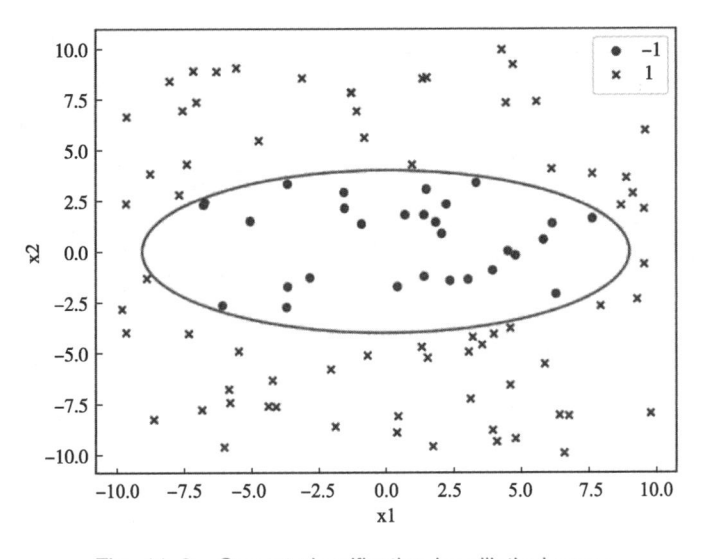

Fig. 11-8　Correct classification by elliptical curve

However, it is relatively difficult to solve the nonlinear classification problem of elliptic curves in Fig. 11-8. Since the linear classification problem is relatively easy to solve, can this dataset be transformed into a linearly classified data through certain nonlinear changes?

Apply nonlinear transformation to the dataset and square the features ($Z = x \times x$). At this point, it can be seen that the converted data can be correctly classified through a straight line. The original elliptical curve has been transformed into the straight line in Fig. 11-9, and the original nonlinear classification problem has been transformed into a linear classification problem.

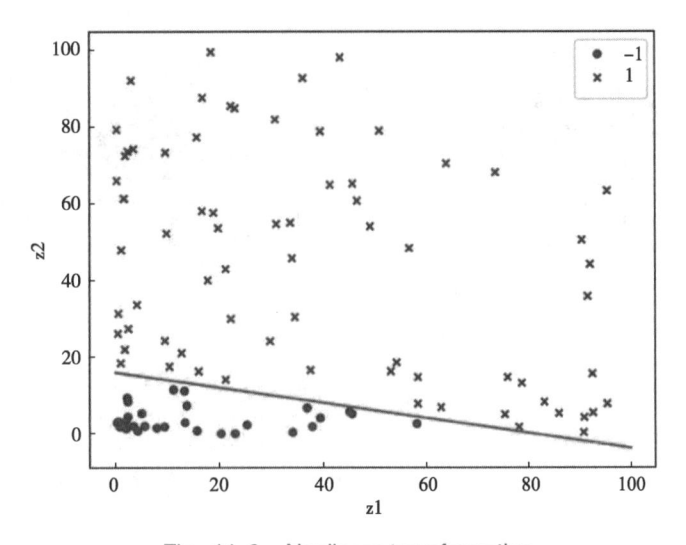

Fig. 11-9   Nonlinear transformation

In the above example, the original space is called the input space X, and the mapped new space is called the feature space H. Kernel functions can be used for transformations. The commonly used kernel functions include linear kernel, Gaussian kernel, and polynomial kernel. Usually, kernel functions are used to convert low-dimensional data into high-dimensional data.

As shown in Fig. 11-10, a kernel function can be used to map the line in $\mathbf{R}^2$ space to a plane in $\mathbf{R}^3$ space. At this point, a plane can be used to separate the two types of data in $\mathbf{R}^3$ space.

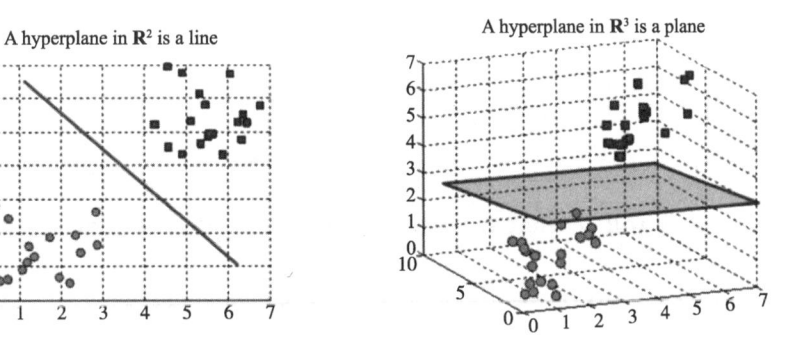

Fig. 11-10   Low dimensional data converted to high dimensional data

The Gaussian kernel function can be used to map a 2D data to a 3D data, and then the hyperplane can be used for separation.

In the classification of cancer tumors, see Fig. 11-11 for the distribution of the "cancer" and "normal" datasets. The sum function (kernel) can be used to map the dataset to a three-dimensional space, where a decision surface can be found to separate the cancer and benign data.

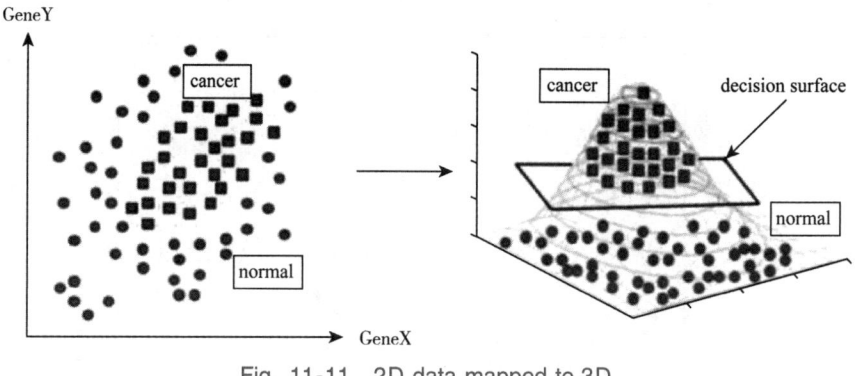

Fig. 11-11   2D data mapped to 3D

2. Code implementation

Use the make_moons( ) function to generate a nonlinear data, with the following code:

```
1.  from sklearn.datasets import make_moons
2.  X, y =make_moons(n_samples =100, noise =0.15, random_state =42)
3.
4.  def plot_dataset(X, y, axes):
5.      plt.plot(X[:, 0][y = =0], X[:, 1][y = =0], "bs")
6.      plt.plot(X[:, 0][y = =1], X[:, 1][y = =1], "g^")
7.      plt.axis(axes)
8.      plt.grid(True, which = 'both')
9.      plt.xlabel(r" $ x_1 $ ", fontsize =20)
10.     plt.ylabel(r" $ x_2 $ ", fontsize =20, rotation =0)
11.
12. plot_dataset(X, y, [-1.5, 2.5, -1, 1.5])
13. plt.show()
```

Run the code and it will generate two crescent datasets interleaved with each other, as shown in Fig. 11-12.

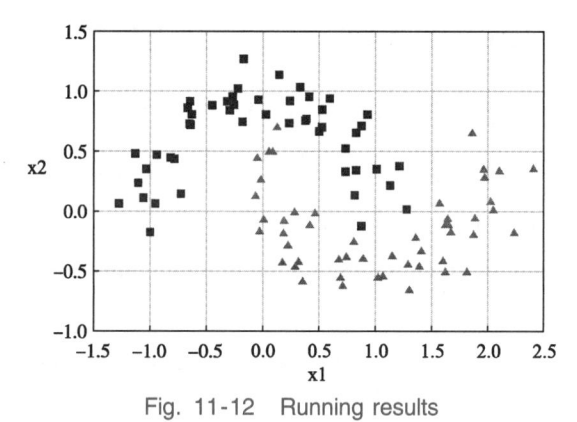

Fig. 11-12   Running results

Define a nonlinear support vector machine with the following code:

```
1.   from sklearn.datasets import make_moons
2.   from sklearn.pipeline import Pipeline
3.   from sklearn.preprocessing import PolynomialFeatures
4.
5.   polynomial_svm_clf = Pipeline((
                    ("poly_features",PolynomialFeatures(degree =3)),
6.                  ("scaler",StandardScaler()),
7.                  ("svm_clf",LinearSVC(C =10,loss = "hinge"))
8.                  ))
9.
10. polynomial_svm_clf.fit(X,y)
```

Here, a Pipeline is used to generate a LinearSVC that can visualize the classification results with the code as follows:

```
1.   def plot_predictions(clf,axes):
2.       x0s =np.linspace(axes[0],axes[1],100)
3.       x1s =np.linspace(axes[2],axes[3],100)
4.       x0,x1 =np.meshgrid(x0s,x1s)
5.       X =np.c_[x0.ravel(),x1.ravel()]
6.       y_pred =clf.predict(X).reshape(x0.shape)
7.       plt.contourf(x0,x1,y_pred,cmap =plt.cm.brg,alpha =0.2)
8.
9.   plot_predictions(polynomial_svm_clf,[ -1.5,2.5, -1,1.5])
10. plot_dataset(X,y,[ -1.5,2.5, -1,1.5])
```

After running the code, it can be seen that the support vector machine can effectively separate the two types of data points, with the visualization results as shown in Fig. 11-13.

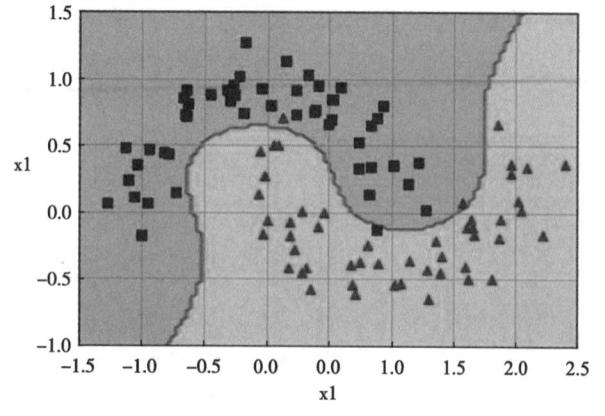

Fig. 11-13   Running results from nonlinear support vector machine

11.4   Differences Between Linear Separable, Linear Inseparable, and Nonlinear SVM

If the data is linearly separable, there must be a unique hyperplane to completely separate the samples and maximize the interval. At this time, the classifier is a linearly separable support vector machine. However, in reality, there are few completely linear separable cases. As shown in Fig. 11-14, due to the existence of two samples, it is actually impossible to have a hyperplane (straight line in 2D) to completely separate the data. Moreover, only a small number of samples lead to linear inseparability (a nonlinear support vector machine is required when completely or mostly linearly inseparable), so the dataset can be approximately regarded as linearly separable. In fact, there is still an infinite hyperplane that can segment the dataset and it is sufficient to select a hyperplane that can ensure the maximum interval and the minimum number of misclassifications. This is the so-called linear inseparable support vector machine based on soft interval maximization. However, when the data is completely linearly inseparable, there is no such hyperplane, and only a hypersurface can be used, as shown in Fig. 11-15. Nonetheless, nonlinear problems are often difficult to solve and need to be transformed into linear problems by mapping the original space to a new space, so that the data can be made linearly separable. As required, the kernel function must be used with nonlinear support vector machines adopted for calculations.

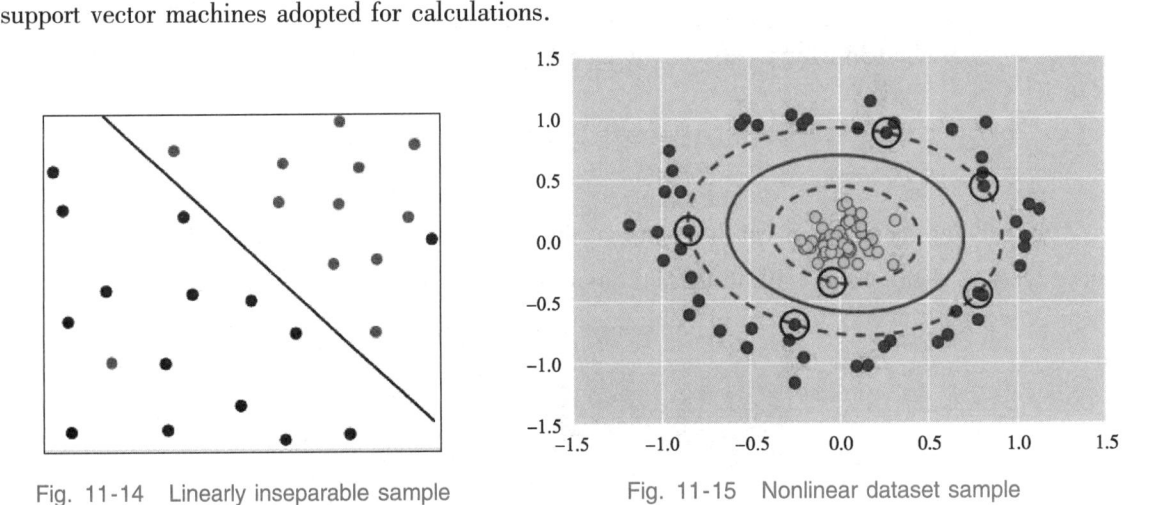

Fig. 11-14   Linearly inseparable sample   Fig. 11-15   Nonlinear dataset sample

Unit 12

Use Bayesian Algorithm to Classify Spams

With the rapid development and widespread application of the Internet, the widespread use of e-mail has brought considerable convenience to production and daily life, but the emergence of spams has brought considerable troubles. In response to the problem of spams, this unit takes the Bayesian algorithm as the theoretical basis, applies theory to engineering practice, and designs and implements a spam filtering system based on the Bayesian algorithm.

12.1 Introduction to Bayes Algorithm Principles

1. Bayes rule

Bayes theorem is such a theorem about the conditional probability (or marginal probability) of the random events A and B. Where, $P(A \mid B)$ is the probability of A occurring in the case of B, also known as Bayes rule, represented as follows:

$$P(A|B) = \frac{P(B|A)P(A)}{P(B)}$$

Derived from the joint probability formula:

$$P(A,B) = P(B|A)P(A) = P(A|B)P(B)$$

Where, $P(B)$ is called prior probability; $P(B|A)$ is called posterior probability; $P(B,A)$ is called joint probability.

2. Bayes algorithm

Bayes algorithm, also known as naive Bayes algorithm. Naiveness: Independent characteristic conditions. Bayesian: Based on Bayes theorem. It belongs to the generative model of supervised learning, simple to implement, free of iteration, and supported by a solid mathematical theory (i. e. Bayes theorem). It can perform well in a large number of samples and is not suitable for scenarios where the feature conditions of the input vector are related.

As a classification method based on Bayes theorem and independence assumption of conditional features, the naive Bayes algorithm can calculate the probability of classification through features and select the case with a large probability. Moreover, as a machine learning classification (supervised learning) method based on probability theory, it is widely used in the field of sentiment classification.

12.2 Three Commonly Used Models of Bayes Algorithm

1. Polynomial model

Repeated words are considered to appear repeatedly, as follows:

P (("acting", "invoice", "value added tax", "invoice", "regular", "invoice") | S) = P ("acting" | S) P ("invoice" | S) P ("value added tax" | S) P ("invoice" | S) P ("regular" | S) P ("invoice" | S) = P ("acting" | S) $P3$ ("invoice" | S) P ("value added tax" | S) P ("regular" | S)

When making statistics and judgments, attention shall be paid to the number of repetitions.

2. Bernoulli model

Treat repeated words as if they only appear once, as follows:

P (("acting", "invoice", "value added tax", "invoice", "regular", "invoice") | S) = P ("acting" | S) P ("invoice" | S) P ("value added tax" | S) P ("regular" | S)

This method is more simplified and convenient. Of course, it loses the word frequency information, so the effect will be much worse.

3. Hybrid model

When calculating sentence probability, the number of occurrences of repeated words is not taken into account, but when calculating the probability P ("word" | S) of words, the number of occurrences of repeated words will be taken into account.

The Bayes algorithm is still effective in situations with limited data and can handle multi-class problems. However, it is more sensitive to the method for preparing input data, suitable for those application scenarios where the data type is nominal data.

12.3 | Spam Classification

Spams used to be a headache that had long plagued e-mail operators and users. According to statistics, most of the e-mails received by users are spams. The traditional spam filtering methods mainly include keyword and checksum methods. The keyword filtering is implemented based on specific words, like spam keywords such as "invoice", "loan", "interest rate", "prize winning", "certificate handling", "lottery", "number", "money", "payment" and "luck", etc. But this method is not ideal and easy to be avoided. Firstly, these keywords may also be present in normal e-mails, which can easily lead to misjudgment.

The classification of spams had not achieved a good effect until the use of the "Bayesian" algorithm was proposed, and as the number of e-mails increases, the Bayesian classification would become more and more effective. The classification method used is to determine whether an e-mail is a spam through multiple words, but the probability is difficult to estimate. Through Bayes rule, it can be transformed into calculating the probability of these words appearing in the spam.

1. Read text data

Read text files in Jupyter Notebook and get a dataset with the code as follows:

```
1.  import numpy as np
2.  import pandas as pd
3.  import seaborn as sns
4.  import matplotlib.pyplot as plt
5.  import seaborn as sns
6.  # display Chinese
7.  plt.rcParams['font.sans-serif'] = ['SimHei']
8.  plt.rcParams['axes.unicode_minus'] = False
9.  df = pd.read_table("../input/SMSSpamCollection", sep = ' \t ',
       names = [' label', 'sms_message'])
10. df.head()
```

See Fig. 12-1 for the code running results. In the dataset, the columns are currently unnamed, indicating two columns. The first column has two values: "ham", indicating that the e-mail is not a spam; "spam", indicating that the e-mail is a spam. The second column is the text content of the classified information.

	label	sms_message
0	ham	Go until jurong point, crazy.. Available only ...
1	ham	Ok lar... Joking wif u oni...
2	spam	Free entry in 2 a wkly comp to win FA Cup fina...
3	ham	U dun say so early hor... U c already then say...
4	ham	Nah I don't think he goes to usf, he lives aro...

Fig. 12-1   Code running results

2. Prepare data

In data preprocessing, we have roughly understood the structure of the dataset. Now, we convert the labels into binary variables, where 0 represents "ham" (non-spam) and 1 represents "spam" (spams), which is more convenient for calculation. Why is this step a must? The reason is the way Scikit-learn processes input. Scikit-learn only processes numerical values, so if the label value is left as a string, Scikit-learn will convert it on its own (more precisely, the string label will be converted to an unknown floating-point value).

If the label is kept as a string, the model can still make predictions, but problems may occur later when the performance indicators (such as accuracy and recall scores) are calculated. Therefore, in order to avoid unexpected traps later on, it is best to convert the classification values to integers and then introduce them into the model.

Introductions: Use the mapping method to convert the values in the "Label" column to numerical values, as follows: { 'ham': 0, 'spam': 1 }. In this way, the "ham" value will be mapped to 0 and the "spam" value to 1. In addition, to know the size of the dataset being processed, use "shape" to output the number of rows and columns, as shown in the following code:

```
1.  print(df.shape)
2.  df['label'] =df.label.map({'ham':0, 'spam':1})
3.
4.  df['label'].head()
```

3. Data conversion

The Bag of Words (BoW) dataset contains a large amount of text data (5572 rows of data). Most machine learning algorithms require the input to be digital data, while e-mails are usually texts. Here follows the concept of Bag of Words (BoW). It is used to indicate that the problem to be addressed has a "large number of words" or a lot of text data. The basic concept of BoW means that a paragraph of text is taken out to calculate the frequency of words appearing in the text. **Note:** BoW treats every word equally, and the order in which words appear is not important.

We can convert a collection of documents into a matrix, where each document is a row and each word (token) is a column. The corresponding (row, column) value is the frequency at which each word (token) appears in this document.

Assuming there are four documents as follows:

```
['Hello, how are you! ',
'Win money, win from home.',
'Call me now',
'Hello, Call you tomorrow? ']
```

The goal is to convert this set of texts into a frequency distribution matrix, as shown in Fig. 12-2.

	are	call	from	hello	home	how	me	money	now	tomorrow	win	you
0	1	0	0	1	0	1	0	0	0	0	0	1
1	0	0	1	0	1	0	0	1	0	0	2	0
2	0	1	0	0	0	0	1	0	1	0	0	0
3	0	1	0	1	0	0	0	0	0	1	0	1

Fig. 12-2 Frequency distribution matrix

As can be seen from Fig. 12-2, the document is numbered in rows, with each word being a column name and the corresponding value being the frequency of the word appearing in the document. Next, we need to convert the document. To manage this step, we need to use the "Count Vector" method, which functions as follows:

① It will tokenize strings (dividing them into individual words) and set an integer ID for each token.

② It will calculate the number of occurrences of each token.

4. Generate dataset

By using the train_test_split method in Scikit-learn, the dataset can be divided into train and test sets. The following variables can be used to split the data subject to the code below:

```
1.  from sklearn.model_selection import train_test_split
2.  X_train, X_test, y_train, y_test = train_test_split(
3.                          df['sms_message'],df['label'],
4.                          random_state =1)
5.  print('Number of rows in the total set: {}'.format(df.shape[0]))
6.  print('Number of rows in the training set: {}'.format(X_train.shape[0]))
7.  print('Number of rows in the test set: {}'.format(X_test.shape[0]))
```

Next, we need to apply the Bag of Words process to the dataset. The data has been split, and the next goal is to convert the data into the expected matrix format. For this purpose, use CountVectorizer() as before. Two steps need to be completed:

Firstly, we need to fit the training data (X_train) on CountVectorizer() and return the matrix.

Secondly, we need to convert the test data (X_test) to return the matrix.

X_train is the training data of the 'sms_message' column in the dataset and this data will be used to train the model. The code is as follows:

```
1.  from sklearn.feature_extraction.text import CountVectorizer
2.  count_vector = CountVectorizer()
3.
```

```
4.  # Fit the training data and then return the matrix
5.  training_data = count_vector.fit_transform(X_train)
6.
7.  # Transform testing data and return the matrix.
8.  testing_data = count_vector.transform(X_test)
```

Scikit-learn can be implemented with multiple naive Bayesian methods, so that there is no need performing calculations from scratch and the dataset can be predicted only by using the sklearn. naive_bayes methods of Scikit-learn.

This classifier is suitable for classifying discrete features (such as word count text classification). On the other hand, the Gaussian naive Bayes algorithm is more suitable for continuous data because it assumes that the input data is subject to a Gaussian (normal) distribution. The code is as follows:

```
1.  from sklearn.naive_bayes import MultinomialNB
2.  naive_bayes = MultinomialNB()
3.  naive_bayes.fit(training_data, y_train)
4.  predictions = naive_bayes.predict(testing_data)
```

In the evaluation model, the test set has already been predicted, and the next goal is to evaluate the effectiveness of the model. Various measurement indicators can be applied:

① The accuracy rate measures the probability of the classifier making correct predictions, namely, the ratio of the number of correct predictions to the total number of predictions (the number of test data points).

② The precision rate refers to the probability that an e-mail classified as spam is actually spam, which is the ratio of the true positive case (classified as spam and actually spam) to all positive cases (all classified as spam, regardless of whether the classification is correct or not). In other words, it is the ratio result of the following formula:

$$[TP/(TP + FP)]$$

③ The recall rate (sensitivity) represents the proportion of the e-mails that are actually spam and classified as spam, that is, the ratio of true positive cases (classified as spam and actually spam) to all e-mails that are spam, in other words, the ratio result of the following formula:

$$[TP/(TP + FN)]$$

④ F1 score. For skewed classification distribution problems (where the dataset belongs to skewed classification), for example, if there are 100 e-mails, only 2 are spam, and the remaining 98 are not, then the accuracy itself is not a good indicator. 90 e-mails are classified as non-spam (including 2 spam e-mails, but classified as non-spam, so they belong to false negative cases) and 10 e-mails are classified as spam (all 10 e-mails are false positive cases), which will still achieve a

relatively high accuracy score. For such situations, precision and recall are very practical. F1 scores can be obtained through these two indicators, namely the weighted average of the precision and recall scores. This score ranges from 0 to 1, and 1 represents the best potential F1 score.

All four indicators, if used, can ensure that the model achieves a good effect. Such four indicators range from 0 to 1, and that the score is made as close as possible to 1 can indicate how the model is effective. The code is as follows:

```
1.  from sklearn.metrics import accuracy_score, precision_score,
    recall_score, f1_score
2.  print (' Accuracy score: ', format (accuracy _ score (y _ test,
    predic - tions)))
3.  print (' Precision score: ', format (precision _score (y _test,
    predic - tions)))
4.  print ('Recall score: ', format(recall_score(y_test, predictions)))
5.  print ('F1 score: ', format(f1_score(y_test, predictions)))
```

Run the code and see Fig. 12-3 for the results. It will output the Accuracy, Precision, Recall, and F1 scores of the model.

```
Accuracy score:   0.9885139985642498
Precision score:   0.9720670391061452
Recall score:   0.9405405405405406
F1 score:   0.9560439560439562
```

Fig. 12-3   Accuracy, Precision, Recall, and F1 scores of the model

Compared with other classification algorithms, a major advantage of the naive Bayes algorithm is its ability to handle a large number of features. In spam classification, there are thousands of different words, each of which is treated as a feature. In addition, even if there are unrelated features, it will have good effects and is not easily affected by such features. Another major advantage is its relative simplicity. The naive Bayes algorithm can be used directly and rarely requires parameter adjustment, unless the distribution data is usually known. It will rarely overfit the data. Another important advantage is its fast training and prediction speed compared to the amount of data it can process. In short, the naive Bayes algorithm is a very practical algorithm.

Unit 13

Use Word Vector Word2Vec Algorithm to Generate Poetry Automatically

The natural languages used in daily life cannot be directly understood by computers. When it is necessary to process these natural languages, specific means need to be used for analysis or preprocessing. Word vectors can be available by using one-hot coding to process texts, but due to the sparsity of data in analyzing unique numbering of texts, the Word2Vec algorithm can solve this problem by mapping each word to a vector, representing the relationship between words.

13.1 Overview of Word2Vec Algorithm

Word to Vector (Word2Vec), namely, from word to vector. Word2Vec uses a layer of neural networks to map one-hot word vectors to distributed word vectors. Word2Vec is a word embedding model proposed by Google in 2013. It is actually a shallow neural network model. There are only two network structures, namely CBOW and Skip-gram. Inclusively, the goal of CBOW is to predict the generation probability of the current word based on the words that appear in the context, while Skip-gram is to predict the generation probability of each word in the context based on the current word.

13.2 Implementation of Word2Vec Algorithm

When processing natural languages, words or characters are usually vectorized, for example,

You can use one- hot encoding to convert words into vectors, for example, "I love Bejing Tiananmen", and after word segmentation, perform one-hot encoding accordingly, resulting in:

"I": $[1,0,0,0]$

"Love": $[0,1,0,0]$

"Beijing": $[0,0,1,0]$

"Tiananmen": $[0,0,0,1]$

However, if there are n words, any of them is encoded with 1 only and the n-1 bit is 0, this will result in very sparse data (with particularly many zeros and very few ones) and high storage consumption.

So, distributed representation was proposed. What is distributed representation? Its idea is to map each word to a shorter word vector through training. What is the dimension of this shorter word vector? Usually, you need to specify it yourself during training.

Fig. 13- 1 shows four different words, which can be represented by a variable dimension length (only the first four dimensions are depicted in the figure).

When using the word embedding model, there can be some relationship between words. For example, the word vector of king minus the word vector of man, plus the word vector of woman, will be equal to the word vector of queen.

king	queen	woman	priness
0.99	0.99	0.02	0.98
0.99	0.05	0.01	0.02
0.05	0.93	0.999	0.94
0.7	0.6	0.5	0.1
...	...	...	...

Fig. 13-1　Four different words represented by variable dimension length

If a heat map is used, the similarity of word vectors can be intuitively seen, as shown in Fig. 13-2.

The vector generated by "king – man + woman" is not exactly equivalent to "queen", but "queen" is the word closest to it in this set, as shown in Fig. 13-3.

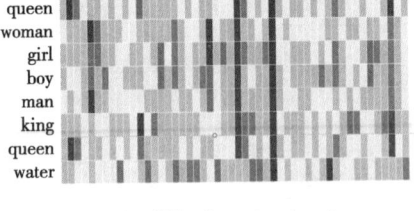

Fig. 13-2　Word vector heat map

Fig. 13-3　King- man + woman vector similarity

13.3　Word2Vec Model

The most typical example of natural language processing is the function of predicting the next

word in the smartphone input method. This is a feature that billions of people use hundreds of times a day. It can be considered that this model received two words (I called) and recommended a set of words ("phone" is the one most likely to be chosen), as shown in Fig. 13-4. The Word2Vec model is actually a simplified neural network, input is feature, output is predict, as shown in Fig. 13-5. It can score all the words it knows (there may be thousands to millions of words in the model vocabulary) based on likelihood, and the input method program will select the one with the highest score and recommend it to the user.

Fig. 13-4   Intelligent input via input method

The output of the natural language processing model is the probability score of the words known by the model. Usually the probability is expressed as a percentage, but in reality, a score of 40% is represented as 0.4 when a vector group is output, as shown in Fig. 13-6.

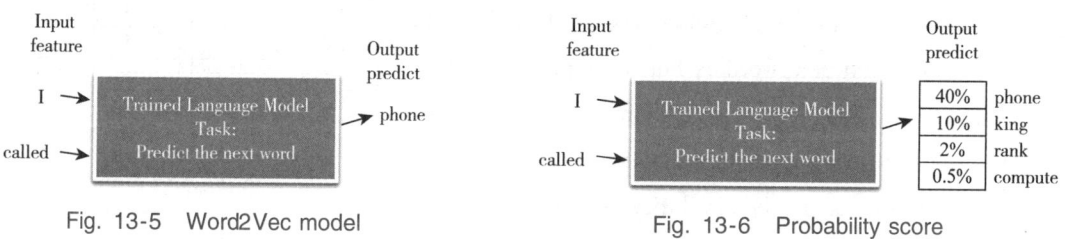

Fig. 13-5   Word2Vec model

Fig. 13-6   Probability score

The Word2Vec model is a simplified neural network that includes an input layer, a trained language model, and an output layer with a softmax( ) function, as shown in Fig. 13-7.

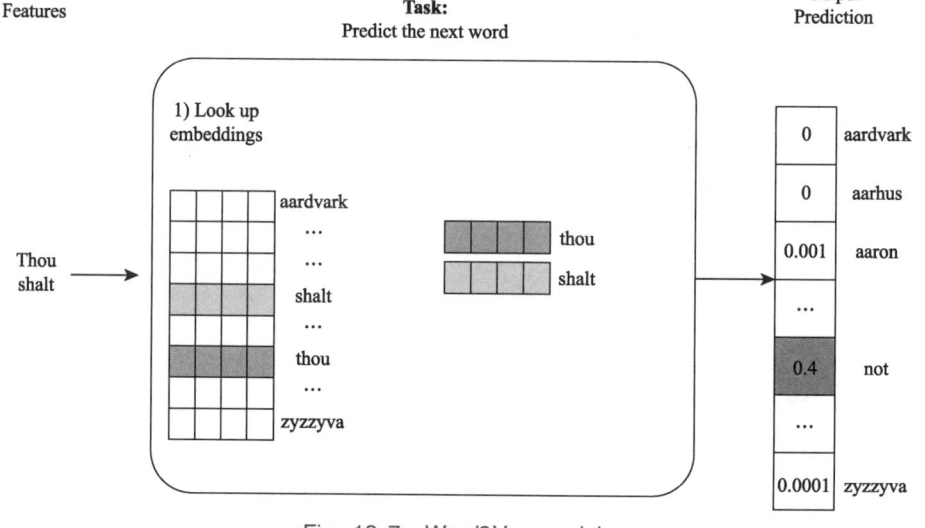

Fig. 13-7   Word2Vec model

For the sentence "I like deep learning and NLP", based on this sentence, we can build a glossary with a size of 6. Suppose 300 features are used to represent a word. Remember that the weight matrix above is w (6,300), and use the unique heat code wt to represent the matrix (300, 1). Multiply the two matrices $wt \times w$ and the hidden neural network will output a 6×1 dimensional matrix, as shown in Fig. 13-8.

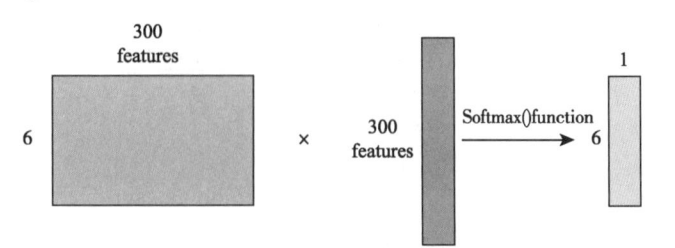

Fig. 13-8 Hidden Layer Neural Network Output

13.4 Generate Word2Vec Data

Firstly, obtain a large amount of text data, and then establish a window that can slide along the text (for example a window containing three words). By using this sliding window, a large amount of sample data can be generated for the training model, as shown in Fig. 13-9.

Fig. 13-9 Window established with text sliding

Make the first two words as features only, and let the third word as label only to generate the first sample in the dataset, as shown in Fig. 13-10.

Fig. 13-10 First sample in the dataset

Slide the window to the next position and generate the second sample in the dataset, as shown in Fig. 13-11.

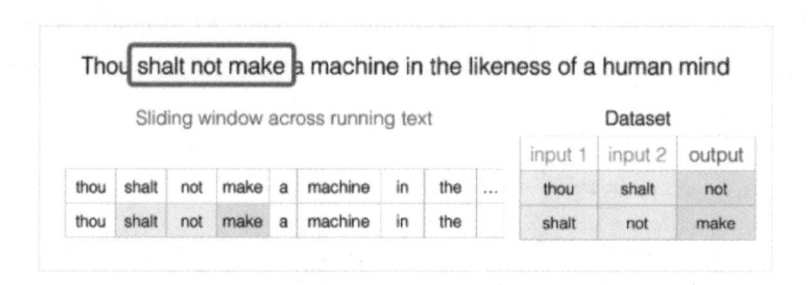

Fig. 13-11   Second sample in the dataset

After sliding on all datasets, a larger dataset will be obtained, as shown in Fig. 13-12.

Thou shalt not make	a machine in	the likeness of a human mind

Sliding window across running text   Dataset

thou	shalt	not	make	a	machine	in	the	...		input 1	input 2	output
thou	shalt	not	make	a	machine	in	the			thou	shalt	not
thou	shalt	not	make	a	machine	in	the			shalt	not	make
thou	shalt	not	make	a	machine	in	the			not	make	a
thou	shalt	not	make	a	machine	in	the			make	a	machine
thou	shalt	not	make	a	machine	in	the			a	machine	in

Fig. 13-12   Larger dataset

13.5　Train the Word2Vec Model

If a corpus is slightly larger, there may be too many possible results, which can be time-consuming to calculate. At this point, two words can be input to see if they correspond to the input and output, which is equivalent to a dichotomy task. However, at this point, the labels constructed from the training set are all 1, which makes it impossible to perform good training. In this case, some negative examples can be added, as shown in Fig. 13-13. The "input word" represents the words input, the "output word" represents the

input word	output word	target
not	thou	1
not		0
not		0
not	shalt	1
not	make	1

Fig. 13-13   Model to be updated

words output, and the target represents the results with 0 and 1 representing positive and negative samples.

By using neural network backpropagation to calculate updates, not only the weight parameter matrix is updated, but also the input data is updated. Finally, the model is trained, and then the model output uses sigmoid to work out a probability, calculate the loss (error), and update the model based on the loss, as shown in Fig. 13-14.

input word	output word	target	input · output	sigmoid()	Error
not	thou	1	0.2	0.55	0.45
not	aaron	0	-1.11	0.25	-0.25
not	taco	0	0.74	0.68	-0.68

Fig. 13-14 Model finally trained

Advantages of Word2Vec: Due to its consideration of the context, Word2Vec performs better than previous word vector methods; compared to the previous word vector method, it has fewer dimensions, so it is faster; moreover, it is quite versatile and can be used in various tasks for processing natural languages.

Disadvantages of Word2Vec: Due to the one-to-one relationship between words and vectors, the problem of polysemy cannot be solved; in addition, Word2Vec is a static approach that, although highly versatile, cannot be dynamically optimized for specific tasks.

13.6 Code implementation of Word2Vec Algorithm

1. Define configuration files

In Jupyter Notebook, set the configuration file for Word2Vec, including sliding window size, word vector dimension, etc. The code is as follows:

```
1.  class CONF:
2.      path = 'Poetry. txt'
3.      window =16 # sliding window size
4.      min_count =60 # filter low-frequency words
5.      size =125 # Word vector dimension
```

```
6.        topn =14 # generate openness of poetry
7.        model_path = 'Word2Vec'
```

2. Define the model

Define a word vector model and complete the model initialization function. The code is as follows:

```
1.    class Model:
2.      def_init_ (self, window, topn, model):
3.          self.window = window
4.          self.topn = topn
5.          self.model = model # word vector model
6.          self.chr_dict = model.wv.index_to_key # dictionary
7.

8.      @ classmethod
9.      def initialize(cls, config):
10.         """Model initialization"""

11.         if exists (config.model_path):
12.            # model reading
13.            model = Word2Vec.load(config.model_path)
14.         else:
15.            # corpus reading
16.            with open (config.path, encoding = 'utf - 8 ') as f:
17.               ls_of_ls_of_c = [list(line.strip()) for line in f]
18.            # model training and saving
19.               # window represents the maximum distance between the
    current word and the predicted word in a sentence
20.               # min_count can truncate the dictionary. The words with
    the word frequency less than min_count results will be discarded,
    5 by default
21.               model = Word2Vec(ls_of_ls_of_c, window = config.window,
    min_count = config.min_count)
22.               model.save(config.model_path)
23.         return cls(config.window, config.topn, model)
24.

25.      def poem_generator(self, title, form):
```

```
26.          """generation of poetry"""
27.          def clean(lst): return [t[0] for t in lst if t[0] not in [',', '。']]
28.          # title completion
29.          if len(title) < 4:
30.              if not title:
31.                  title + = choice(self.chr_dict)
32.              for _ in range(4 - len(title)):
33.                  # similar_by_key calculation similarity
34.                  similar_chr = self.model.wv.similar_by_key(title
         [-1], self.topn // 2)
35.                  similar_chr = clean(similar_chr)
36.                  # choice returns a list item, selects a data item at
         random, and returns a random item
37.                  char = choice([c for c in similar_chr if c not in title])
38.                  title + = char
39.          # text generation
40.          poem = list(title)
41.          for i in range(form[0]):
42.              for _ in range(form[1]):
43.                  predict_chr = self.model.predict_output_word(
44.                      poem[-self.window:], max(self.topn, len
         (poem) +1))
45.                  predict_chr = clean(predict_chr)
46.                  char = choice([c for c in predict_chr if c not in
         poem[len(title):]])
47.                  poem.append(char)
48.              poem.append(',' if i % 2 == =0 else '。')
49.          length = form[0] * (form[1] +1)
50.          return '《% s》' % ''.join(poem[: - length]) + ' \n' + ''.join
         (poem[ - length:])
```

3. Model training

Load text data, train and save the model. Call the model to achieve the function of writing poetry automatically, and the code is as follows:

```
1.   def main (config = CONF):
2.       """main function"""
```

```
3.     form = {'five-character quatrain': (4,5), 'seven-character
       quatrain': (4,7), 'couplet': (2,9)}
4.  m = Model.initialize(config)
5.  while True:
6.      title = input ('Input title:').strip()
7.      poem = m.poem_generator(title, form['five-character quatrain'])
8.      print(' \033[031m% s \033[0m' % poem) # red
9.      poem = m.poem_generator(title, form['seven-character quatrain'])
10.     print(' \033[033m% s \033[0m' % poem) # yellow
11.     poem = m.poem_generator(title, form['couplet'])
12.     print(' \033[036m% s \033[0m \n' % poem) # purple
```

After the code runs, we need to enter a title, for example, Huatian. The program will generate five-character quatrains, seven-character quatrains and couplets automatically, specifically as follows:

Input Title: Huatian

Five-character quatrain: Cudrania land grain without mulberry, field rice garden with apricot. Cropland seedlings with silkworm raisers, cultivation for millet, vegetable and wheat.

Seven-character quatrain: Rice field plowed and hoed for wheat and vegetables, land seedlings growing to cover mulberry planters. Cudrania surrounded with millet, silkworm and vegetables, children's meal from good food due to fertile jujube humus.

Couplet: Meicun wheat garden fenced land for apricot, grain and mulberry, cudrania field borders for cultivating silkworms and jujube seeds.